# Current Topics in Microbiology 144 and Immunology

Editors

R. W. Compans, Birmingham/Alabama · M. Cooper,
Birmingham/Alabama · H. Koprowski, Philadelphia
I. McConell, Edinburgh · F. Melchers, Basel
V. Nussenzweig, New York · M. Oldstone,
La Jolla/California · S. Olsnes, Oslo · H. Saedler,
Cologne · P. K. Vogt, Los Angeles · H. Wagner, Ulm
I. Wilson, La Jolla/California

# Transforming Proteins of DNA Tumor Viruses

Edited by
R. Knippers and A. J. Levine

With 85 Figures

Springer-Verlag
Berlin Heidelberg New York
London Paris Tokyo Hong Kong

Rolf Knippers, Dr.
Fakultät für Biologie,
Universität Konstanz,
D-7750 Konstanz, FRG

Arnold J. Levine, Ph. D.
Department of Biology,
Princeton University,
Princeton, N.J. 08544-1014 USA

ISBN 3-540-50909-7 Springer-Verlag Berlin Heidelberg New York
ISBN 0-387-50909-7 Springer-Verlag New York Berlin Heidelberg

# Preface

It is surprising, and even disappointing, that there have been very few meetings and published volumes resulting from these meetings that focus attention upon all of the groups of DNA tumor viruses. Historically, separate meetings were held each year for the adenovirus-SV40-polyoma researchers, the herpesviruses, hepatitis B virus and the papillomaviruses. It was as if these four virus groups were four fields of study developing independently with a literature and culture of their own. When a virologist crossed the field from the adenovirus group to the herpesvirus or papillomaviruses, he or she was lost to their former group because of the structure of separate meetings and remote literature. This, of course, has resulted from historical accident and is being rectified by the rapid progress made in our understanding of how these viruses contribute to the causation of cancer in animals and humans. It was precisely because of these factors that it was time to hold a meeting and publish its proceedings on the subject of transforming proteins of DNA tumor viruses. For the first time, DNA tumor viruses were defined as all of the virus groups that can contribute to cancer in animals with the exception, unfortunately, of the poxviruses. The purpose of the meeting was to bring together scientists who rarely attend meetings together but actually work on the same problems with different viruses. This was done at Titisee, Federal Republic of Germany, in the Black Forest, at a meeting sponsored by the Boehringer-Ingelheim Foundation, to whom all the participants are indeed grateful. By every criteria, the meeting was a success. The presentations and the informal discussions reinforced what we have begun to appreciate: there are indeed common mechanisms, themes, and strategies employed by different viruses to interact with their host cells. The justification for focusing our attention upon the DNA tumor virus oncogenes are, in the end, the very same reasons why the cellular-encoded oncogenes are important in our understanding of cancer: (1) Some of the DNA tumor viruses – such as the human papillomaviruses 16, 18, 33, hepatitis B virus, Epstein-Barr virus – have been shown to be continuously associated with specific human cancers and very likely contribute to one of several events that cause these cancers. (2) The study of how these virally encoded oncogenes function will provide fundamental information about the control of cell growth. The available evidence demonstrates that some virally encoded oncogene products interact or function with the normal cellular proto-oncogene products and, in so doing, alter their activities. Other viral oncogene products are able to bypass the normal cellular control mechanisms which are regulated by proto-oncogenes and their products. Clearly, the study of the DNA viral oncogenes will inevitably lead us to the cellular-encoded proto-oncogenes.

One of the common themes that ran through the meeting was the number of viral oncogene products that can function as *trans*-activators of viral or cellular genes. The adenovirus E1A gene products have been shown for a long time now to regulate positively or negatively viral and cellular genes at the level of transcription. The SV40 large T antigen and the papillomavirus E7 gene product can positively regulate viral genes at the transcription level as well. While the mechanisms of *trans*-activation are less clear, the Epstein-Barr virus nuclear antigen II (EBNA-II) may also alter the activity of some genes. Perhaps a related observation is the fact that the adenovirus E1A product, the SV40 large T antigen, and the papillomavirus E7 product all bind to the Rb protein, or retinoblastoma susceptibility gene product. The Rb protein is thought to behave as an anti-oncogene. It could be the case that the Rb protein negatively regulates the expression of growth-regulating genes, and its binding to the DNA tumor virus oncogene product then inactivates this function. The analogous observation that the SV40 large T antigen and the adenovirus E1B 55K tumor antigen both bind to the cellular p53 protein has gained a new significance in the past year or so. Several lines of evidence suggest that p53, like the Rb protein, can act to regulate negatively (anti-oncogene) transformation and tumorigenesis. Thus, the SV40 large T antigen is shown in this volume to encode two domains (functional and structural) that are required for the transformation of different cell types. These domains can be localized by a genetic analysis between amino acid residues 1–120 and 325–625 (out of 708 amino acids). The Rb protein binds to the 1–120 domain, and p53 binds to the 325–625 domain. Similarly, the adenovirus E1A protein binds the Rb protein, while the adenovirus E1B 55K binds the p53 protein, indicating that separate proteins, not separate domains, carry out the same interactions in adenovirus-infected cells. The E1A and E1B 55K functions, like the two T-antigen domains, are both required for transformation of cells by this virus.

Yet another similar theme of viruses interacting with their hosts emerges from the roles of the SV40 large T antigen and the EBV latent membrane protein, or LMP. B cells that carry the EBV genome in an episomal state (not integrated in the genome of the host) synthesize the LMP gene product and insert it into their plasma membrane. There it is recognized by the cytotoxic T cells (CTL) of the host as foreign, resulting in a CTL response. Based upon this, the LMP antigen was first detected and termed the lymphocyte-determined membrane antigen, or LYDMA. Thus, LMP, a major transforming gene product of EBV, is also its tumor-specific transplantation antigen (TSTA). The case is very similar with SV40. The large T antigen, the major transforming gene product, is also the virally encoded TSTA for the transformed cell.

A third common theme that ran through the meeting presentations involved the functional analogies between DNA virus-encoded oncogene products and cellular-encoded oncogene products. The cellular proto-oncogene products have been shown to function as peptide hormones (such as platelet-derived growth factor), receptors for hormones (such as the epidermal growth factor (EGF) receptor), proteins that appear to mediate or transduce signals from the plasma membrane to the nucleus (*ras, src*, G-proteins, protein kinases), and in the nucleus, transcriptional *trans*-activators (*jun*, AP-1, *fos* complex). The DNA tumor viruses encode oncogene products that act as growth factors or hormones (poxviruses, EGF-like hormone), that could alter the transduction of signals (polyoma middle T antigen-*src* complex), and, as indicated previously, that act as nuclear *trans*-activators at the transcriptional level. Clarifying the interactions and functional homologies between viral- and cellular-encoded oncogenes was a theme of the meeting and this volume.

Like any good meeting, and in any good collection of papers derived from such a meeting, what we do not know or what is yet to be found became as clear as what we think we do now know. Does hepatitis B virus encode an oncogene product? remains an open question. Can hepatocellular carcinoma, so closely associated with a chronic infection by this virus, be explained by (1) immunological destruction of infected liver cells and constant regeneration of such cells, increasing the risk of proto-oncogene mutations, (2) the use of the hepatitis B virus polymerase by cellular DNA which, in turn, may have a high error frequency increasing the risk of mutation rate, or (3) a yet to be found (the x-gene?) oncogene that can act directly. Second, what are the cellular targets for the DNA viral *trans*-activators? Which genes and gene products are apparently regulated by Rb and p53? Third, of the multiple functions of the viral proteins detailed in this volume, which function plays a critical role in the transformation process? And, why do almost all of the groups of DNA-containing viruses have some representatives that can cause cancer in animals? What is inherent in the evolution and development of such viruses that has led to this result? Fourth, how do we begin to translate our results in vitro, in cell culture, to the disease of cancer in vivo, in animals? It seems likely that most cancers develop in multiple independent steps (with time) and that the DNA tumor viruses may contribute one or more of these events on the pathway to cancer by providing exogenous genetic information. Some of these events might be measurable only in vivo, in cells found in a three-dimensional organ, on a specific extracellular matrix, expressing cellular genes and products that are not expressed in vitro. How do we develop the tools to detect such potential events? Clearly, the increased use of transgenic animals will play a role in bridging this gap in our knowledge.

When this meeting and these proceedings were being planned, we particularly saw the need to bring together individuals studying a wide variety of diverse DNA tumor viruses. After all, these researchers were often asking the same questions, using common techniques and approaches, and even generating similar answers with different virus groups. Once the diverse terminology, the jargon of a virus group, was put away, and the individual viral oncogenes and their functions were isolated and discussed, new insights and interactions emerged. The utility of the meeting became clear, and the contacts between groups stimulated new experiments, collaborations, and interests. For this, the editors are indebted to the participants, speakers, authors of these papers, and the Boehringer-Ingelheim Foundation. The combination of these individuals, groups and the DNA tumor viruses themselves made the event and these proceedings as great as could have been expected.

A. J. Levine, R. Knippers
Princeton, New Jersey, USA
Konstanz, Germany
December 1988

# Table of Contents


**Part I: The Transforming Protein of Simian Virus 40:
Large T antigen and Its Interactions
with the Cellular Protein p53**

**Part IV: Transformation by Adenoviruses**



**Part V: Herpesviruses: The Cellular and Molecular Biology
of Epstein-Barr Virus**



**Part VI: Hepatitis B Virus and Liver Cancer**

# List of Contributors

You will find the address at the beginning of the respective contribution

ADDISON, C.
AMSTERDAM, A.
ARTHUR, A.
BAICHWAL, V. R.
BEARD, P.
BIESINGER, B.
BOS, J. L.
BRAIN, R.
BRUGGMANN, H.
CHENG, S. H.
CHITTENDEN, T.
COUKOS, W. J.
COURTNEIDGE, S. A.
CRIPE, T. P.
DEPPERT, W.
DESROSIERS, R. C.
DiMAIO, D.
DOERFLER, W.
EFRAT, S.
ESPINO, P. C.
FANNING, E.
FLECKENSTEIN, B.
FREY, A.
FUCHS, P. G.
GRAESSMANN, A.
GRAESSMANN, M.
GRASSMANN, R.
GREEN, M. R.
GRIFFIN, B. E.
GRIMALDI, M.
HAI, T.
HAMMERSCHMIDT, W.
HANAHAN, D.
HARREY, R.
HÖSS. A.
HOFSCHNEIDER, P. H.
HORWITZ, B. H.

HOWLEY, P. M.
IFTNER, T.
JENKINS, J. R.
JESSBERGER, R.
KIESSLING, U.
KLAUSING, K.
KLEINHEINZ, A.
VON KNEBEL-DOEBERITZ, M.
KNIPPERS, R.
KOSHY, R.
LEE, K. A. W.
LEVINE, A. J.
LICHTENBERG, U.
LILLIE, J. W.
LOEBER, G.
LUBBE, L.
MARTIN, K. J.
MICHALOVITZ, D.
MOAREFI, I.
MODROW, S.
MÜNGER, K.
MURTHY, S. C. S.
OFFRINGA, R.
OREN, M.
PARSONS, R.
PFISTER, H.
PHELPS, W. C.
PIWNICA-WORMS, H.
PLATZER, M.
PRAKASH, S. S.
ROBERTS, T. M.
RUDGE, K.
SCHAAK, J.
SCHALLER, H.
SCHEFFNER, M.
SCHEIDTMANN, K. H.
SCHLICHT, H. J.

SCHNEIDER, J.
SETTLEMAN, J.
SHENK, T.
SMITH, A. E.
STAHL, H.
STEINMEYER, T.
STRAUSS, M.
STÜRZBÄCHER, H.-W.
SUGDEN, B.
TEGTMEYER, P.

TIMMERS, H. T. M.
TRIMBLE, J.
TUREK, L. P.
VAN DAM, J. A. F.
VAN DEN HEUVEL, S. J. L.
VAN DER EB, A. J.
WESSEL, R.
YEE, C. I.
ZANTEMA, A.

# Part I: The Transforming Protein of Simian Virus 40: Large T Antigen and Its Interaction with the Cellular Protein p 53

# Introduction

The genome of simian virus 40 (SV40) is a circular, double-stranded DNA composed of 5243 base pairs. The genome is divided into two functionally distinct genetic regions of about equal size: The "early" region codes for two regulatory proteins, and the "late" region codes for structural proteins of the virus coat. SV40 DNA is packaged in vivo as chromatin, composed of about 25 nucleosomes, forming a nucleoprotein complex; it is usually referred to as the viral minichromosome because it resembles cellular chromatin in composition and general architecture.

Depending on the cellular environment, SV40 can adopt two different life styles. In its natural host, the African green monkey cell, SV40 multiplies manifold, produces thousands of progeny virus particles, and eventually causes the death of the infected cell. SV40 also infects other mammalian cells, for example, rodent cells. In these cases, the viral genome is replicated poorly, if at all, and is integrated into the genome of the infected cell. Integration occurs at many sites of the host cell genome, and the circular viral genome is cut at random sequences before or during integration. If the early genes of the viral DNA remain intact, two proteins can be expressed, the small and the large tumor antigen, about 17K and 90K in size, respectively. The larger of the two proteins induces profound changes of the phenotype of the infected rodent cells. They acquire the growth characteristics of an "immortalized" or "established" cell line and are eventually transformed to tumorigenicity. Animals bearing tumors of SV40-transformed cells recognize the "early" viral proteins as foreign antigens, hence their designation as tumor antigens (reviewed in TOOZE 1981).

The large tumor antigen (T antigen) is one of the most intensely studied eukaryotic proteins because it performs an amazingly large variety of functions which are of interest to biochemists studying transcription and DNA replication of eukaryotes as well as to cell biologists and tumor biologists (for reviews: RIGBY and LANE 1983; LIVINGSTON and BRADLEY 1987; STAHL and KNIPPERS 1987).

T antigen is a phosphoprotein of 708 amino acids with several functional domains. It is a DNA binding protein with a high affinity to DNA sequences containing the pentanucleotide GAGGC in defined arrangements (Fig. 1). The DNA binding property of T antigen is discussed later in more detail by FANNING et al. These authors also present a tentative model of the structure of the T-antigen DNA binding domain. Just outside of this domain, but within a region that is in close contact to DNA in T-antigen-DNA complexes, is a sequence of amino acids that would theoretically form a Zinc finger loop, a motif frequently found in DNA

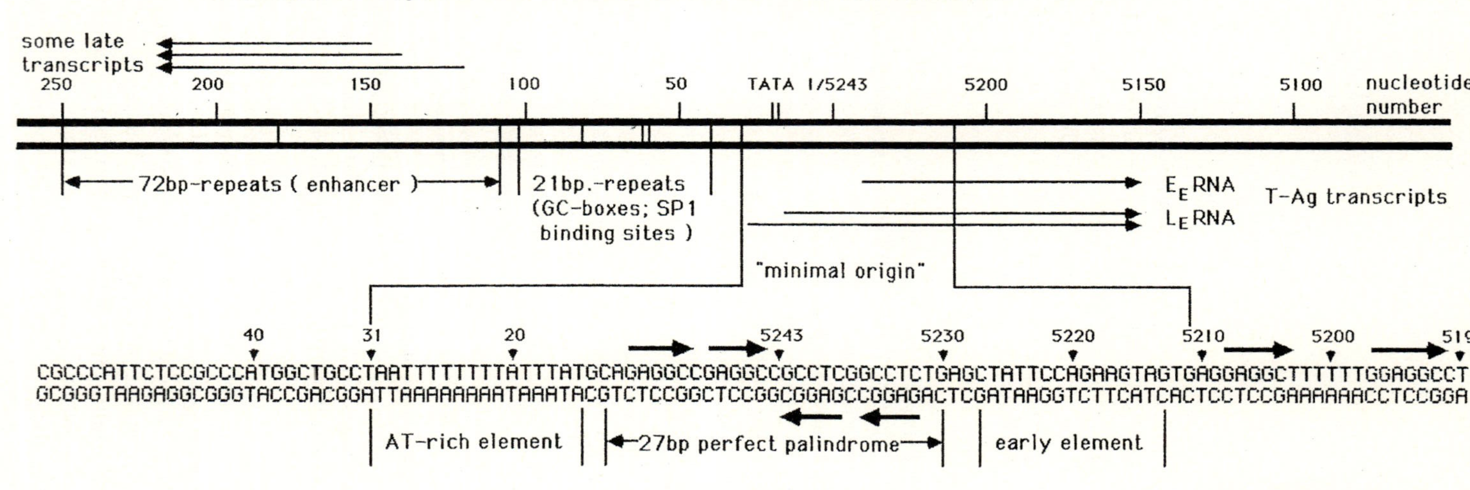


**Fig. 1.** T-antigen binding sites in the SV40 genome. The *upper part* of the figure shows a schematic representation of the genetic control region of the SV40 genome. The nucleotide numbering system is given in Tooze (1981). The *TATA*-box, the *GC-boxes*, and the *enhancer* for the early coding region are indicated as are the start sites for the immediate early and late early transcripts ($E_E$ *RNA*, $L_E$ *RNA*). The start sites for late transcription are less well-defined. In this figure only the start sites of some late transcripts are shown. The nucleotide sequence, shown in the *lower part* of the figure, includes the two most important T-antigen binding sites. *Binding site I* contains two tandemly oriented GAGGC pentanucleotides separated by an AT-rich spacer (Ryder et al. 1985). T antigen bound to site I autoregulates its expression probably by repression of transcription (Rio and Tjian 1983). *Binding site II* contains four pentanucleotides, arranged as two pairs oriented in opposite directions. This binding site is part of the viral "minimal" origin of replication (Deb et al. 1986). T antigen bound to site II initiates viral DNA replication (Dean et al. 1987; Wold et al. 1987; Stahl et al. 1988). (The figure is reproduced from an article by Stahl and Knippers 1987)

binding proteins (review: KLUG and RHODES 1987). LOEBER et al. show in their contribution that this amino acid sequence is important for the function of T antigen. The interaction of T antigen with DNA appears to be regulated by protein phosphorylation as summarized later by KLAUSING and KNIPPERS.

A second domain of T antigen includes the active center of an ATP-hydrolyzing activity (ATPase), which is most likely a part of an interesting biochemical function of T antigen, namely its ability to unwind double-stranded DNA. Bound to the origin of replication, T antigen initiates the unwinding of the viral genome as a first step for the establishment of replication forks. STAHL and his associates (SCHEFFNER et al.) show below that T antigen unwinds not only the origin of DNA but, under proper conditions, double-stranded DNA of any sequence. This property could be important for the effects of T antigen on the cellular genome.

As discussed below, specific DNA binding and the unwinding of double-stranded DNA are important functions of T antigen in productively infected cells. Genetic evidence has shown that T antigen also induces cellular gene expression and cellular DNA replication in the infected cells. These properties are only poorly understood. One possibility is that T antigen induces or activates cellular transcription factors. This will be considered by BEARD and BRUGGMANN in their contribution.

Even less is known in molecular terms about the role that T antigen plays in transformed cells. More than 90% of the T antigen in transformed (as well as in productively infected) cells is located in the nucleus, and a small but significant fraction is found at the cytoplasmic membrane. KALDERON et al. (1984) have analyzed a class of SV40 mutants producing a T antigen that accumulates in the cytoplasm and cannot be transported to the nucleus. These mutants are actively able to transform established mouse or rat cell lines but fail to transform primary cells (LANFORD et al. 1985), suggesting that (a high concentration of) nuclear T antigen is required for immortalization, but not for transformation to tumorigenicity. In fact, replicative T-antigen functions such as specific DNA binding and DNA unwinding are dispensable for the transforming activity, which is linked to a domain in the *N*-terminal fifth of the protein, separated from the DNA binding and the ATPase domains (see Fig. 1 in FANNING et al., this volume).

An important property of T antigen appears to be its ability to associate with cellular proteins including DNA polymerase and a transcription factor (references in STÜRZBECHER et al., this volume). Recently, DE CAPRIO et al. (1988) described an interaction of T antigen with the protein coded for by the retinoblastoma susceptibility (*Rb*) gene. Based on cell biological and epidemiological evidence it is believed that the Rb protein may be involved in the negative regulation of cell proliferation. An obvious model predicts that the T antigen neutralizes this function and causes a restrictive growth of the affected cells. The Rb protein appears to bind to the *N*-terminal region of T antigen.

Another intriguing cellular protein, p53, binds to a more centrally located section of T antigen.

It was discovered more than a decade ago that much of the T antigen expressed in SV40-transformed cells is not free but associated with the cellular p53 protein. This protein appears also to be involved in the regulation of cell proliferation, probably in initiation of DNA replication. Moreover, p53 was shown to be an oncogene product in its own right (for reference, see STÜRZBECHER et al., this volume). For these reasons, p53 has attracted much attention in recent years, and many laboratories are investigating the functional and structural properties of the p53-T-antigen complex.

STÜRZBECHER et al. briefly summarize below the current knowledge about the genetic and biochemical properties of the p53 protein and present their data showing that rodent p53 inhibitis a replicative function of T antigen. DEPPERT and STEINMAYER provide evidence showing that the stabilization of p53 may not necessarily be related to its association with T antigen but could equally well be due to a metabolic property of the transformed cell. A possible link is suggested by the observation of SCHEIDTMANN that T antigen may induce a function in transformed rodent cells which causes the phosphorylation of p53. Finally, OREN et al. (MICHALOVITZ et al.) present new evidence showing that an overproduction of p53 increases the transformation activity of T antigen, and that p53 can cooperate with a *ras* oncogene product for the transformation of primary cells to tumorigenicity.

Studies with transgenic mice have clearly supported the concept that T antigen is necessary and sufficient for transformation. The results obtained with plasmid constructs carrying a T-antigen coding region linked to tissue-specific promoter-enhancers have shown that T-antigen expression induces a hyperplasia of the affected tissue. Only some of the hyperplastic cells, however, develop into solid tumors. In their contribution to this volume, EFRAT and HANAHAN show that the expression of T antigen above a certain level is necessary for tumorigenicity via hyperplasia. Suboptimal levels of T antigen do not induce tumor growth, even though stabilization or increased expression of p53 accompanies the synthesis of T antigen.

Both p53 and the Rb protein are nuclear proteins. It remains to be shown then why T-antigen mutants which cannot be transported into the nucleus are able to transform established cell lines. Clearly, for an investigation of this and related questions it would be useful to possess methods which allow an experimentally controlled regulation of T-antigen expression. One possible way to achieve this goal would be the inhibition of gene action by intracellular antisense-RNA. GRAESSMANN and GRAESSMANN describe below their initial experience following this rationale; the method may become a valuable tool for further experiments in this field.

# References

Dean FB, Bullok P, Murakami Y, Wobbe CA, Weissbach L, Hurwitz J (1987) Simian virus 40 (SV40) DNA replication: SV40 large T antigen unwinds DNA containing the SV origin of replication. Proc Natl Acad Sci USA 84: 16–20

Deb S, De Lucia A, Baur CP, Koll A, Tegtmeyer P (1986) Domain structure of the simian virus 40 core origin of replication. Mol Cell Biol 6: 1663–1670

De Caprio JA, Ludlow JW, Figge J, Shew JY, Huang CM, Lee WH, Marsilio E, Paudra E, Livingston DM (1988) SV40 large tumor antigen forms a specific complex with the product of the retino blastoma gene. Cell 54: 275–283

Kalderon D, Richardson WD, Markham AF, Smith AE (1984) Sequence requirements for nuclear location of simian virus large T antigen. Nature 311: 33–38

Klug A, Rhodes D (1987) Zinc fingers: a novel protein motif for nucleic acid recognition. Trends Biochem Sci 12: 464–469

Lanford RE, Wong C, Bute JS (1985) Differential ability of a T antigen transport-defective mutant of simian virus 40 to transform primary and established rodent cells. J Virol 5: 1043–1050

Livingston D, Bradley M (1987) The simian virus 40 large T antigen. A lot packaged into a little. Mol Biol Med 4: 63–80

Rigby P, Lane D (1983) The structure and function of SV40 large T antigen. Adv Viral Oncol 3: 31–57

Rio DC, Tjian R (1983) SV40 T antigen binding site mutations that affect autoregulation. Cell 32: 1227–1240

Ryder K, Vakalopouloi E, Metz R, Mastrangelo I, Hough P, Tegtmeyer P, Fanning E (1985) Seventeen base pairs of region I encode a novel tripartite binding signal for SV40 T antigen. Cell 42: 539–548

Stahl H, Knippers R (1987) The simian virus 40 large tumor antigen. Biochim Biophys Acta 910: 1–10

Stahl H, Scheffner M, Wiekowski M, Knippers R (1988) DNA unwinding function of the SV40 large tumor antigen. Cancer Cells 6: 102–115

Tooze J (ed) (1981) DNA tumor viruses: molecular biology of tumor viruses, pt 2. Cold Spring Harbor Laboratory, Cold Spring Habor NY

Wold MS, Li JJ, Kelly TJ (1987) Initiation of simian virus 40 DNA replication in vitro: large-tumor-antigen- and origin-dependent unwinding of the template. Proc Natl Acad Sci USA 84: 3643–3647

# Structure and Function of SV 40 Large T Antigen Communication Between Functional Domains

E. Fanning[1], J. Schneider[1], A. Arthur[1], A. Höss[1], I. Moarefi[1], and S. Modrow[2]

## 1 Introduction

The large T antigen of simian virus 40 (SV40) is a phosphoprotein involved in both negative and positive transcriptional control of viral genes as well as cellular genes (reviewed by Rigby and Lane 1983; DePamphilis and Bradley 1986; Stahl and Knippers 1987). Moreover, it is required for initiation of viral DNA replication and can induce cellular DNA synthesis in quiescent cells. It is well-documented that specific binding of T antigen to two major sites in the regulatory region of the viral genome is needed either directly or indirectly for these activities. Although both sites harbor two or more repeats of the pentanucleotide binding signal GAGGC, they differ in function: Binding of T antigen to site I is primarily involved in repression of early transcription, whereas binding to site II is a crucial step in initiation of SV40 DNA replication (DiMaio and Nathans 1982). Induction of cellular RNA and DNA synthesis by T antigen may also require the sequence-specific DNA binding activity of the protein.

T antigen has previously been shown to be differentially phosphorylated (Fanning et al. 1981; Scheidtmann et al. 1982). Much of the protein in both infected and transformed cells is extracted in the form of highly phosphorylated oligomers, whereas a minor fraction, containing the newly synthesized T antigen, is underphosphorylated and monomeric. Despite conflicting data in the literature, most workers now agree that the minor, underphosphorylated T-antigen molecules have a higher specific activity or affinity for DNA binding, particularly to site II in the origin of replication, than the more abundant, highly modified oligomers (Dorn et al. 1982; Fanning et al. 1982; Scheidtmann et al. 1984; Simmons et al. 1986; Runzler et al. 1987; Mohr et al. 1987; Klausing et al. 1988; M. K. Bradley personal communication).

In this contribution, we have created two sets of mutant T antigens to investigate the role of the phosphorylated domains of the protein in origin DNA binding. A set of truncated T antigens was expressed in *Escherichia coli* (*E. coli*) to define the minimal amino acid sequence necessary for origin binding activity. A second set of mutants carried point mutations at each of the previously mapped phosphorylated serines and threonines (Scheidtmann et al. 1982; Schneider and Fanning 1988). Origin binding studies with the mutant proteins demonstrated that the mini-

[1] Institute for Biochemistry, Karlstrasse 23, 8000 Munich 2, FRG
[2] Max von Pettenkofer Institute, Pettenkoferstr. 9a, 8000 Munich 2, FRG

Current Topics in Microbiology and Immunology, Vol. 144
© Springer-Verlag Berlin · Heidelberg 1989

mal origin DNA binding domain is distinct from but interacts with the phosphorylated domains of the protein, and suggest that phosphorylation may modulate the origin binding specificity of the protein, thus defining functional subclasses of T-antigen molecules.


# 2  Experimental Procedures

## 2.1  Expression of Truncated T Antigens in E. coli

The creation of an intron-less T-antigen gene and its expression in bacteria as a *lacZ* fusion protein using the vector pUC9 has been described in detail (ARTHUR et al. 1988). One truncated T antigen, T260, carries a stop codon at position 260, whereas another, Ta, contains 18 C-terminal missense amino acids following the last authentic T-antigen residue at 239. A third one, Tbb, is fused in the *lacZ* reading frame via a novel restriction site, such that the T-antigen coding sequences lack the first 130 residues and those after residue 682 (see Fig. 2B).


## 2.2  Construction of Rat Cell Lines Expressing T Antigens
##      with Point Mutations in the Phosphorylation Sites

Each of the phosphorylated T-antigen residues was replaced by alanine or in two cases cysteine (SV111/112, residue 112) or glutamic acid (SV124E), as described (SCHNEIDER and FANNING 1988) (Fig. 1). One mutant carried an alanine substitution for an unphosphorylated serine residue (SV120). Rat2 cell lines transformed by DNA transfection with mutant and wild-type plasmid DNAs were described previously (SCHNEIDER and FANNING 1988). Each line was named according to the number of the mutant residue in the T antigen.


## 2.3  Origin DNA Binding Assays

T antigens were immunoprecipitated from bacterial lysates using monoclonal Pab419 antibody (HARLOW et al. 1981), except for Tbb, which lacks the Pab419 epitope and was therefore precipitated with Pab1630 (BALL et al. 1984). T antigens were immunoprecipitated from transformed rat cell extracts with Pab108 (GURNEY et al. 1986). Purified immune-complexed T antigens were then incubated with excess end-labeled SV40 DNA fragments to equilibrium; unbound DNA was washed away, and bound DNA was detected by agarose gel electrophoresis and autoradiography as described previously (HINZPETER et al. 1986; VOGT et al. 1986). A *Hin*dIII-*Eco*RI digest of pONwt, which carries a synthetic site I, was used to test binding to site I (RYDER et al. 1985). Site II binding was assayed using *Hin*dIII-cleaved p1097, which lacks a 31-bp sequence spanning site I (HUBER et al. 1985). Binding to intact origin DNA was tested with *Hin*dIII-digested pSVwt DNA (FANNING et al. 1982).

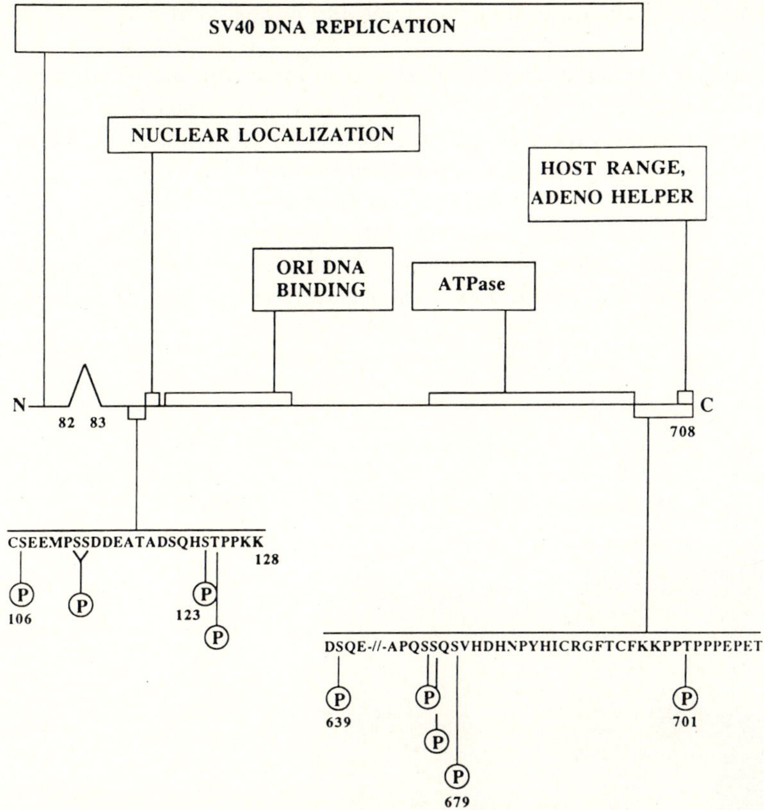



**Fig. 1.** Localization of functional domains and phosphorylation sites of SV40 T antigen. The functional domains represent those sequences found to be necessary and sufficient for each activity (CLARK et al. 1983; KALDERON et al. 1984; COLE et al. 1986; SIMMONS 1986; STRAUSS et al. 1987; ARTHUR et al. 1988)

# 3 Results

## 3.1 Localization of the Origin DNA Binding Domain of T Antigen

Expression of SV40 T antigen in *E. coli* requires the creation of an intron-less copy of the gene and the introduction of bacterial expression signals upstream of the coding sequences. The first requirement was met by generating new restriction sites at the 5′ and 3′ borders of the intron by oligonucleotide-directed mutagenesis, such that ligation of the two sites recreated the authentic T-antigen coding sequences (ARTHUR et al. 1988). The second requirement was satisfied by inserting the T-antigen coding sequences in frame in the polylinker of pUC9, resulting in a *lac*Z-T-antigen fusion protein expressed in *E. coli* under *lac* control (ARTHUR et al. 1988). Using this expression system, the full-length protein and a series of truncated T antigens were produced and tested for their ability to bind specifically to the SV40 origin region.

Figure 2B presents a schematic diagram of several of these truncated T antigens. Tbb lacks the *N*-terminal 130 residues of T antigen, as a novel restriction site was created to insert the coding sequences into the polylinker (ARTHUR et al. 1988). A new stop codon generated at T-antigen residue 260 resulted in the truncated T260 protein. A still shorter fusion peptide Ta was generated by using a restriction site at residue 239 to insert the coding sequence into the polylinker. Each of the constructions was shown by Western blotting of bacterial extracts to produce polypeptides with the predicted length and T antigen-specific monoclonal antibody

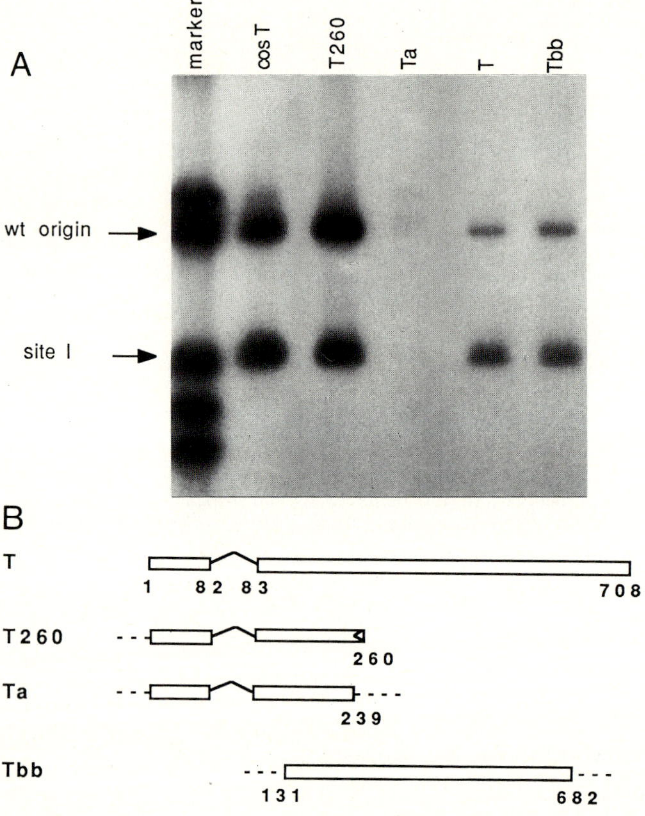

Fig. 2. Origin DNA binding of truncated antigens expressed in *E. coli*. **A** T-antigen polypeptides expressed in *E. coli* and a control T antigen from an SV40-transformed monkey line COS1 (GLUZMAN 1981) were assayed for specific binding to a mixture of labeled DNA fragments containing intact wild-type (*wt*) origin DNA and *site I* DNA. *Marker*, 4% of the input DNA mixture. Approximately equal molar amounts of each T antigen were used, as determined by Western blotting (not shown), but different exposure times were used: 5 h (marker, COST, T260, Ta) or 1 h (T, Tbb). **B** T-antigen coding sequences expressed in *E. coli* are diagrammed as *bars*. *Numbers below the bars* indicate the first or last T antigen residue present in the fusion proteins. *Dotted lines* indicate missense or *lacZ* residues resulting from the constructions (ARTHUR et al. 1988)

epitopes (ARTHUR et al. 1988). None of the T-antigen polypeptides produced in *E. coli* were detectably phosphorylated (ARTHUR, unpublished data).

Specific DNA binding activity of the bacterial T antigens was assayed by immunoprecipitating the soluble proteins from bacterial lysates and incubating the purified immune complexes with excess end-labeled SV40 DNA to equilibrium (Fig. 2A). Specifically bound DNA retained by the immune complexes was a measure of the binding activity of the proteins. The results demonstrate that: (a) full-length T antigen, as well as T antigens lacking either the N-terminal 130 residues or the C-terminal residues after 259 bind specifically and efficiently to the SV40 regulatory region and to the isolated site I (Fig. 2A). (b) The full-length protein binds well to the isolated site II in the origin of replication, whereas the activity of Tbb and T260 at site II is reduced (data not shown; it should be noted that the level of DNA binding varies with the monoclonal antibody bound to the peptide; ARTHUR et al. 1988). (c) Further truncation of T antigen to residue 239 led to a complete loss of sequence-specific binding detectable with this assay (Fig. 2A). Therefore the minimal sequence responsible for site I and site II origin-binding activity must be localized in the region between residues 131 and 259, though it remains to be tested whether the 129-amino acid peptide alone would be sufficiently stable for origin binding (see: Note added in proof). (d) Since apparently unphosphorylated bacterial T antigens and peptides lacking either of the known phosphorylation site clusters have origin binding activity, phosphorylation of T antigen is probably dispensable for origin binding activity.

## 3.2 DNA Binding Properties of T Antigens Carrying Mutations in the Phosphorylated Domains

Oligonucleotide-directed mutagenesis was used to replace each of the known phosphorylated serines and threonines in the large T antigen by an evolutionarily conserved amino acid residue unable to undergo phosphorylation, creating a set of mutant proteins (SCHNEIDER and FANNING 1988). The replication and cell transformation properties of the mutants have already been described (SCHNEIDER and FANNING 1988), and detailed mapping of the phosphorylation sites in the mutant proteins will be presented elsewhere (SCHEIDTMANN et al., manuscript in preparation). DNA binding studies of the mutant proteins expressed in transformed rat cells have revealed novel patterns of DNA binding specificity (Fig. 3). With one exception, SV677, the mutant proteins bind to T-antigen binding sites in the genomic control region with much the same activity as T antigen from wild type-transformed rat cells (Fig. 3A). SV677 T-antigen binding activity is severely reduced on this template, though is is detectable. This result is puzzling because SV677 replicated and produced plaques with at least the frequency of wild-type SV40 (SCHNEIDER and FANNING 1988). Therefore, origin binding of the mutant proteins was also assayed using the isolated site I and site II templates (Fig 3A, right and 3B). SV677 T antigen was unable to bind to the site I template in this assay, but bound as much site II origin DNA as the wild-type protein, correlating with its ability to initiate SV40 DNA replication.

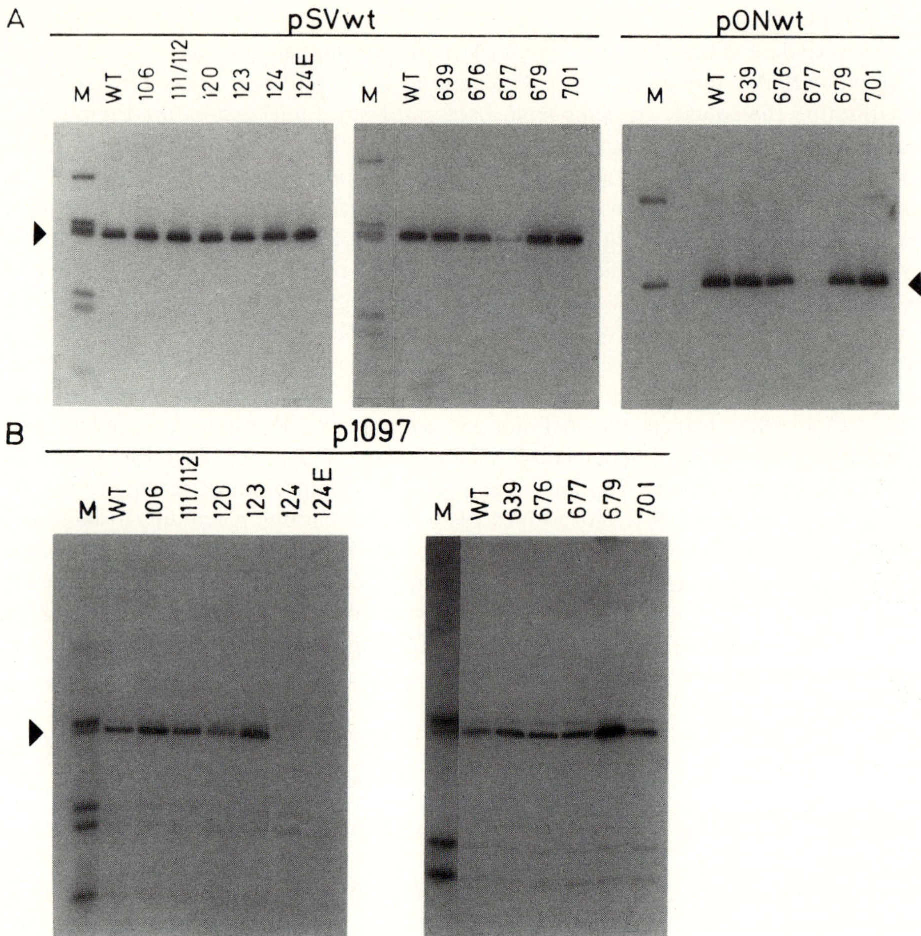

**Fig. 3A, B.** Origin DNA binding of T antigens carrying point mutations in the phosphorylated domains. Equal amounts of T antigen immunoprecipitated from cell lines transformed by each point mutant and wild-type (*WT*) SV40 were assayed for specific binding to intact origin DNA (*pSVwt*) and site I DNA (*pONwt*) **(A)** and to site II DNA (*p1097*) **(B)** as indicated. Fragments containing T-antigen binding sites are indicated by *arrows*. *M*, marker lane containing 5% of the input DNAs used in the binding reactions

Binding studies on the isolated site II template revealed three other mutant proteins with novel properties. SV124 and SV124E T antigens were unable to bind site II DNA in this assay (Fig. 3B), though they bound normally to intact origin DNA (Fig. 3A) and to isolated site I DNA (not shown). Both these T antigens were replication defective (SCHNEIDER and FANNING 1988). Finally, SV679 T antigen reproducibly bound more site II origin DNA than did an equivalent amount of wild-type T antigen (Fig. 3B). Consistent with this result, SV679 replicated about two- to three-fold better than wild-type SV40 (SCHNEIDER and FANNING 1988).

# 4 Discussion

We have used two sets of mutant SV40 T antigens to define a minimal origin DNA binding domain and to study its interactions with the phosphorylated domains at the N- and C-termini of the protein. The origin binding domain of the protein is localized within residues 131 to 259 (Fig. 2). Numerous point mutations that disrupt origin binding activity (PAUCHA et al. 1986), as well as two pseudorevertants of origin region mutants (MARGOLSKEE and NATHANS 1984) lie within this region. It is noteworthy that a putative zinc finger structure predicted for the papovavirus T antigens (BERG 1986) (SV40 T-antigen residues 302 to 320) is not included in this domain, although it is included in the smallest DNA binding peptide generated by tryptic cleavage of T antigen (SIMMONS 1986). The conserved $Cys_2$-$His_2$ motif may affect DNA binding in a more subtle fashion (LOEBER et al., this volume) or have other functions, such as protein-protein interactions (FRANKEL and PABO 1988).

Secondary structure predictions for the sequence 131 to 259 reveal a prominent alpha helix beginning at or slightly before residue 166 and extending to residue 178 (Fig. 4a). The alpha helix is the core DNA recognition motif in the helix-turn-helix DNA binding domains of a number of bacterial repressor proteins (reviewed by PABO and SAUER 1984). The predicted helix of T antigen includes several residues conserved in the corresponding helix of other papovavirus T antigens (Glu 166, Lys 167, Leu 171, Tyr 172, Lys 178) in addition to the pseudorevertants mapping at residues 157 and 166, suggesting that residues within the helix may interact directly with the GAGGC pentanucleotide binding signal in the major groove of the DNA. However, several point mutations that inactivate origin binding map at residues 147 to 153 or in the C-terminal half of this origin binding domain, implying that other specific amino acid residues are also necessary for either DNA recognition or structural stability (PAUCHA et al. 1986 and references therein).

The helix is predicted to carry several lysine residues on one side, whereas the other side of the helix is rich in uncharged and hydrophobic amino acids (Fig. 4b). We speculate that the lysine-rich face of the helix interacts with the pentanucleotide. The β-sheet and β-turn structures predicted for sequences flanking the alpha helix are also conserved among the papovavirus T antigens. They may interact with and possibly stabilize the helix-DNA interaction, together playing a role functionally analogous to that of the second helix in the bacterial repressor DNA binding domains.

These considerations may be visualized in a preliminary working model for the T-antigen origin binding motif, dubbed the dog-bone model (Fig. 4c). In this representation, the lysine-rich face of the helix would project out of the page and the hydrophobic face into the page, interacting with the β sheets, here depicted as anti-parallel. The N-terminal β-sheet structures have been omitted for simplicity. This structure bears some resemblance to a structure predicted for the zinc finger motif, although it carries none of the characteristic Cys and His residues (BERG 1988). The model depicted here does not account for the C-terminal half of the minimal origin binding domain. It is possible that this region is also involved in DNA recognition. Alternatively, as in bacterial repressor proteins, it could mediate T antigen-T antigen interactions required for high affinity binding (RYDER et al. 1985).

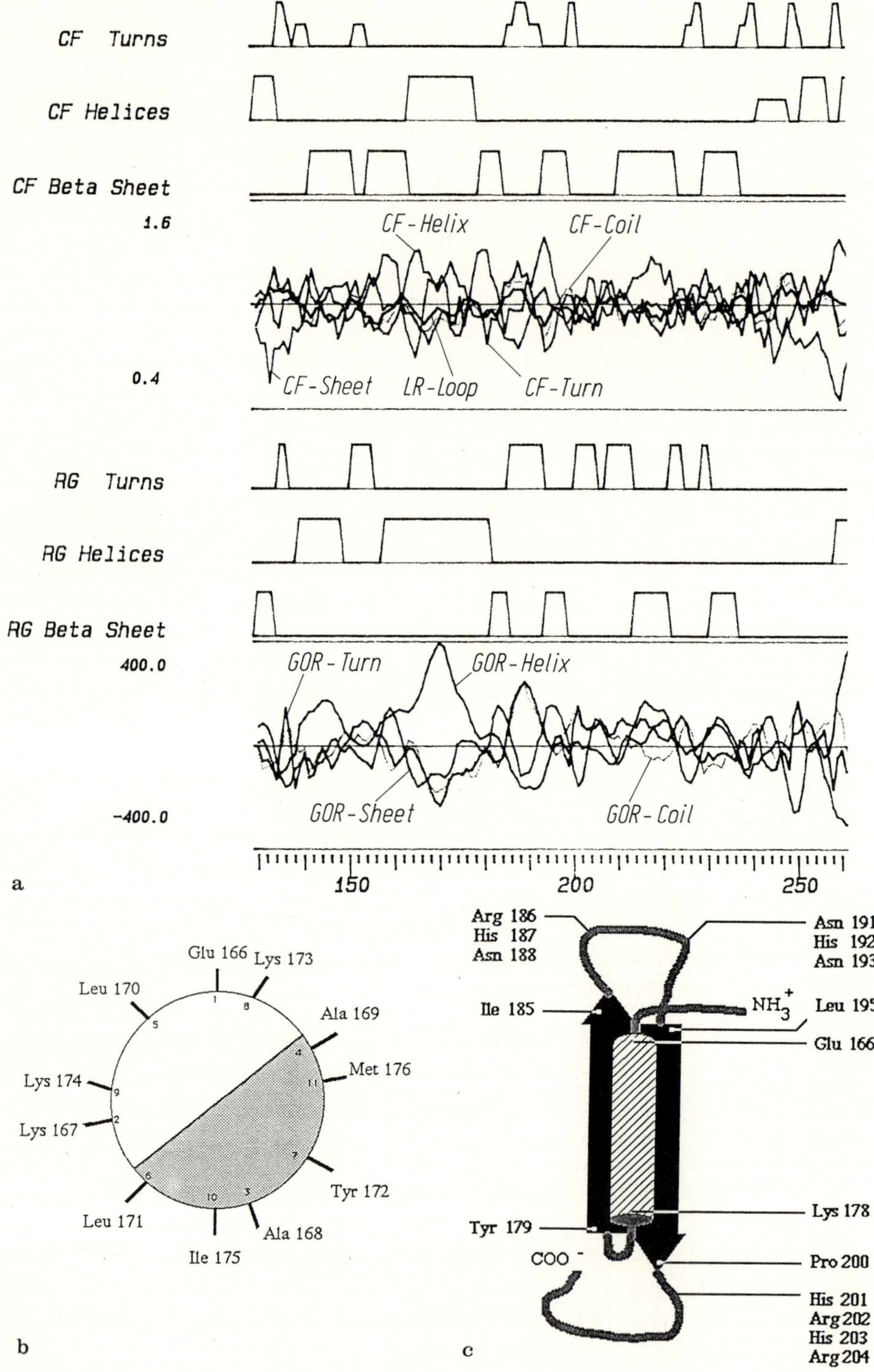
CF Turns
CF Helices
CF Beta Sheet
1.6
CF-Helix
CF-Coil
0.4
CF-Sheet
LR-Loop
CF-Turn
RG Turns
RG Helices
RG Beta Sheet
400.0
GOR-Turn
GOR-Helix
GOR-Sheet
GOR-Coil
-400.0
150
200
250
a
Glu 166
Lys 173
Leu 170
Ala 169
Lys 174
Met 176
Lys 167
Tyr 172
Leu 171
Ala 168
Ile 175
b
Arg 186
His 187
Asn 188
Asn 191
His 192
Asn 193
Ile 185
NH3+
Leu 195
Glu 166
Tyr 179
Lys 178
COO-
Pro 200
His 201
Arg 202
His 203
Arg 204
c

Although efforts are underway to test the essential features of the dog-bone model, an understanding of T-antigen-DNA recognition will ultimately require detailed three-dimensional structural data.

The origin binding domain of each monomeric mass of T antigen interacts with a pentanucleotide GAGGC present in multiple copies in both of the major binding sites (RYDER et al. 1986) and references therein). Site I carries at least two copies of the pentanucleotide arranged in tandem and separated by a 7-bp spacer specifically required for binding of T antigen (RYDER et al. 1985), wherease site II harbors four pentanucleotides arranged in a palindrome (reviewed by RIGBY and LANE 1983). Protein-protein interactions between bound T-antigen subunits are probably required for high affinity binding. We report here four mutations in the phosphorylated domains of T antigen that differentially affect binding of the protein to one or the other type of site. Given the localization of the minimal origin binding domain of the protein within residues 131 to 259, these results demonstrate that in the folded protein the phosphorylated regions of the protein communicate with the origin binding domain. These interactions apparently modulate the protein-protein contacts between T antigen monomers, distinguishing site II binding from site I binding.

The communication between the origin binding domain and phosphorylated domains is probably controlled by the phosphorylation state of the protein. If this conjecture is correct, one would predict that the phosphorylation site mutants which show altered binding to site I or II when expressed in mammalian cells should lose their altered DNA binding specificity when expressed in an unphosphorylated form, for example in bacteria. Preliminary results show no major differences in DNA binding between mutant and wild-type T antigen when expressed in *E. coli* (A. ARTHUR and I. MOAREFI, personal communication). These results and others (SIMMONS et al. 1986; MOHR et al. 1987; KLAUSING et al. 1988) suggest that differential phosphorylation of T antigen could determine functional subclasses of T-antigen molecules able to bind to site I and hence repress early transcription, or to site II in order to initiate new rounds of SV40 DNA replication. Phosphorylation of T antigen would thus represent a sensitive and dynamic mechanism for regulating the course of a lytic viral infection.

*Acknowledgements.* We thank Silke Dehde and Birgit Posch for invaluable technical assistance, Ronald Mertz and Dorit Weigand for oligonucleotide synthesis, all of our colleagues who made vectors, bacterial strains, cells, and monoclonal antibodies available to us, and Maria Ihmsen for excellent secretarial assistance. The financial support of the Bundesministerium für Forschung und Technologie (Genzentrum), Deutsche Forschungsgemeinschaft and Fonds der Chemischen Industrie is gratefully acknowledged.

*Note added in Proof* A peptide containing the 129 amino acids from T antigen residues 131–259 is relatively stable and binds specifically to SV40 origin DNA (Hoess and Arthur, unpublished).

---

◀ **Fig. 4a–c.** Speculative model for the structure of the T-antigen DNA binding motif. **a** Secondary structures predicted for T-antigen residues 130 to 260 using the program PROT/PLOT (WOLF et al. 1987). **b** "Helical wheel" representation of T-antigen residues 166 to 176, predicted to form a prominent alpha helix. The hydrophobic face of the helix is *shaded*. **c** Schematic diagram of the predicted recognition helix and C-terminal β-pleated sheets and turns

# References


Arthur A, Höss A, Fanning R (1988) Expression of SV40 T antigen in *Escherichia coli*: localization of T-antigen origin DNA-binding domain to within 129 amino acids. J. Virol 62: 1299–2006

Ball RK, Siegl B, Quellhorst G, Brandner G, Braun DG (1984) Monoclonal antibodies against simian virus 40 large T antigen: epitope mapping, papovavirus cross-reaction and cell surface staining. EMBO J 3: 1485–1491

Berg J (1986) Potential metal-binding domains in nucleic acid binding proteins. Science 232: 485–486

Berg J (1988) Proposed structure for the zinc-binding domains from transcription factor IIIA and related proteins. Proc Natl Acad Sci USA 85: 99–102

Clark R, Peden K, Pipas J, Nathans D, Tjian R (1983) Biochemical activities of T-antigen proteins encoded by simian virus 40 A gene deletion mutants. Mol Cell Biol 3: 220–228

Cole CN, Tornow J, Clark R, Tjian R (1986) Properties of the simian virus 40 (SV40) large T-antigens encoded by SV40 mutants with deletions in gene A. J Virol 57: 539–546

DePamphilis ML, Bradley MK (1986) Replication of SV40 and polyoma virus chromosomes. In: Salzman NP (ed) The papovaviridae, vol 1. Plenum, New York, pp 99–246

DiMaio D, Nathans D (1982) Regulatory mutants of simian virus 40. Effect of mutations at a T antigen binding site on DNA replication and expression of viral genes. J Mol Biol 156: 531–548

Dorn A, Brauer D, Otto B, Fanning E, Knippers R (1982) Subclasses of simian virus 40 large tumor antigen. Partial purification and DNA binding properties of two subclasses of tumor antigen from productively infected cells. Eur J Biochem 128: 53–62

Fanning E, Nowak B, Burger C (1981) Detection and characterization of multiple forms of simian virus 40 large T antigen. J Virol 37: 92–102

Fanning E, Westphal K-H, Brauer D, Görlin D (1982) Subclasses of simian virus 40 large T antigen: differential binding of two subclasses of T antigen from productively infected cells to viral and cellular DNA. EMBO J 1: 1023–1028

Frankel AD, Pabo CO (1988) Fingering too many proteins. Cell 53: 675

Gluzman Y (1981) SV40-transformed simian cells support the replication of early SV40 mutants. Cell 23: 175–182

Gurney R, Tamowsky S, Deppert W (1986) Antigenic binding sites of monoclonal antibodies specific for SV40 large T. J Virol 57: 1168–1172

Harlow E, Crawford LV, Pimand DC, Williamson NM (1981) Monoclonal antibodies specific for simian virus 40 tumor antigens. J Virol 39: 861–869

Hinzpeter M, Fanning E, Deppert W (1986) A new sensitive target-bound DNA binding assay for SV40 large T antigen. Virology 148: 159–167

Huber B, Vakalopoulou E, Burger C, Fanning R (1985) Identification and biochemical analysis of DNA replication-defective large T antigen from SV40-transformed cells. Virology 146: 188–202

Kalderon D, Richardson WD, Markham AF, Smith AE (1984) Sequence requirements for nuclear location of simian virus 40 large T antigen. Nature 311: 33–38

Klausing K, Scheidtmann K-H, Baumann EA, Knippers R (1988) Effects of in vitro dephosphorylation on DNA-binding and DNA helicase activities of SV40 large tumor antigen. J Virol 62: 1258–1265

Margolskee R, Nathans D (1984) Simian virus 40 mutant T antigens with relaxed specificity for the nucleotide sequence at the viral origin of DNA-replication. J Virol 49: 386–393

Mohr IJ, Stillman B, Gluzman Y (1987) Regulation of SV40 DNA replication by phosphorylation of T antigen. EMBO J 6: 153–160

Pabo CO, Sauer RT (1984) Protein-DNA recognition. Annu Rev Biochem 53: 293–321

Paucha E, Kalderon D, Harvey R, Smith AE (1986) Simian virus 40 origin DNA-binding domain on large T antigen. J Virol 57: 50–64

Rigby P, Lane D (1983) The structure and function of SV40 large T antigen. Adv Viral Oncol 3: 31–57

Runzler R, Thompson S, Fanning E (1987) Oligomerization and origin DNA-binding activity of simian virus 40 large T antigen. J Virol 61: 2076–2083

Ryder K, Vakalopoulou E, Mertz R, Mastrangelo I, Hough P, Tegtmeyer P, Fanning E (1985) Seventeen base pairs of region I encode a novel tripartite binding signal for SV40 T antigen. Cell 42: 539–548

Ryder K, Silver S, DeLucia A, Fanning E, Tegtmeyer P (1986) An altered DNA conformation in origin region I is a determinant for the binding of SV40 large T antigen. Cell 44: 719–725

Scheidtmann K-H, Echle B, Walter G (1982) Simian virus 40 large T antigen is phosphorylated at multiple sites clustered in two separate regions. J Virol 44: 116–133

Scheidtmann K-H, Hardung M, Echle B, Walter G (1984) DNA-binding activity of SV40 large T antigen correlates with a distinct phosphorylation state. J Virol 50: 1–12

Schneider J, Fanning E (1988) Mutations in the phosphorylation sites of SV40 T antigen alter its origin DNA binding specificity for sites I or II and affect SV40 DNA replication activity. J Virol 62: 1598–1605

Simmons DT (1986) DNA-binding region of the simian virus 40 tumor antigen. J Virol 57: 776–785

Simmons D, Chou W, Rodgers K (1986) Phosphorylation downregulates the DNA binding activity of simian virus 40 T antigen. J Virol 60: 888–894

Stahl H, Knippers R (1987) The SV40 large tumor antigen. Biochim Biophys Acta 910: 1–10

Strauss M, Argani P, Mohr I, Gluzman Y (1987) Studies on the origin-specific DNA-binding domain of SV40 large T antigen. J Virol 61: 3326–3330

Vogt B, Vakalopoulou E, Fanning E (1986) Allosteric control of SV40 T antigen binding to viral origin DNA. J Virol 58: 765–772

Wolf H, Modrow S, Motz M, Jameson B, Hermann G, Förtsch B (1988) An integrated family of amino acid sequence analysis programs. CABIOS 4: 187–191

# A Genetic Analysis of the Zinc Finger
# of SV40 Large T Antigen

G. Loeber, R. Parsons, and P. Tegtmeyer

## 1 Introduction

Zinc finger structures are common to a wide variety of nucleic acid binding proteins of both prokaryotic and eukaryotic origin. Examples of these proteins are transcription factors for RNA polymerase II (Kadonaga et al. 1987) and RNA polymerase III (Miller et al. 1985), steroid hormone receptors (Green and Chambon 1987), developmentally regulated genes (Chowdhury et al. 1987; Rosenberg et al. 1986), and single-stranded DNA binding proteins such as the gene 32 protein of bacteriophage T4 (Giedroc et al. 1986). The zinc finger proteins have in common the sequence motif Cys-$X_{2-5}$-Cys-$X_{2-12}$-His/Cys-$X_{2-5}$-His/Cys in which the symbol X may be a variety of amino acids. The histidines and cysteines are thought to interact specifically with a zinc ion to form a core structure with two minor loops and a major loop, the "finger". Klug and Rhodes (1987) have proposed that the finger comes into close contact with the major groove of DNA.

Simian virus 40 (SV40) large T antigen has a single zinc finger motif between amino acids 302 and 320 of the 708-amino acid protein. T antigen has a number of related functions during the lytic cycle of the virus in permissive host cells. These functions include binding to specific sequences in the origin of DNA replication (Tjian 1978; DeLucia et al. 1983), opening of the dsDNA in the origin of DNA replication (Dean et al. 1987; Wold et al. 1987), a helicase activity (Stahl et al. 1986), association with cellular proteins such as DNA polymerase-alpha (Smale and Tjian 1986) and p53 (Lane and Crawford 1979), repression of its own synthesis (Tegtmeyer et al. 1975), stimulation of the expression of late viral genes (Brady et al. 1984), induction of cellular DNA synthesis (Chou and Martin 1975), and other functions. In non-permissive cells, such as rodent cells, T antigen induces and maintains cellular transformation (Tooze 1980).

Many T-antigen functions can be correlated with specific domains on the protein. For example, a region which is sufficient for binding to specific recognition sequences in the origin of DNA replication maps between the amino acids 131 and 247 (Paucha et al. 1986; Strauss et al. 1987). The zinc finger sequence is located outside this domain, but still inside the region which is required for optimal binding to origin DNA (Simmons 1986). The zinc finger domain also overlaps the regions required for the helicase activity (Stahl et al. 1986) and those responsible for

Department of Microbiology, State University of New York at Stony Brook, Stony Brook, NY 11794, USA

Current Topics in Microbiology and Immunology, Vol. 144
© Springer-Verlag Berlin · Heidelberg 1989

DNA polymerase-alpha (Smale and Tjian 1986) and p53 binding (Gannon and Lane 1987). Other possible activities of the finger sequence include the specific opening of the SV40 origin of replication, the oligomerization of T antigen, or perhaps other functions not yet identified.

Because the zinc finger sequence is highly conserved among all polyomaviruses, this structural element is likely to have an important function. We have undertaken a systematic mutational analysis of the zinc finger region by inducing single amino acid exchanges using oligonucleotide-directed mutagenesis. Analysis of mutant function in lytic and transforming infections has allowed us to identify distinct domains in the zinc finger region.

# 2 Construction of the Mutants

The zinc finger sequence of SV40 T antigen, between amino acids 302 and 320, corresponds to bases 3858 through 3914 in the viral genome (Tooze 1980). Base substitution mutants in this region of the large T-antigen gene were constructed by oligonucleotide-directed mutagenesis using a uracil-containing, single-stranded DNA as template (Kunkel 1985). This strategy exploits a plasmid vector (Bluescript SK +; Stratagene, La Jolla) with a single-stranded phage (M 13) origin of DNA replication as well as a normal plasmid origin of DNA replication (Col E1). Thus, DNA can be isolated either as double-stranded DNA for functional studies or as single-stranded DNA with positive polarity for oligonucleotide-directed mutagenesis. Since this vector system does not have the stringent size limitations of other vectors derived from single-stranded filamentous phages, we could use the entire SV40 genome as a target for mutagenesis. Using oligomers with 1- to 2-bp mismatches, we found that 50%–90% of the screened clones carried the desired mutation.

## 2.1 Choice of Amino Acid Substitutions

To compare the importance of various amino acid positions and to identify possible functional domains in the zinc finger region, we chose to make amino acid exchanges of similar severity in each position. As a guide for the choice of appropriate substitutions, we used the amino acid score matrix of Staden (1982), derived from the data of Dayhoff (1969). This matrix is based on a comparison of the sequences of 72 families of functionally related proteins to establish a hierarchy of the functional interchangeability of various amino acids. Except for some special situations, the mutations introduced into the protein were of intermediate severity. Figure 1 shows the positions of the changes in the amino acid sequence of the finger.

## 2.2 Viability of the Zinc Finger Mutants

To test whether the mutants retained their ability to produce viable virus, the mutant viral DNAs were cleaved out of the plasmid with the enzyme *Bam*H1, and the total

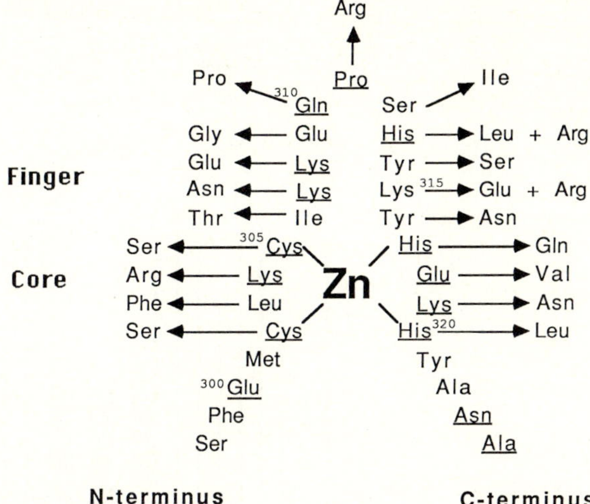


**Fig. 1.** Amino acid alterations in the zinc finger sequence of SV40 (strain 776) large T antigen. Amino acids are arranged in the zinc binding structure that has been proposed for other fingers (KLUG and RHODES 1987). The coordinate binding of zinc by cysteines and histidines forms two minor loops in a "core" structure and a major loop in the "finger" structure. Amino acid positions are *numbered*, and the mutations that we have used are shown. The *underlined amino acids* are highly conserved among polyomaviruses

DNA was religated in a large volume using T4 DNA ligase. Approximately 5 % of the DNA was converted to closed, circular SV40 DNA capable of efficient replication in permissive cells. The mixture of religated DNAs was transfected into confluent CV1 cells in 3-cm petri dishes using the DEAE transfection method (MCCUTCHAN and PAGANO 1968), and an agar overlay was added. The plates were incubated at 37 °C, and the plaques induced by the mutants were visualized by adding a neutral red overlay after 1 week. Under the conditions used approximately 200 plaques/ng of total religated SV40 wild-type (WT) DNA were obtained. Figure 2 shows a summary of the results of the plaque assays. The mutants can be divided into three groups depending on their ability to form plaques on CV-1 cells.

The first group is by far the largest. Mutants of this group do not have any effect on viral replication; their plaque number and plaque size are equivalent to those of the WT virus. This group includes the mutants located in both minor loops of the zinc finger and the N-terminal half of the major loop. A remarkable finding is that a change of the proline at position 311 to an arginine at the "tip" of the finger has no effect on the viability, although the proline is highly conserved among primate papovaviruses, and a computer-generated structural model for the T antigen predicts a turn in the protein at this site (DEPAMPHILIS and BRADLEY 1986). These findings suggest that these regions have a spacer function. However, mutations more severe than the ones tested here could have an effect on viability.

The second group consists of the mutants 302-S, 305-S, 313-L, 317-Q, and 320-L, which failed to induce any visible cytopathic effect even after prolonged incubation at 37 °C for 8 weeks. Except for the mutation at amino acid 313, all of these mutations change the putative zinc-binding amino acids of the core. To investigate

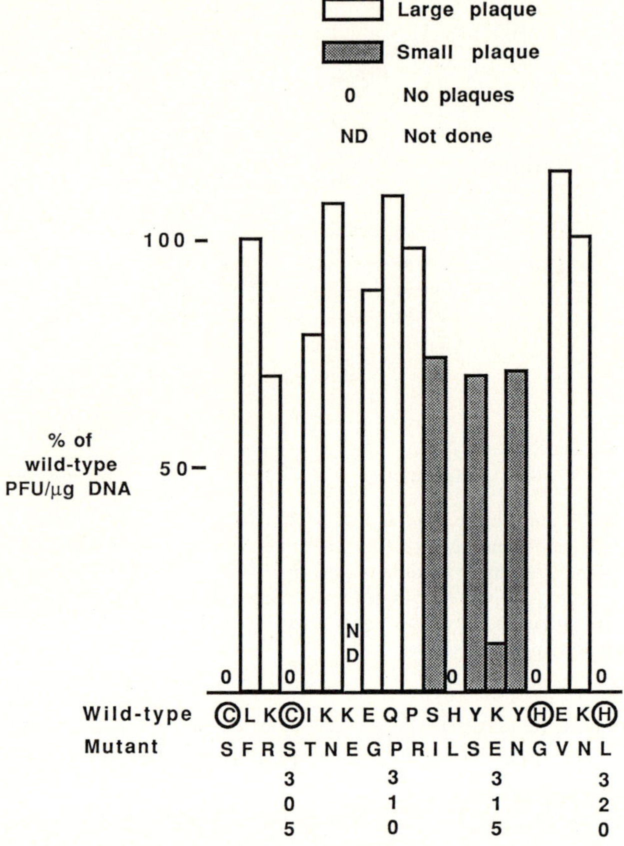


**Fig. 2.** Viral replication of the zinc finger mutants. Wild-type and mutant amino acids are shown in the single letter code; the corresponding position in the protein is indicated *below* the sequences. The *height* of the column represents the ability of a mutant to induce plaques on permissive CV-1 cells compared with WT SV40 DNA under identical conditions. The *bar codes* are defined in the figure

whether the failure of mutant 313-L to replicate can be compensated by the introduction of a more similar amino acid at that position, we exchanged an arginine for the histidine at this position. Although this maintains the basic charge at amino acid 313, the new mutant, 313-R, is also completely unable to replicate. Thus, the lack of viability of the mutants at position 313 is not a simple charge effect.

The third group consists of the mutants between positions 312 and 316. These mutants replicate, but the number of plaques induced is slightly reduced. Furthermore, the plaques appear later after transfection than WT plaques and grow to a smaller size. As estimated from plaque size, the viability of mutant 316-N is reduced only slightly. Mutations at positions 312 and 314 (312-I and 314-S) cause a moderate decrease in replication efficiency; plaques induced by these mutants have approximately 50% of the diameter of WT plaques. Mutant 315-E exhibited a more drastic reduction in viability. Plaques induced by 315-E first appeared about

4 weeks after transfection and did not grow to more than pinpoint size after 8 weeks. Since this mutant changes a lysine to a glutamic acid, another mutant was constructed maintaining the positive charge at this position by a change to arginine. It replicated at WT levels. Thus, the reduction in the growth of 315-E is probably a charge effect.

The clustering of mutants with reduced ability to grow at the C-terminus of the major loop of the zinc finger suggests that this region might be an active site for a specific function of the finger. At present, we are investigating the biochemical defects of the mutants in this active site and in the putative zinc-binding sequences. We are also constructing additional mutants in this region to investigate possible spacer functions within the finger.

## 3 DNA Replication In Vivo

The results of the plaque assay have identified mutants with impaired or no ability to grow in CV-1 cells. To localize these defects, which could affect either early functions directly involved in DNA replication or other functions in the growth cycle, we transfected mutant DNA into CV-1 cells. Episomal DNA was extracted after 48 h, and the Mbol assay was used to discriminate unreplicated input DNA from replicated DNA (DeLucia et al 1986). No replicated bands could be detected after transfection with DNA from mutants which affect the "core" zinc finger sequence (302-S, 305-S, 317-Q, 320-L). A mutation of the histidine at position 313 of the finger also blocked DNA replication completely (data not shown). Replication of mutant 315-E DNA was markedly reduced compared with WT DNA (data not shown). A quantitation of the replicated bands relative to input DNA and a comparison with WT replication showed a more than tenfold decrease in replication ability. These data correlate with the reduced growth rate observed in the plaque assays. Thus, the defect in viral growth can be localized to a function prior to or during DNA replication.

## 4 Transformation of Nonpermissive Rodent Cells

To test the ability of the mutants to transform nonpermissive rodent cells, plasmid DNA was transfected into C57/Black mouse embryo fibroblast (MEF) cells using the $CaPO_4$ technique (Graham and van der Eb 1973). Between 1 and 5 µg of plasmid DNA were added to the cells, which had been seeded into 75-cm$^2$ flasks the previous day at a density of $10^5$ cells/flask. The precipitate was kept on the cells overnight, and the cells were kept at 37 °C and fed twice a week. After 3–5 weeks, the cells were fixed and stained, and the foci were scored according to number and size. Figure 3 summarizes the results of these experiments, showing that the mutants can be divided into two groups depending on their ability to transform.

The first group consists of all the mutants which are able to replicate, including those which replicate only at a reduced rate. Mutants from this group induce

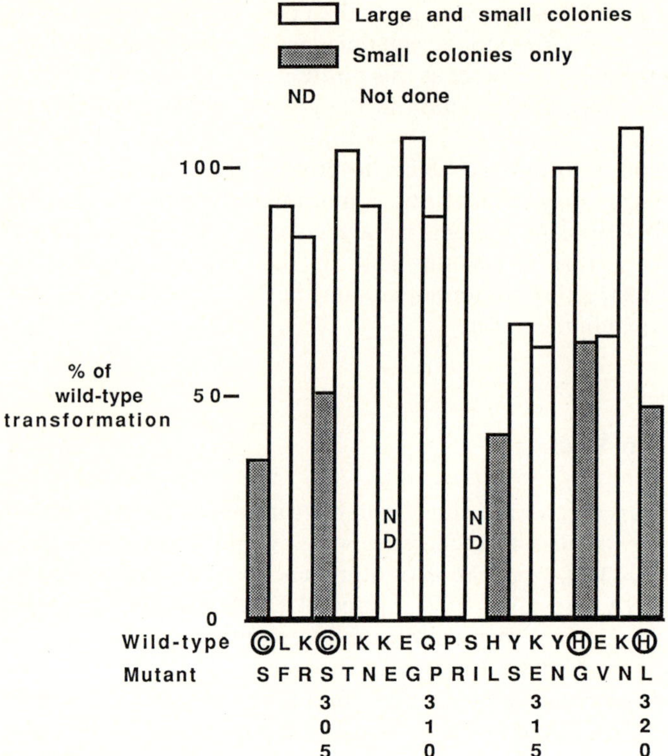



**Fig. 3.** Transformation of primary mouse embryo fibroblast (MEF) cells. Wild-type and mutant amino acids are shown in the single letter code; the corresponding position in the protein is indicated beneath. The *height* of the column represents the number of foci induced by mutant DNA compared with WT DNA. The *bar codes* are defined in the figure

transformed foci in the MEF cultures with WT efficiency. The foci also have a WT morphology. We consider the small differences in the number of foci among the mutants to be insignificant, since even the transformation rate of WT DNA differs somewhat in duplicate samples. Foci induced by these mutants were cloned and expanded to stable transformed cell lines. The transformed cells were positive for T-antigen expression as seen in immunofluorescent antibody staining. The fact that the highly replication-deficient mutant 315-E is able to induce transformed foci at WT level suggests that the intrinsic replication function of the major loop of the zinc finger is not required for the neoplastic transformation.

The second group consists of those mutants which fail to replicate in permissive cells, the core zinc finger mutants 302-S, 305-S, 317-Q, and 320-L, and the mutant 313-L located in the major loop of the finger. These mutants are able to transform MEF cells to a certain degree, but the number of foci is slightly lower. Furthermore, the foci induced by these mutants are significantly smaller and appear later after transfection. As seen in immunofluorescence antibody staining, most of these foci were positive for T-antigen expression. Mutant 313-R, which maintained the positive charge at that position in the finger, showed the same deficiency in inducing foci as 313-L. Attempts to clone single foci induced by these mutants and expand them to stable transformed cell lines have failed to date.

Since the replication defect seen in the small plaque mutant 315-E did not affect the transformation ability at all, it is highly unlikely that this replication function of the protein is responsible for the transformation defect. It seems more likely that a major change in structure, related to zinc binding, affects the larger regions of T antigen required for stable transformation and immortalization. It is also possible that the zinc finger structure is required for the stability of the protein and that an alteration causes a higher susceptibility to proteolytic digestion and thus reduces the level of available active protein in the transformed cell.

The fact that mutations of the histidine at position 313 near the "tip" of the proposed finger cause the same effects as mutations of cysteines and histidines in the "core" zinc binding region of the finger is remarkable. The phenotype of this mutant and the fact that a histidine is essential suggest that this position is involved in the same structural framework as the other cysteines and histidines. It is possible that this histidine might also interact with zinc in a coordinate manner. Coordination of a zinc ion with five ligands in a metastable intermediate has been reported in the active center of carboxypeptidase A, where zinc is held in coordinate bonds. CHRISTIANSON et al. (1987) have proposed that the ligands of carboxypeptidase A may switch during the catalytic reaction. Studies of the biochemical properties of this putative zinc-binding site in T antigen are in progress.

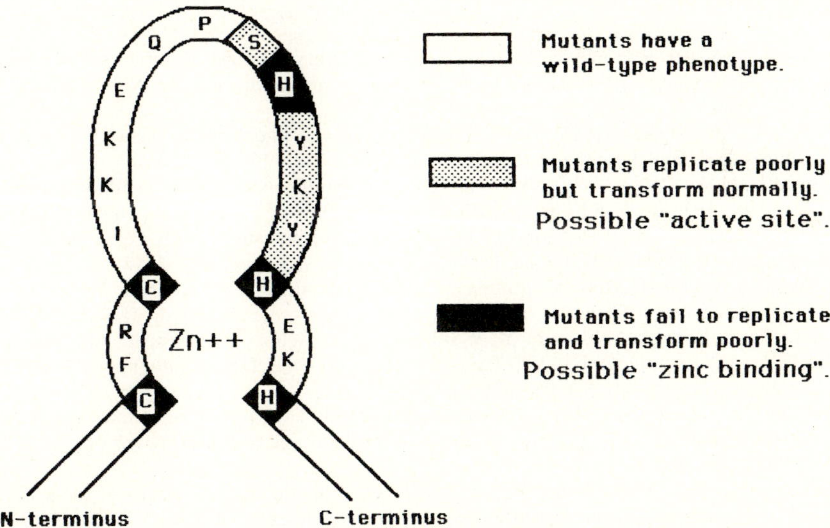


Fig. 4. Zinc finger of SV40 large T antigen. Amino acids 302 through 320 are given in the single letter code. The cysteines at positions 302 and 305 and the histidines at positions 317 and 320 are thought to form coordinate bonds with a zinc ion. The *bar codes* are defined in the figure

# 5 Conclusions

The mutational analysis of the zinc finger region of SV40 large T antigen has shown that it can be divided into three domains characterized by distinct roles in DNA replication and in transformation. Figure 4 shows a model for the zinc finger. Domain 1 consists of the amino acids forming the two minor loops and the N-terminal part of the major loop. Mutations in this area do not affect either replication or transformation. These findings suggest a spacer function. Domain 2 is shown in grey shading in the model and spans the C-terminus of the major loop. Mutations here have no effect on transformation but reduce the viability of the virus and DNA replication in vivo to different degrees. These findings suggest a defect in an active center important for a specific function involved in DNA replication. Domain 3 is represented by the "core" zinc finger amino acids and a single histidine in the major loop. This domain is shown in black in Fig. 4. Mutations in this domain abolish the viability of the virus completely and have a distinctive effect on transformation functions of T antigen. Thus, the five cysteines and histidines may play an important role in determining the overall structure of the protein.

*Acknowledgements.* This investigation was supported by PHS grants CA-18808, CA-38146, and 5-T32 CA-09176 awarded by the National Cancer Institute, DHHS, and by a postdoctoral fellowship Lo 375/1-1 to G.L. from the Deutsche Forschungs-gemeinschaft.

# References

Brady J, Bolen JB, Radonovich MF, Salzman N, Khoury G (1984) Stimulation of simian virus 40 late gene expression by simian virus 40 large tumor antigen. Proc Natl Acad Sci USA 81: 2040–2044

Chou JY, Martin RG (1975) DNA infectivity and the induction of host DNA synthesis with temperature-sensitive mutants of simian virus 40. J Virol 15: 145–150

Chowdhury K, Deutsch U, Gruss P (1987) A multigene family encoding several "finger" structures is present and differentially active in mammalian genomes. Cell 48: 771–778

Christianson DW, David PR, Lipscomb WN (1987) Mechanism of carboxypeptidase A: hydration of a ketonic substrate analogue. Proc Natl Acad Sci USA 84: 1512–1515

Dayhoff MO (1969) Atlas of protein sequence and structure. National Biomedical Research Foundation, Silver Springs

Dean FB, Bullock P, Murakami Y, Wobbe R, Weissbach L, Hurwitz J (1987) Simian virus 40 (SV40) DNA replication: SV40 large T antigen unwinds DNA containing the SV40 origin of replication. Proc Natl Acad Sci USA 84: 16–20

DeLucia AL, Lewton BA, Tjian R, Tegtmeyer P (1983) Topography of simian virus 40 A protein-DNA complexes: arrangement of pentanucleotide interaction sites at the origin of replication. J Virol 46: 143–150

DeLucia AL, Deb S, Partin K, Tegtmeyer P (1986) Functional interactions of the simian virus 40 core origin of replication with flanking regulatory sequences. J Virol 57: 138–144

DePamphilis ML, Bradley MK (1986) Replication of SV40 and polyoma virus chromosomes. In: Salzman N (ed) The papovaviridae. Plenum, New York, pp. 100–246

Gannon, JV, Lane DP (1987) P53 and DNA polymerase A compete for binding to SV40 T antigen. Nature 329: 456–458

Giedroc DP, Keating KP, Williams KR, Konigsberg WH, Coleman JE (1986) Gene 32 protein, the single stranded DNA binding protein from bacteriophage T4, is a zinc metalloprotein. Proc Natl Acad Sci USA 83: 8452–8456

Graham FL, Van der Eb AJ (1973) A new technique for the assay of infectivity of human adenovirus 5 DNA. Virology 52: 456–467

Green S, Chambon P (1987) Oestradiol induction of a glucocorticoid responsive gene by a chimeric receptor. Nature 325: 75–78

Kadonaga JT, Carner KR, Masiarz FR, Tjian R (1987) Isolation of a cDNA encoding transcription factor Sp1 and functional analysis of the DNA binding domain. Cell 51: 1079–1090

Klug A, Rhodes D (1987) "Zinc fingers": a novel protein motif for nucleic acid recognition. Trends Biochem Sci 12: 464–469

Kunkel TA (1985) Rapid and efficient site-specific mutagensis without phenotypic selection. Proc Natl Acad Sci USA 82: 488–492

Lane DP, Crawford LV (1979) T antigen is bound to a host protein in SV40-transformed cells. Nature 278: 261–263

McCutchan JH, Pagano JS (1968) Enhancement of the infectivity of simian virus 40 desoxy-ribonucleic acid by diethylaminoethyl-dextran. J Natl Cancer Inst 41: 351–357

Miller J, McLachlan AD, Klug A (1985) Repetitive zinc binding domains in the protein transcription factor IIIA from *Xenopus* oocytes. EMBO J 4: 1609–1614

Paucha E, Kalderon D, Harvey RW, Smith AE (1986) Simian virus 40 origin DNA-binding domain on large T-antigen. J Virol 57: 50–64

Rosenberg UB, Schroeder C, Preiss A, Kienlin A, Cote S, Riede I, Jaeckle H (1986) Structural homology of the product of the *Drosophila* Krueppel gene with *Xenopus* transcription factor IIIA. Nature 319: 336–339

Simmons DT (1986) DNA-binding region of the simian virus 40 tumor antigen. J Virol 57: 776–785

Smale ST, Tjian R (1986) T-antigen-DNA polymerase A complex implicated in simian virus 40 DNA replication. Mol Cell Biol 6: 4077–4087

Staden R (1982) An interactive graphics program for comparing and aligning nucleic acid and amino acid sequences. Nucleic Acids Res 10: 2951–2961

Stahl H, Droege P, Knippers R (1986) DNA helicase activity of SV40 large tumor antigen. EMBO J 5: 1939–1944

Strauss M, Argani P, Mohr IJ, Gluzman Y (1987) Studies on the origin-specific DNA-binding domain of simian virus 40 large T antigen. J Virol 61: 3326–3330

Tegtmeyer P, Schwartz M, Collins JK, Rundell K (1975) Regulation of tumor antigen synthesis by simian virus 40 A gene. J Virol 16: 168–178

Tjian R (1978) The binding site on SV40 DNA for a T antigen-related protein. Cell 13: 165–179

Tooze J (ed) (1980) DNA tumor viruses: the molecular biology of tumor viruses, 2nd edn. Cold Spring Harbor Laboratory, Cold Spring Harbor

Wold MS, Li JJ, Kelly TJ (1987) Initiation of simian virus 40 DNA replication in vitro: large-tumor-antigen and origin-dependent unwinding of the template. Proc Natl Acad Sci USA 84: 3643–3647

# Effects of Amino Acid Phosphorylation
on the DNA Binding Properties of Large T Antigen

K. KLAUSING and R. KNIPPERS

SV40 large T antigen can be post-translationally modified by N-terminal acetylation (PAUCHA et al. 1978), acylation (KLOCKMANN and DEPPERT 1983), poly (ADP-ribosyl)ation (GOLDMAN et al. 1981), and phosphorylation (SCHEIDTMANN et al. 1982). Phosphorylation of T antigen has recently attracted considerable attention as it appears to be involved in the regulation of some of the activities that T antigen performs in lytically infected cells.

A total of ten amino acids, eight serine and two threonine residues, can be phosphorylated. They are distributed in two clusters, an N-terminal and a C-terminal cluster, each one including four phosphoserines and one phosphothreonine (Fig. 1) (SCHEIDTMANN et al. 1982). Studies with nuclear transport mutants of T antigen suggest that the two threonines, Thr 124 and Thr 701, are probably phosphorylated by cytoplasmic protein kinases whereas the serines may be phosphorylated later, some time after the entry of T antigen into the nucleus (SCHEIDTMANN et al., 1984a; FISCHER-FANTUZZI et al. 1986).

To assess the role that phosphorylation may play in regulating the biochemical activities of T antigen several research groups have used alkaline calf intestine phosphatase to dephosphorylate T antigen in vitro. This treatment leads to a complete dephosphorylation of all phosphoserines (with the exception of Ser 639 which is only partially dephosphorylated) but not of Thr 124 and of Thr 701 (GRÄSSER et al. 1987; KLAUSING et al. 1988) (Fig. 1).

T antigen treated with alkaline phosphatase is several times more active than untreated T antigen as an initiator of SV40 DNA replication in vitro (MOHR et al. 1987; GRÄSSER et al. 1987). This increase in replication activity is not due to a stimulation of the DNA-unwinding activity of T antigen since dephosphorylated T antigen unwinds the double helical section of a partially double-stranded DNA substrate with the same efficiency as untreated T antigen (KLAUSING et al. 1988). It rather appears that T antigen treated with alkaline phosphatase interacts at a several-fold higher affinity than untreated T antigen with the specific binding site in the viral origin of replication (MOHR et al. 1987; KLAUSING et al. 1988) (Fig. 2). The same treatment does not affect the general, unspecific DNA binding property of T antigen. Also, the binding of dephosphorylated T antigen to binding site I, which is not part of the minimal origin of replication, is only slightly stimulated (Fig. 2). Thus, T antigen dephosphorylated by alkaline phosphatase is specifically directed to the origin of replication.

Fakultät für Biologie, Universität Konstanz, 7750 Konstanz, FRG

Current Topics in Microbiology and Immunology, Vol. 144
© Springer-Verlag Berlin · Heidelberg 1989

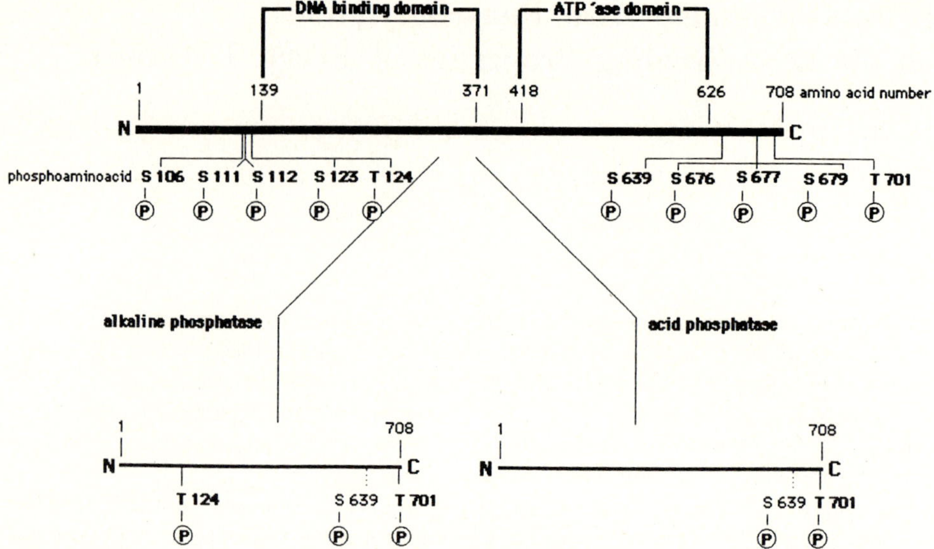



**Fig. 1.** Phosphoamino acids in SV40 large T antigen. *Upper part*, the polypeptide of 708 amino acids is represented as a *horizontal line* with the N-terminus on the *left (N)* and the C-terminus on the *right (C)*. The DNA binding domain and the ATPase domain have been defined in the Introduction to this part (see also: FANNING et al., this volume). The location of the phosphoamino acids has been determined by SCHEIDTMANN et al. (1982). *Lower part*, alkaline phosphatase dephosphorylates the phosphoserines (with the exception of Ser 639 which is only partially dephosphorylated). Acid phosphatase additionally removes the phosphate group of Thr 124 (KLAUSING et al. 1988)

Treatment with acid phosphatase removes not only the phosphate groups from all serines but also from amino acid Thr 124, leaving Thr 701 (and to some extent Ser 639) as the only remaining phosphoamino acid (Fig 1) (KLAUSING et al. 1988). Surprisingly, T antigen desphosphorylated by acid phosphatase has the DNA binding properties of untreated T antigen, i.e., a preference for the binding site II in the viral origin of replication cannot be detected (Fig. 2).

It appears therefore that phosphorylation has subtle effects on the DNA binding properties of T antigen. The phosphorylation of Thr 124 appears to be important for an effective interaction of T antigen with the origin of replication. However, the additional phosphorylation of neighboring serine residues serves to downregulate the specific origin binding activity.

This conclusion is consistent with other observations. First, OREN et al. (1980) have shown that newly synthesized T antigen binds more tightly to the SV40 origin than "old" T antigen, the form that accumulates in the nucleus; newly synthesized T antigen is underphosphorylated compared with the storage form of nuclear T antigen (BURGER and FANNING 1983; SCHEIDTMANN et al. 1984b).

Second, site-specific mutagenesis leads to the conclusion that Thr 124 is important for the replicative function of T antigen. Replacement of Thr 124 by either Ile (KALDERON and SMITH 1984; PAUCHA et al. 1986) or Ala (SCHNEIDER and FANNING 1988) eliminates the replicative activity of the T antigen. SCHNEIDER and FANNING (1988) demonstrated that the loss of replicative activity was due to the inability

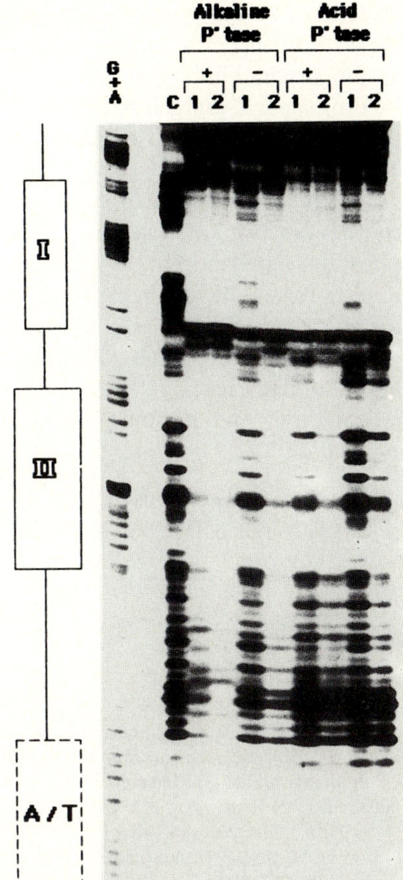



**Fig. 2.** Footprints of T antigen bound to the SV40 DNA origin region: Results of an experiment using $^{32}$P-end labeled SV40 origin DNA after limited DNase I digestion (GALAS and SCHMITZ 1978) and electrophoresis on a sequencing gel. *Left lane (G + A)*, the labeled DNA was chemically cleaved according to MAXAM and GILBERT (1980) to identify binding sites I and II. *Lane C*, control lane, DNase I-digested origin DNA without bound T antigen. *Lanes 1 and 2*, 50 and 100 ng T antigen, respectively, per 1 ng labeled DNA. *Minus (−) lanes*, T antigen incubated in dephosphorylation buffer without phosphatase; *plus (+) lanes*, T antigen dephosphorylated with alkaline or acid phosphatase as described by KLAUSING et al. (1988)

of their Ala mutant T antigen to bind to the SV40 origin whereas its binding to site I was not affected. Interestingly, a Thr 124-Glu mutant had reduced but clearly detectable replicative activity, suggesting that a negative charge at residue 124 may be important for T-antigen-origin interaction (SCHNEIDER and FANNING 1988).

As shown in Fig. 1, the clusters of phosphorylated amino acids do not overlap with the two or three known functional domains of T antigen. In particular, the *N*-terminal cluster of phosphoamino acids is clearly not within the DNA binding region which begins at amino acid residue 139 (PAUCHA et al. 1986) and may reach as far as residue 371 (SIMMONS 1986, 1988, FANNING et al., this volume). Thus, direct effects of phosphate groups on protein-DNA interaction appear to be unlikely. This conclusion is supported by the observation that the extent of phosphorylation does not affect the affinity of T antigen to binding site I.

It should be recalled that the architecture of binding site I differs from that of binding site II (see Introduction to this part). Binding site I consists of two tandem GAGGC-recognition pentanucleotides, separated by a 7-bp AT-rich spacer, whereas binding site II includes two pairs of pentanucleotides, oriented in opposite directions. A monomer mass of T antigen appears to bind to each pentanucleotide (MASTRANGELO et al. 1985) implying that the arrangement of T-antigen molecules at the origin binding site II must be quite different from their arrangement at binding site I. T-antigen molecules aligned at the origin probably form specific protein-protein contacts which could be affected by the phosphorylation of certain amino acid side chains.

In summary, the available data suggest the following order of events with regard to the function of T-antigen phosphorylation. Newly synthesized T antigen is phosphorylated at Thr 124 and Thr 701 in the cytoplasm and transported to the nucleus where it is directed to the origin of replication. T antigen unwinds the origin region as a first step to establishing replication forks. During or at the end of the replication cycle the serine residues of T antigen also become phosphorylated, and T antigen loses its specific affinity for the origin region. It accumulates in the infected cell and performs functions which are not necessarily related to DNA replication.

*Acknowledgements.* This work was supported by the Deutsche Forschungsgemeinschaft through SFB 156. K.K. is a recipient of a BMFT fellowship.

# References

Burger C, Fanning E (1983) Specific binding activity of T antigen subclasses varies among different SV40 transformed cell lines. Virology 126: 19–31

Fischer-Fantuzzi L, Scheidtmann KH, Vesco C (1986) Biochemical properties of a transforming nonkaryophilic T antigen of SV40. Virology 153: 87–95

Galas DJ, Schmitz A (1978) DNase I footprinting: a simple method for the detection of protein-DNA binding specificity. Nucleic Acids Res 5: 3152–3170

Goldman N, Brown M, Khoury G (1981) Modification of SV40 T antigen by poly ADP-ribosylation. Cell 24: 567–572

Grässer FA, Mann K, Walther G (1987) Removal of serine phosphates from simian virus large T antigen increases its ability to stimulate DNA replication in vitro but has no effect on ATPase and DNA binding. J Virol 61: 3373–3380

Kalderon D, Smith AE (1984) In vitro mutagenesis of a putative DNA binding domain of SV40 large T. Virology 139: 109–134

Klausing K, Scheidtmann KH, Baumann EA, Knippers R (1988) Effect of in vitro dephosphorylation on DNA binding and DNA helicase activity of simian virus 40 large tumor antigen. J Virol 62: 1258–1265

Klockmann U, Deppert W (1983) Acylation: A new post-translational modification specific for plasmamembrane-associated simian virus 40 large T antigen. FEBS Lett 151: 257–259

Mastrangelo IA, Hough PVC, Wilson VG, Wall JS, Hainfeld JF, Tegtmeyer P (1985) Monomers through trimers of large T antigen in region I and monomers through tetramers in region II bind to SV40 origin as stable structures in solution. Proc Natl Acad Sci USA 82: 3626–3630

Maxam A, Gilbert W (1980) Sequencing end-labeled DNA with base-specific chemical cleavages. Methods Enzymol 65: 499–560

Mohr IJ, Stillman B, Gluman Y (1987) Regulation of SV40 DNA replication by phosphorylation of T antigen. EMBO J 6: 153–160

Oren ME, Winocour E, Prives C (1980) Differential affinities of simian virus 40 large tumor antigen to DNA. Proc Natl Acad Sci USA 77: 220–224

Paucha E, Mellor A, Harvey R, Smith AE, Hewick RM, Waterfield MD (1978) Large and small tumor antigens from simian virus 40 have identical amino termini mapping at 0.65 map units. Proc Natl Acad Sci USA 75: 2165–2169

Paucha E, Kalderon D, Harvey RW, Smith AE (1986) Simian virus 40 origin DNA-binding domain on large T antigen. J Virol 57: 50–64

Scheidtmann KH, Echle B, Walter G (1982) Simian virus 40 large T antigen is phosphorylated at multiple sites clustered in two separate regions. J Virol 44: 116–133

Scheidtmann KH, Hardung M, Echle B, Walter G (1984a) DNA-binding activity of simian virus 40 large T antigen correlates with a distinct phosphorylation state. J Virol 50: 1–12

Scheidtmann KH, Schickedanz J, Walter G, Lanford RL, Butel JS (1984b): Differential phosphorylation of cytoplasmic and nuclear variants of simian virus 40 large T antigen encoded by simian virus 40-adenovirus 7 hybrid viruses. J Virol 50: 636–640

Schneider J, Fanning E (1988) Mutations in the phosphorylation sites of simian virus 40 (SV40) T antigen alter its origin DNA binding specificity for sites I or II and affect SV40 DNA replication activity. J Virol 62: 1598–1605

Simmons DT (1986) DNA-binding region of the simian virus 40 tumor antigen. J Virol 57: 776–785

Simmons DT (1988) Geometry of the simian virus 40 large tumor antigen-DNA complex as probed by protease digestion. Proc Natl Acad Sci USA 85: 2086–2090

# SV40 T Antigen Catalyzed Duplex DNA Unwinding

M. Scheffner, R. Wessel, and H. Stahl

## 1 Introduction

Large tumor antigen (T antigen) combines several functional domains on one poly-peptide chain of 708 amino acids. One domain is responsible for a specific and an unspecific interaction of T antigen with DNA (see: Fanning et al., Loeber et al., this volume) and a second domain contains an enzymatic activity that breaks the hydrogen bonds of double-stranded (ds) DNA at the expense of ATP (Stahl et al. 1986; for review, see also: Rigby and Lane 1983; Stahl and Knippers 1987).

The DNA-unwinding activity of T antigen was originally discovered as a T-antigen function operating at the forks of replicating SV40 chromatin. We have shown that T antigen is specifically associated with replication forks in replicating SV40 chromatin (Stahl and Knippers 1983; Stahl et al. 1985). In support of this fact an essential role of T antigen during the elongation phase of SV40 DNA replication was found, as monoclonal antibodies directed against several domains of T antigen were able to inhibit DNA replication in an in vitro SV40 chromatin replication system (Stahl et al. 1985; Wiekowski et al. 1987). The function of T antigen during the elongation of viral DNA replication could be more specifically defined by an uncoupling of DNA synthesis and fork movement in an in vitro system (Dröge et al. 1985). Under these conditions a function of T antigen for the separation of parental DNA strands was detected (Wiekowski et al. 1987). This function is most likely performed by its DNA helicase activity (Stahl et al. 1986).

In addition to its elongation function T antigen is also essential during the initiation process of SV40 DNA replication: It opens up the duplex DNA structure at the origin of SV40 DNA to allow the assembly of the DNA replication machinery, which is composed of cellular proteins (Wold et al. 1987; Dodson et al. 1987).

Biochemical studies of the unwinding activity have to reflect the different DNA structures with which T antigen has to deal in vivo. Here we present a characterization of the T-antigen helicase with respect to the DNA substrate requirements during the initiation process. Knowledge about which DNA structures are suitable as an entry site for the T-antigen helicase may allow us to understand better its function in the viral infection cycle as well as in SV40-transformed cells.

---

Fakultät für Biologie, Universität Konstanz, 7750 Konstanz, FRG

## 2 Results and Discussion

### 2.1 Unwinding of Closed Circular DNA

As we have shown before T antigen can efficiently start its helicase activity at partially dsDNA substrates with a 3'-single-stranded (ss) region (WIEKOWSKI et al. 1988). The single strand has to be at least 5–10 nucleotides long. The structure of this DNA substrate most probably resembles a replication fork with T antigen bound to the template on the leading strand. The translocation of T antigen on partially dsDNA templates and at the forks of replicating minichoromosomes is inhibited by several T-antigen-specific monoclonal antibodies (STAHL et al. 1986; WIEKOWSKI et al. 1987). Some of the inhibiting antibodies do not interfere with the ATPase center of T antigen, demonstrating the complex nature of the process by which T antigen is translocated along a substrate DNA strand. Translocation proceeds over a distance of at least several hundred base pairs. During this process T antigen remains stably bound to the DNA template. Salt concentration as high as 1 $M$ NaCl fail to dissociate active T antigen from its substrate. We have reported that small, fully dsDNA fragments (30–40 nucleotides long) could not be unwound, implying that T antigen, like all known DNA helicases, needs an ssDNA region to initiate DNA unwinding (STAHL et al. 1988). On the other hand, these results could also be interpreted to mean that several T-antigen molecules participate in DNA unwinding,

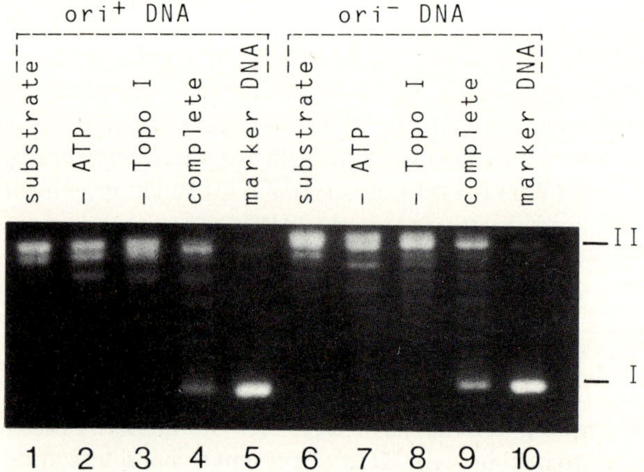


**Fig. 1.** T antigen unwinds closed circular DNA. Reaction mixtures contained 1.5 µg T antigen, 0.1 µg relaxed substrate DNA, and 0.1 µg topoisomerase I. The mixtures were incubated for 90 min as described before (STAHL et al. 1986). The reactions were stopped by the addition of SDS (1 %), EDTA (50 m$M$), and tRNA (5 µg). After phenol and chloroform extraction the DNA was ethanol precipitated and redissolved in TE buffer. The products were electrophoresed on 0.8 % agarose gels in 40 m$M$ Tris, 5 m$M$ Na acetate, 2 m$M$ EDTA, pH 8.3, and ethidium bromide stained. *Lanes marked ori⁻* contained pAT DNA and *ori⁺* contained pAT DNA with the SV40 origin sequence (bp 5092 to 160; see TOOZE 1980). As marker DNA we used the respective native plasmids before treatment.

and hence the 30-bp duplex DNA fragment might be too small to serve as an appropriate DNA-substrate.

We have found that T antigen is able to unwind closed circular DNA-molecules (Fig. 1). Relaxed plasmids were incubated at 37 °C with T antigen, ATP, and DNA topoisomerase I as described (STAHL et al. 1986).

The reaction products were analyzed by agarose gel electrophoresis. The relaxed input DNA was supercoiled in the presence of topoisomerase I, and the generation of supercoiled DNA forms was completely dependent upon the presence of T antigen and ATP (Fig. 1, lanes 1 and 2). The reaction showed no dependence upon other proteins like the *Escherichia coli* single-strand-specific, binding protein (SSB). Moreover, the reaction does not require the SV40 origin sequence (ori$^+$ DNA) since the rapidly migrating DNA topoisomers were generated efficiently with the pAT plasmid (ori$^-$ DNA). The unwinding reaction was very sensitive to salt, but at 100 m$M$ sodium acetate ori$^-$ DNA unwinding could be more efficiently inhibited than ori$^+$ DNA (data not shown, but see STAHL et al. 1988).

## 2.2 Influence of Antibodies on Plasmid Unwinding

We tested the influence of T-antigen-specific monoclonal antibodies (PAb) on the formation of topoisomers in a reaction like that shown in Fig. 1. To our surprise some antibodies such as PAb-101, 108, KT3, and 1623, which had previously been shown to inhibit strand separation of partially dsDNA substrates by T antigen (STAHL et al. 1986), stimulated the plasmid unwinding reaction whereas others, like

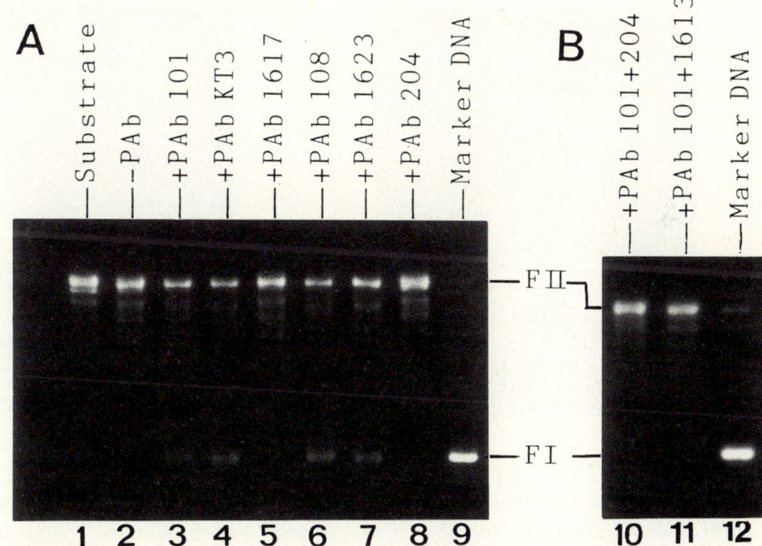


**Fig. 2A, B.** Influence of monoclonal antibodies on T-antigen unwinding. Reaction mixtures were as described in Fig. 1 except that T antigen was preincubated with 10 µg of the indicated monoclonal antibody (PAb) (for source of antibodies and some of their characteristics see STAHL et al. 1986; WIEKOWSKI et al. 1987). As a substrate we used relaxed SV40 DNA

PAb204 and 1613, completely inhibited the reaction (Fig. 2). The latter antibodies have been shown before to block the ATPase activity of T antigen (CLARK et al. 1981; WIEKOWSKI et al. 1987). Antibodies PAb204 and 1613 also inhibit the antibody-stimulated plasmid unwinding (Fig. 2B). The stimulating effect of the antibodies PAb101, 108, KT3, and 1623 was not dependent on which epitope on the T-antigen protein they bind to or on the presence of the SV40 origin sequence. This may confirm the hypothesis mentioned above that several T-antigen molecules are essential for the duplex DNA opening step as binding of T antigen (mono- to tetramers) to each arm of a monoclonal antibody could well result in stabilization and proper orientation of closely adjacent molecules, facilitating protein interactions which are most probably essential fot the duplex DNA opening reaction.

To verify that the faster migrating topoisomers resulted from unwinding of the DNA, the reaction products were analyzed by gel electrophoresis in the presence of chloroquine, an intercalating agent (SHURE et al. 1977). Native, supercoiled SV40 DNA (form I) migrated more slowly in the presence of 3.5 μ$M$ chloroquine due to the neutralization of superhelical turns in compensation for the decrease in the twist of the DNA which results from the intercalation of the drug (Fig. 3, lane 2). In contrast, fully relaxed SV40 DNA migrated more rapidly due to the

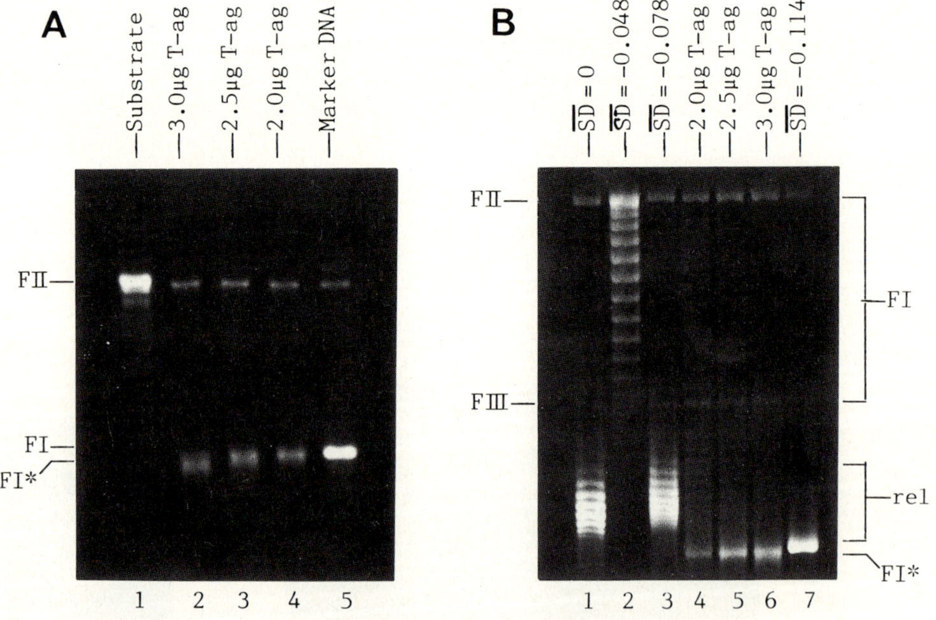

**Fig. 3A, B.** Extent of DNA unwinding. Reaction mixtures were the same as in the previous experiments, except that the indicated amounts of T antigen were preincubated with 10 μg of PAb101. After incubation, reaction mixtures were divided into two parts and analyzed by agarose gel electrophoresis without (**A**) and in the presence of (**B**) 3.5 μ$M$ chloroquine. In **B** lanes 1–3 and lane 7 show SV40 DNA with a superhelical density (SD) of 0, 0.048, 0.078, and 0.114, respectively, as markers. *FI** indicates the position of strongly underwound SV40 DNA; *rel,* the position of (partially) relaxed SV40 DNA; and *FI, FII,* and *FIII* the positions of native, nicked circular, and linear SV40 DNA under the respective separation conditions

introduction of positive superhelical turns (Fig. 3, lane 1). Plasmids that were exposed under unwinding conditions to limited amounts of T antigen migrated more slowly (Fig. 3, lane 4), more or less like supercoiled DNA containing approximately 36 negative superhelical twists (Fig. 3, lane 3). After incubation with increasing T-antigen concentrations, most of the DNA migrated similarly to or even more rapidly than supercoiled DNA containing about 58 superhelical twists (Fig. 3, Fl*, lane 7).

Taken together, the data presented so far clearly demonstrate that T antigen is able to unwind closed circular DNA molecules. This reaction is not dependent on the SV40 origin sequence or other factors like SSB. Our results are in good agreement with reports by others (WOLD et al. 1987). However, the T-antigen-dependent formation of negative supercoiled topoisomers could be due to several reactions related to an alteration of the topological state of circular DNA, namely: wrapping of the DNA around T-antigen molecules, a change in the number of base pairs for each turn of the DNA helix, and duplex DNA opening.

## 2.3 Unwinding of Linear Duplex DNA

To discriminate between these possibilities we tested the T-antigen-catalyzed unwinding of fully double-stranded linear DNA fragments, measuring the extent of DNA strand separation (Fig. 4A). As SV40 origin sequences appear to be unnecessary for T-antigen-induced DNA unwinding, we used linearized pAT plasmid as a substrate. As can be seen (Fig. 4A), T antigen is able to unwind the 2.8-kbp long, linear DNA, producing ss DNA. However, in contrast to the results obtained using closed circular DNA as substrate, we found that the unwinding of linearized plasmid DNA was stimulated by SSB (data not shown). It is possible

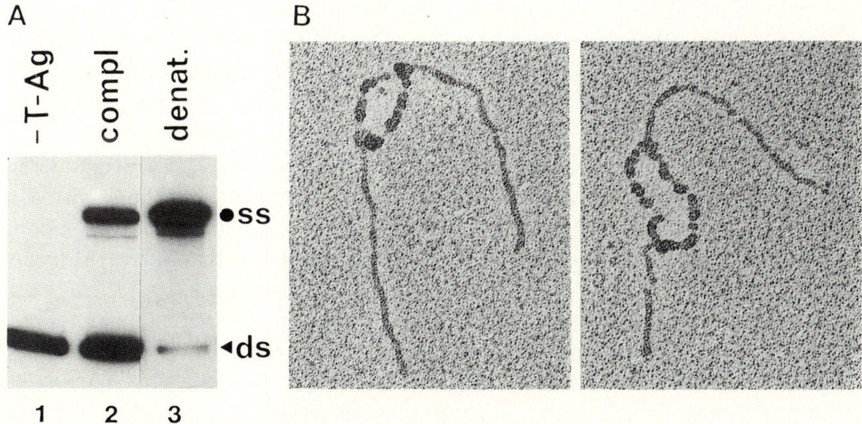



**Fig. 4A, B.** Unwinding of linear pAT DNA: 100 ng T antigen was incubated with 10 ng EcoRI-linearized, $^{32}$P end-labeled pAT DNA under helicase reaction conditions. In **A** the reaction mixture was incubated for 60 min analyzed by acrylamide gel electrophoresis and autoradiography (ss marks the position of ss DNA and ds the position of ds pAT DNA). In **B** the reaction was performed for 15 min and subsequently processed for electron microscopy (see text)

though that under these experimental conditions melting at the ends of the linearized plasmid DNA could generate ss regions serving as entry sites for T antigen.

To exclude this possibility the unwinding of the linearized pAT DNA was examined by electron microscopy. To enrich partially unwound DNA molecules, the incubation time of the unwinding reaction was reduced to about 15 min. Under these conditions only 5% of the input DNA was converted to single strands (for an estimate of the efficiency of the unwinding reaction during longer incubation times, see Fig. 4A). Unwinding was stopped by the addition of 0.22 $M$ NaCl, known to remove unspecifically bound T antigen from DNA. The mixture was passed over a sepharose CL-4B column, and after fixation with glutaraldehyde (0.1%) the molecules were spread for electron microscopy. Partially unwound DNA molecules with ss regions of various lengths as well as completely ssDNA molecules were observed. About 20% of partially unwound DNA molecules showed ss bubbles which were located within various internal sections of the DNA, indicating a sequence-unspecific start of the unwinding reaction. Characteristic molecules are shown in Fig. 4B. Note that the single strands were coated with SSB and thus appeared condensed in length (about 2.7 times, see KORNBERG 1982) and thickened relative to duplex DNA. The electron micrographs show typical bubbled DNA structures after exposure to the T-antigen helicase. This was never observed in a control reaction performed without T antigen (data not shown). This result demonstrates that T antigen is able to perform a duplex DNA opening reaction in the absence of the SV40 origin sequence.


## 2.4 T-Antigen Interaction with dsDNA in the Presence of ATP

Previous reports on the role of T antigen during the initiation of SV40 DNA replication suggested that the opening of the origin sequences is probably preceded by a sequence-specific binding of T antigen, resulting in an ATP-dependent formation of a specialized nucleoprotein structure (DEAN et al. 1987; DEB and TEGTMEYER 1987). Since we expected a similar mechanism for the sequence-independent duplex DNA opening reaction, the influence of ATP and nonhydrolyzable ATP analogues on T-antigen binding to ds-DNA was examined. T antigen was incubated at 37 °C with nick-translated, linear pAT DNA. The resulting DNA protein complexes were digested with a large excess of DNase I and isolated by binding to nitrocellulose filters (for method, see DEB and TEGTMEYER 1987). Under helicase reaction conditions but in the absence of ATP, T antigen bound only poorly as demonstrated by counting the radioactivity bound to the filter. But in the presence of ATP (as well as of nonhydrolyzable ATP analogues), the binding efficiency increased at least tenfold (data not shown; SCHEFFNER et al., 1989). We prepared T-antigen-pAT DNA complexes for electron microscopy exactly as described in Fig. 4 except that the protein-DNA complexes formed in the presence of ATP were not treated with 0.22 $M$ NaCl before fixation (Fig. 5). In accordance with the biochemical data long stretches of DNA complexed by T antigen were found. These complexes were not seen in the absence of ATP. The T-antigen-DNA binding appears to be cooperative as long clusters of bound T-antigen molecules could be detected on the DNA (Fig. 5A). Moreover, under the helicase reaction conditions T antigen apparently formed protein-DNA aggregates

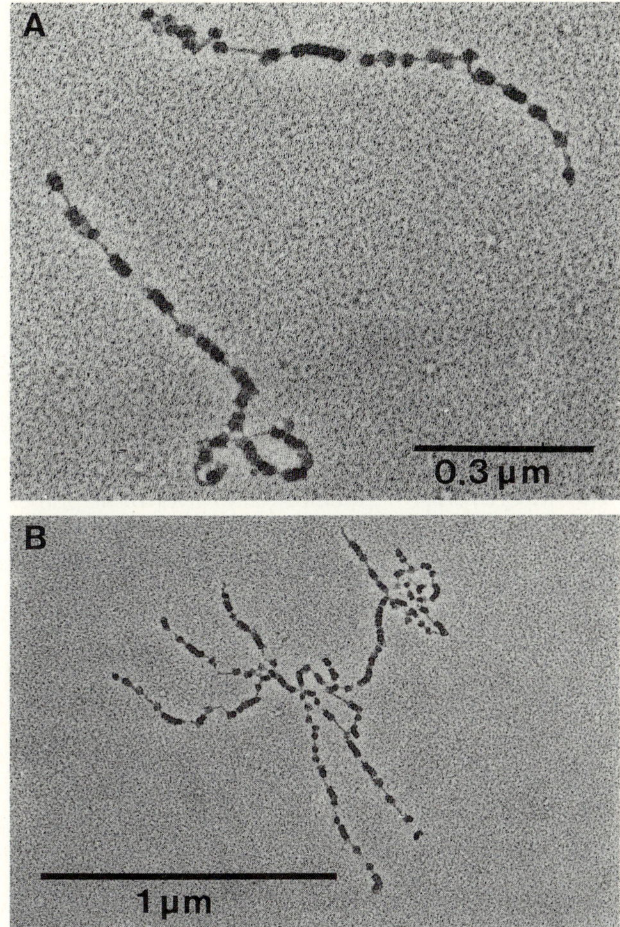

**Fig. 5A, B.** T-antigen binding to dsDNA in the presence of ATP. Experimental conditions were the same as in Fig. 4 except that the salt treatment of the DNA was omitted before fixation with glutaraldehyde

(Fig. 5B) comparable to that formed by bacterial recA protein in an early step of the recA-catalyzed reaction (synapsis; for review, see GRIFFITH and HARRIS 1988). This similarity between T antigen and recA may not be accidental as both proteins share a similar region in their amino acid sequences and have several properties in common (SEIF 1982), for example, a sequence-independent DNA-unwinding activity.

## 3 Conclusions

Duplex DNA opening is most probably one of the first steps of the SV40 DNA replication process. We show here that this reaction can be catalyzed by purified T antigen. First, T antigen is able to unwind closed circular DNA molecules as demonstrated by the introduction of superhelical twists in the presence of topoisomerase I. Second, T antigen can separate the strands of ds DNA fragments by starting from internal sites. In contrast to SV40 DNA replication, duplex DNA opening is not dependent on the SV40 origin sequence, at least under the conditions essential for optimal DNA helicase activity (e.g., low salt). Initiation of duplex DNA opening requires a minimum of 50–60 base pairs as smaller dsDNA fragments are not unwound by T antigen. We interpret this to indicate that several T-antigen molecules are needed for the start of the unwinding reaction. This hypothesis is confirmed by the fact that several T-antigen-specific monoclonal antibodies that bind to different epitopes strongly stimulate the duplex DNA opening reaction probably by holding adjacent T-antigen molecules in an optimal configuration. Moreover, T-antigen binding to dsDNA is strongly stimulated by ATP or nonhydrolyzable ATP analogues. Under these conditions, T antigen appears to bind cooperatively to DNA.

*Acknowledgements*. We thank R. KNIPPERS for helpful discussions and critical reading of the manuscript and T. KAPITZA for his excellent technical assistance. This research was supported by the Deutsche Forschungsgemeinschaft through SFB 156.

# References

Clark R, Lane DP, Tjian R (1981) Use of monoclonal antibodies as probes of simian virus 40 T antigen ATPase activity. J Biol Chem 256: 11854–11858

Dean FB, Dodson M, Echols H, Hurwitz J (1987) ATP-dependent formation of a specialized nucleoprotein structure by simian virus 40 (SV40) large T antigen at the SV40 replication origin. Proc Natl Acad Sci USA 84: 8981–8985

Deb SP, Tegtmeyer P (1987) ATP enhances the binding of simian virus 40 large T antigen to the origin of replication. J Virol 61: 3649–3654

Dodson M, Dean FB, Bullock P, Echols H, Hurwitz J (1987) Unwinding of duplex DNA from the SV40 origin of replication by T antigen. Science 238: 964–967

Dröge P, Sogo JM, Stahl H (1985) Inhibition of DNA synthesis by aphidicolin induces supercoiling in simian virus 40 replicative intermediates. EMBO J 4: 3241–3246

Griffith JD, Harris LD (1988) DNA strand exchanges. CRC Crit Rev Biochem 23: 43–86

Kornberg A (1982) Supplement to DNA replication. Freeman, San Francisco

Rigby PWJ, Lane DP (1983) Simian virus 40 large T antigen. Adv Viral Oncol 3: 21–57

Seif R (1982) New properties of simian virus 40 large T antigen. Mol Cell Biol 2: 1463–1471

Scheffner M, Wessel R, Stahl H (1989) Sequence independent duplex DNA opening reaction catalyzed by SV40 large tumor antigen. Nucleic Acids Res 17: 93–106

Shure M, Pulleyblank ED, Vinograd J (1977) The problems of eukaryotic and prokaryotic DNA packaging and in vivo conformation posed by superhelix density heterogeneity. Nucleic Acids Res 4: 1183–1205

Stahl H, Knippers R (1983) SV40 large T antigen on replicating viral chromatin. Tight binding and localization on the viral genome. J Virol 47: 65–76

Stahl H, Knippers R (1987) The simian virus 40 large tumor antigen. Biochem Biophys Act 910: 1–10

Stahl H, Dröge P, Zentgraf HW, Knippers R (1985) A large-tumor-antigen-specific monoclonal antibody inhibits DNA replication of simian virus 40 minichromosomes in an in vitro elongation system. J Virol 54: 473–482

Stahl H, Dröge P, Knippers R (1986) DNA helicase activity of SV40 large tumor antigen. EMBO J 5: 1939–1944

Stahl H, Scheffner M., Wiekowski M, Knippers R (1988) DNA unwinding function of the SV40 large tumor antigen. In: Kelly T, Stilman B (eds) Cancer cell, vol 6. Cold Spring Harbor Laboratory, Cold Spring Harbor

Tooze J (1980) DNA tumor viruses. Molecular biology of tumor viruses, 2. Cold Spring Harbor Laboratory, Cold Spring Harbor

Wiekowski M, Dröge P, Stahl H (1987) Monoclonal antibodies as probes for a function of large T antigen during the elongation process of simian virus 40 DNA replication. J Virol 61: 411–418

Wiekowski M, Schwartz MW, Stahl H (1988) Simian virus 40 large T antigen DNA helicase. J Biol Chem 263: 436–442

Wold MS, Li JJ, Kelly TJ (1987) Initiation of simian virus 40 DNA replication in vitro: large-tumor-antigen- and origin-dependent unwinding of the template. Proc Natl Acad Sci USA 84: 3643–3647

# Control of Transcription in Vitro
# from Simian Virus 40 Promoters by Proteins
# from Viral Minichromosomes

P. Beard and H. Bruggmann

## 1 Introduction

The simian virus (SV40) genome comprises two transcription units containing the early genes and, in the other orientation, the late genes. Before viral DNA replication starts, transcription is mostly of the early genes. After the viral DNA begins to replicate, the late promoter is activated, and transcription is predominantly of the late genes (De Pamphilis and Wassarman 1982; Tooze 1980). The SV40 early gene product, large T antigen, is responsible for repression of early transcription and, indirectly or directly, for the increase in late transcription (Brady et al. 1984; Hartzell et al. 1984; Keller and Alwine 1984).

Most of the DNA sequences involved in control of late transcription probably function by interacting with regulatory proteins. It seemed to us that such proteins might be associated with transcribed SV40 chromosomes in the late phase of the lytic cycle. We showed earlier (Beard and Nyfeler 1982; Tack and Beard 1985) that the activity of the late promoter in SV40 chromosomes is determined by a combination of a stable nucleoprotein structure and *trans*-acting factors. Here, we describe a transcriptional regulatory factor (the late-promoter-activating factor, LAF) obtained from SV40 chromosomes which activates transcription from the major late SV40 RNA start site in vitro.

## 2 Methods and Experimental Procedures

SV40 chromosomes were extracted from infected CV-1 cells 36 h after infection essentially as described previously (Beard and Nyfeler 1982; Tack and Beard 1985). Proteins associated with viral minichromosomes were prepared by absorption to hydroxylapatite and batchwise elution.

Nuclear extracts of HeLa cells were prepared as described by Dignam et al. (1983). Transcription reactions and analysis of RNA products by denaturation with glyoxal and electrophoresis were as previously described (Tack and Beard 1985).

---

Swiss Institute for Experimental Cancer Research, 1066 Epalinges/Lausanne, Switzerland

Current Topics in Microbiology and Immunology, Vol. 144
© Springer-Verlag Berlin · Heidelberg 1989

# 3 Results

## 3.1 The Effect of SV40 Chromosomal Proteins on the Activity of Viral Promoters in Vitro

When SV40 DNA is transcribed by nuclear extracts of HeLa cells the early promoter is in vitro up to tenfold stronger than the late promoter. With isolated SV40 chromosomal templates on the other hand, the late promoter is three- to tenfold more active than the early (BEARD and NYFELER 1982; TACK and BEARD 1985). We tested the effect of added SV40 chromosomal proteins on the pattern of transcription of purified SV40 DNA. *Pst*I-cut SV40 DNA was incubated with HeLa nuclear extract in the absence or presence of different protein fractions. Figure 1 shows a map of the run-off transcripts expected. We identified an RNA of about 2 kb as a run-off transcript from the early promoter to the *Pst*I site at position 3204, and RNAs of about 1.7 and 1.8 kb as run-off transcripts from the major late start site and from a minor late start site at position 170, respectively, to the *Pst*I site at position 1988 (TACK and BEARD 1985).

Transcription of the DNA alone gave principally the 2-kb early RNA (Fig. 2). Addition of a 1 *M* KCl/5 *M* urea eluate from hydroxylapatite, in increasing amounts, resulted in a complete switch in the transcription pattern: Transcription from the early promoter was abolished, while that from the major late start site was stimulated by a factor of about 10. Transcription from the minor late start site was also stimulated. The magnitude of the effect on transcription depended on the amount of extract with an optimum of 10 µl extract in a 50 µl reaction. Inhibition was observed with 15 µl of extract.

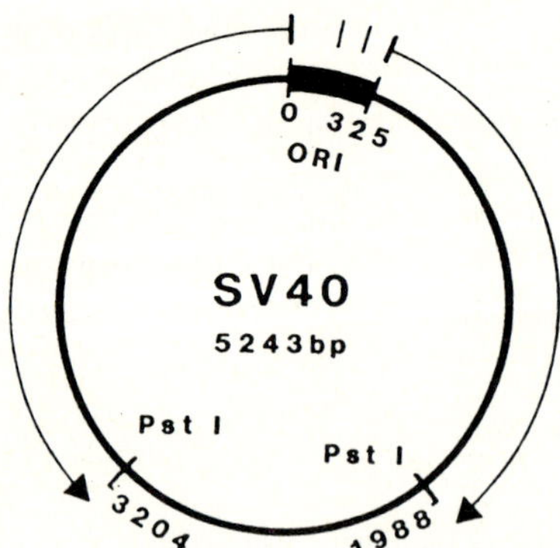


**Fig. 1.** Map of the SV40 genome showing the predicted lengths of run-off transcripts initiated at the early or late promoter and terminated at the *Pst*I cleavage sites

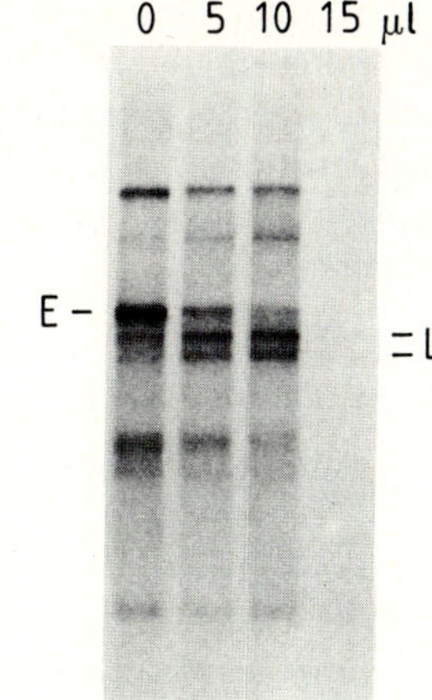


**Fig. 2.** Effect of adding varying amounts of SV40 chromosomal proteins on transcriptional specificity. *Pst*I-cut SV40 DNA was transcribed and the products analyzed by agarose gel electrophoresis and autoradiography. The positions of early and late transcripts are shown by *E* and *L*. Standard transcription reactions were supplemented with 0, 5, 10, or 15 µl of a 1 *M* KCl/5 *M* urea eluate of hydroxylapatite-bound SV40 chromosomes with 0.75 µg protein per µl

## 3.2 T Antigen and the Transcriptional Switch

A 1 *M* KCl/5 *M* urea eluate of hydroxylapatite-bound SV40 chromosomes was depleted of T antigen by incubating it with T antigen-specific immunoglobulin immobilized on fixed *Staphylococcus aureus*. Polyacrylamide gel electrophoresis, transfer, and probing with T antigen-specific antibodies showed that the depletion was effective (Fig. 3, right panel, lane C). When the T-antigen-depleted eluate was tested for its effect on transcription in vitro, it no longer abolished early transcription,

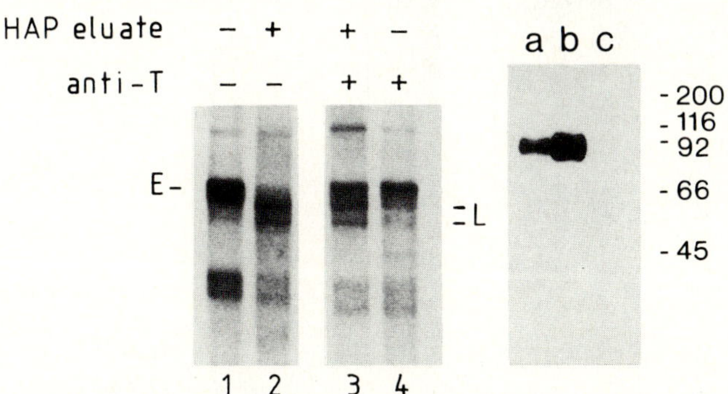


**Fig. 3.** T antigen and transcriptional regulation by proteins from SV40 chromosomes: Removal of T antigen from a 1 *M* KCl/5 *M* urea eluate of hydroxylapatite-bound SV40 chromosomes using antibodies bound to *Staphylococcus aureus*. *Left*, effects on transcription. The hydroxylapatite (*HAP*) *eluate* (or dialysis buffer alone) was depleted of T antigen by treatment with antibody bound to *Staphylococcus aureus* (or mock depleted). Transcription reactions were as in Fig. 2, the presence or not of the hydroxylapatite eluate and antigen-specific antibody (*anti-T*) being indicated by + or –. Early (*E*) and late (*L*) transcripts are indicated as before. *Right*, detection of T antigen by immunoblotting. Samples loaded on the gel were: lane *a*, the hydroxylapatite eluate; lane *b*, intact SV40 chromosomes as marker for T antigen; lane *c*, the hydroxylapatite eluate depleted of T antigen, corresponding to that added to the transcription reaction in lane 3. The positions of protein molecular weight markers are shown

yet it still stimulated the late (Fig. 3, lane 3). The factor activating the late promoter therefore did not behave like the SV40 T antigen with respect to reaction with T antigen-specific antiserum. In other experiments (not shown) the late activating factor was eluted from hydroxylapatite by 1 *M* KCl, while elution of T antigen required urea. Inhibition of the early promoter, on the other hand, correlated with the presence of T antigen.

## 3.3 Interaction of Proteins Extracted from SV40 Chromosomes with the Viral DNA of the Origin-Promoter Region

DNase I footprinting was used to look at how the proteins which eluted from hydroxylapatite-bound SV40 chromosomes interact with the region of viral DNA between the DNA replication origin and the major late RNA start site. Several proteins from HeLa cells are already known which bind to this region (reviewed by JONES et al. 1988) including Sp1 (BRIGGS et al. 1986) and LSF (KIM et al. 1987), which bind to the guanosine/cytosine-rich 21-bp repeats, and AP-1, AP-2, AP-3,

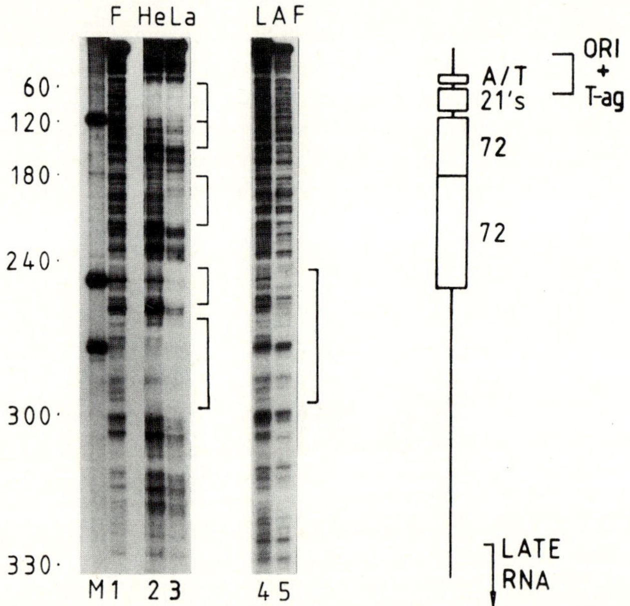


**Fig. 4.** Protection of SV40 origin-promoter region DNA against DNase I by proteins from HeLa cells and from SV40 chromosomes. The *scale on the left* shows the gel positions corresponding to SV40 nucleotide numbers. Lane *M*, size markers of SV40 DNA digested by *Hin*fI with lengths of 237, 109, and 83 nucleotides. Lane *1*, the labeled DNA probe (the *Hpa*II-*Hin*dIII origin fragment recloned, see Methods) was digested with DNase I in the absence of added cellular extract. Lanes *2* and *3*, DNase I protection by proteins in HeLa cell nuclear extract (the amounts of protein added were lane 2, 20 µg; lane 3, 40 µg). Lanes *4* and *5*, DNase I protection by proteins in a 1 *M* KCl/5 *M* urea eluate of hydroxylapatite-bound SV40 chromosomes (the amounts of protein added were lane 4, 4.6 µg; lane 5, 9.2 µg). On the *right* is a map, to the same scale, of the SV40 origin-promoter region

AP-4 (MITCHELL et al. 1987; MERMOD et al. 1988), EBP20 (JOHNSON et al. 1987), octamer binding proteins (BOHMANN et al. 1987; ROSALES et al. 1987 and references therein), and the GT-II proteins (XIAO et al. 1987), which bind to different sites within the SV40 enhancer. Using an *Hpa*II-*Hind*III fragment of SV40 DNA labeled adjacent to the *Hpa*II site, we first looked at DNA binding of factors in the nuclear extract of HeLa cells that was used for transcription (Fig. 4, lanes 2 and 3). This extract protected against DNase I the 21-bp repeats, the origin-proximal part of each copy of the 72-bp repeat, and a region extending from position 240 towards the major late RNA start site. The region of the RNA start site, however, was not protected.

SV40 chromosomes were bound to hydroxylapatite and the proteins eluted with 1 *M* KCl/5 *M* urea as before. The proteins in this eluate did not protect the 21- or the 72-bp repeats (Fig. 4, lanes 4 and 5). They did protect a region between positions 240 and 300 and, occasionally and less strongly, from 305 to 325, that is between the 72-bp repeat and the major late RNA start site at position 325. Although both HeLa cellular and SV40 chromosomal proteins gave a footprint centered on position 270, examination of the footprints showed that several details of the pattern of bands around and within the protected region differed. There was with the SV40 chromosomal extract a frequently cut site at about position 280. Thus the proteins from SV40 chromosomes which recognize this upstream part of the SV40 late promoter are not the same as the proteins from HeLa cells, mentioned above, which bind to SV40 DNA.

# 4 Discussion

Entry to the late phase of the SV40 lytic cycle requires the early viral protein large T antigen. The late promoter activating factor that we detected in this work, LAF, does not appear, however, to be T antigen. Although this result would seem unexpected, in fact it fits with several earlier observations. Temperature-shift experiments using temperature-sensitive mutants in T antigen (ALWINE and KHOURY 1980; COWAN et al. 1973; ROBBINS et al. 1986) suggested that T antigen may be required transiently for establishment of late transcription but that once under way late transcription can continue in its absence. We have shown (TACK and BEARD 1985) that the majority of SV40 chromosomes active in late transcription in vitro do not contain bound T antigen, while we confirmed that replicating chromosomes do (TACK and DE PAMPHILIS 1983). And finally, purified T antigen did not stimulate transcription from the late promoter in vitro, while it did inhibit the early (MYERS et al. 1981). Nevertheless, we do not exclude that T antigen could, in certain circumstances, exert a direct effect on the activity of the late promoter, in particular for RNA start sites closer to the origin of DNA replication (BRADY and KHOURY 1985).

In DNase I footprinting experiments LAF preparations protected a DNA region centered on SV40 nucleotide position 270. Analysis of mutations affecting the late

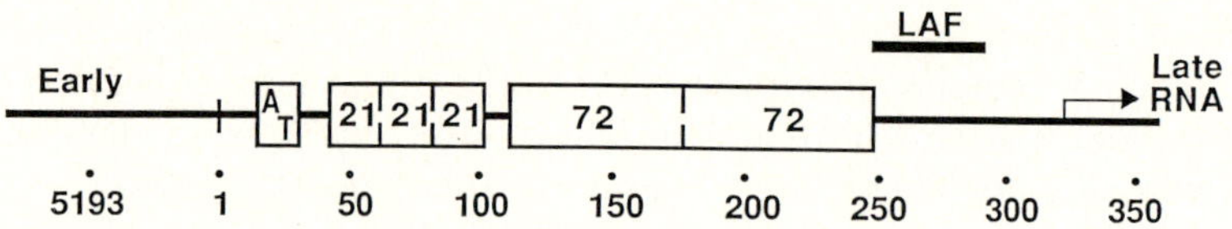

**Fig. 5.** Map of the SV40 promoter. The AT-rich sequence, the GC-rich 21-base pair repeats, the 72-base pair repeats, and the major late RNA start site are shown. The DNA sequence protected by proteins in LAF preparations is indicated by a *bar*

promoter show that this region is important for late promoter activity. Deletions between position 273 and 297 cut late promoter activity by half or more (ERNOULT-LANGE et al. 1987; KELLER and ALWINE 1985). So the simplest interpretation of our findings is that the same factor stimulates the late promoter in the transcription assay and protects against DNase I in the footprinting assay. Figure 5 shows the position of the LAF binding site on SV40 DNA.

Several proteins in the HeLa cell nuclear extract also interact with sequences around SV40 position 270 (Fig. 4; reviewed in JONES et al. 1988). However, LAF is different from these proteins from HeLa cells because the footprint boundaries and sites cleaved within the overall protected regions are different.

The SV40-coded capsid proteins could have an effect on the activity of the late promoter. The tsD class of mutants of SV40, which have altered capsid proteins VP2 and VP3, are affected in the control of transcription. However, the properties of tsD mutants suggest that the function of VP2 or VP3 is to inhibit SV40 transcription late in infection, not to stimulate it (LLOPIS and STARK 1981). So we think LAF is unlikely to be VP2 or VP3.

Functional LAF activity was not found in extracts of uninfected cells (TACK and BEARD 1985). Gel retardation experiments (submitted for publication) show that proteins from SV40-infected CV-1 cells form a complex with a DNA fragment containing the LAF binding site. This complex has a higher stability and a different rate of gel migration when compared with the complex formed with proteins from uninfected cells. LAF therefore appears to be a cellular transcription factor which is modified or induced by SV40 infection. DNA sequences homologous to a sequence within the LAF footprint (CACAGC at nucleotide 270) are found in some cellular promoters, e.g., that of the transferrin receptor gene (at −65; MISKIMINS et al. 1986). Transferrin receptor is expressed at a high level in SV40-transformed cells. Therefore, increases in LAF activity could also be part of the mechanism by which SV40 stimulates expression of some cellular genes.

*Acknowledgements.* We are very grateful to B. HIRT for helpful comments, B. BENTELE for tissue culture, E. HARLOW and H. TÜRLER for T antigen-specific antibodies, M. COCKELL for advice on footprinting, D. GYGAX and A.-M. RODEL for preparing the manuscript, and the Fonds National Suisse de la Recherche Scientifique for financial support.

# References

Alwine JC, Khoury G (1980) Effect of a tsA mutation on simian virus 40 late gene expression: variation between host cell lines. J Virol 33: 920–925

Beard P, Nyfeler K (1982) Transcription of simian virus 40 chromosomes in an extract of HeLa cells. EMBO J 1: 9–14

Bohmann D, Keller W, Dale T, Schöler HR, Tebb G, Mattaj IW (1987) A transcription factor which binds to the enhancers of SV40, immunoglobulin heavy chain and U2 snRNA genes. Nature 325: 268–272

Brady J, Khoury G (1985) *Trans* activation of the simian virus 40 late transcription unit by T-antigen. Mol Cell Biol 5: 1391–1399

Brady J, Bolen JB, Radonovich M, Salzman NP, Khoury G (1984) Stimulation of simian virus 40 late gene expression by simian virus 40 late tumor antigen. Proc Natl Acad Sci USA 81: 2040–2044

Briggs MR, Kadonaga JT, Bell SP, Tjian R (1986) Purification and biochemical characterization of the promoter-specific transcription factor Sp1. Science 234: 42–52

Cowan K, Tegtmeyer P, Anthony DD (1973) Relationship of replication and transcription of simian virus 40 DNA. Proc Natl Acad Sci USA 70: 1927–1930

De Pamphilis ML, Wassarman P (1982) Organization and replication of papovavirus DNA. In: Kaplan A (ed) Organization and replication of viral DNA. CRC, Boca Raton, pp 37–114

Dignam JD, Lebovitz R, Roeder RG (1983) Accurate initiation by RNA polymerase II in a soluble extract from isolated mammalian nuclei. Nucleic Acids Res 11: 1475–1489

Ernoult-Lange M, Omilli F, O'Reilly D, May E (1987) Characterization of the simian virus 40 late promoter: relative importance of sequences within the 72-base-pair repeats differs before and after viral DNA replication. J Virol 6: 167–176

Hartzell SW, Byrne BJ, Subramanian KN (1984) The SV40 minimal origin and the 72 base pair repeats are required simultaneously for efficient induction of late gene expression with large tumor antigen. Proc Natl Acad Sci USA 81: 6335–6339

Johnson PF, Landschulz WH, Graves BJ, McKnight SL (1987) Identification of a rat liver nuclear protein that binds to the enhancer core element of three animal viruses. Genes 1: 133–146

Jones NC, Rigby PWJ, Ziff EB (1988) *Trans*-acting protein factors and the regulation of eukaryotic transcription: lessons from studies on DNA tumor viruses. Genes Dev 2: 267–281

Keller JM, Alwine JC (1984) Activation of the SV40 late promoter: direct effects of T antigen in the absence of viral DNA replication. Cell 36: 381–389

Keller JM, Alwine JC (1985) Analysis of an activable promoter: sequences in the simian virus 40 promoter required for T-antigen-mediated *trans* activation. Mol Cell Biol 5: 1859–1869

Kim CH, Heath C, Bertuch A, Hansen U (1987) Specific stimulation of simian virus 40 late transcription in vitro by a cellular factor binding the simian virus 40 21-base-pair repeat promoter element. Proc Natl Acad Sci USA 84: 6025–6029

Llopis R, Stark GR (1981) Two deletions within genes for simian virus 40 structural proteins VP2 and VP3 lead to formation of abnormal transcriptional complexes. J Virol 38: 91–103

Mermod N, Williams TJ, Tjian R (1988) Enhancer binding factors AP-4 and AP-1 act in concert to activate SV40 late transcription in vitro. Nature 332: 557–561

Miskimins WK, McClelland A, Roberts MP, Ruddle FH (1986) Cell proliferation and expression of the transferrin receptor gene: promoter sequence homologies and protein interactions. J Cell Biol 103: 1781–1788

Mitchell PJ, Wang C, Tjian R (1987) Positive and negative regulation of transcription in vitro: enhancer-binding protein AP-2 is inhibited by SV40 T antigen. Cell 50: 847–861

Myers RM, Rio DC, Robbins AK, Tjian R (1981) SV40 gene expression is modulated by the cooperative binding of T antigen to DNA. Cell 25: 373–384

Robbins PD, Rio DC, Botchan MR (1986) *Trans* activation of the simian virus 40 enhancer. Mol Cell Biol 6: 1283–1295

Rosales R, Vigneron M, Macchi M, Davidson I, Xiao JH, Chambon P (1987) In vitro binding of cell-specific and ubiquitous nuclear proteins to the octamer motif of the SV40 enhancer and related motifs present in other promoters and enhancers. EMBO J 6: 3015–3025

Tack LC, Beard P (1985) Both *trans*-acting factors and chromatin structure are involved in the regulation of transcription from the early and late promoters in simian virus 40 chromosomes. J Virol 54: 207–218

Tack LC, De Pamphilis ML (1983) Analysis of simian virus 40 chromosome-T-antigen complexes: T-antigen is preferentially associated with early replicating DNA intermediates. J Virol 48: 281–295

Tooze J (1980) DNA tumor viruses. Molecular biology of the tumor viruses, pt. 2. Cold Spring Harbor Laboratory, Cold Spring Harbor

Xiao, JH, Davidson I, Ferrandon D, Rosales R, Vigneron M, Macchi M, Ruffenach F, Chambon P (1987) One cell-specific and three ubiquitous nuclear proteins bind in vitro to overlapping motifs in the domain B1 of the SV40 enhancer. EMBO J 6: 3005–3013

# Dissection of the T Antigen/Mouse p53 Complex and Its Inhibitory Effects on Viral Origin-Directed DNA Replication in Vivo and in Vitro

H.-W. STÜRZBECHER, K. RUDGE, R. BRAIN, C. ADDISON, M. GRIMALDI
and J. R. JENKINS

Several lines of evidence suggest that the cellular phosphoprotein p53, overproduced in a variety of neoplastic cell types (reviewed in CRAWFORD 1983; JENKINS and STÜRZBECHER 1988; OREN 1985; ROTTER and WOLF 1985), is an oncogene product, directly involved in the process of malignant transformation. p53 cDNA expression constructs immortalise both adult (JENKINS et al. 1984, 1985) and embryo (JENKINS et al. 1985; ROVINSKI and BENCHIMOL 1988) cells and cooperate with an activated *ras* oncogene in malignant transformation (ELIYAHU et al. 1984; JENKINS et al. 1984; PARADA et al. 1984; ROVINSKI and BENCHIMOL 1988; FINLAY et al. 1988). The potential of wild-type p53 to score in transformation assays is dependent upon powerful heterologous promoter/enhancers driving p53 expression (JENKINS et al. 1984, 1985). Immortalisation and *ras* complementation are distinct activities of p53 and can be separated by deletion mutagenesis (JENKINS et al. 1985). Generation of specific variant p53 species can give rise to protein products with increased stability and enhanced transforming/biological activity (JENKINS et al. 1985), indicating that the p53 gene can be mutationally activated. Near identical arrangements of the endogenous p53 gene coding sequences have been identified in vivo and are associated with malignant transformation by Friend leukaemia virus (ROVINSKI et al. 1987). Subsequently, further in vitro-generated p53 mutants with enhanced activity in cotransformation assays have been identified (FINLAY et al. 1988).

The transforming protein products of two unrelated DNA tumour viruses, the adenovirus 5 E1b 57K protein and the simian virus 40 (SV40) large tumour antigen (T antigen) form tight non-covalent complexes with p53 (LANE and CRAWFORD 1979; LINZER and LEVINE 1979; SARNOW et al. 1982).

Some mutant, but not wild-type, p53 proteins bind to cellular heat-shock-related proteins 72/73 (hsp72/73) (HINDS et al. 1987; JENKINS et al. 1988; STÜRZBECHER et al. 1987), and some of these hsp-binding mutant p53s have an increased half-life and accumulate at elevated levels in transfected cells (JENKINS et al. 1985; STÜRZBECHER et al. 1987, 1988a). However, hsp 72/73 binding does not correlate with either stabilization or oncogenic activation as both stable non-hsp72/73-binding and oncogenically activated non-hsp72/73-binding mutant p53 proteins have been described (JENKINS et al. 1985; STÜRZBECHER et al. 1987, 1988a). Structural requirements for hsp72/73 binding are clearly distinct from those implicated in T-antigen binding and depend on p53 protein conformation (JENKINS et al. 1988; STÜRZBECHER et al.

Marie Curie Research Institute, The Chart, Oxted, Surrey RH8 OTL, UK

Current Topics in Microbiology and Immunology, Vol. 144
© Springer-Verlag Berlin · Heidelberg 1989

1988a). Mouse p53s competent to bind T antigen react with the monoclonal p53-selective antibody PAb 246 (MILNER and COOK 1986; YEWDELL et al. 1986), whereas p53 capable of association with hsp72/73 does not express this epitope (FINLAY et al. 1988; JENKINS et al. 1988; STÜRZBECHER et al. 1987, 1988a; TAN et al. 1986), again arguing for at least two different conformations of p53. Non-primate p53s competent to bind to T antigen profoundly suppress SV40 origin-dependent DNA replication in permissive monkey COS cells (BRAITHWAITE et al. 1987; STÜRZBECHER et al. 1988b), and mouse p53 is capable of displacing DNA polymerase α from T-antigen-Pol α complexes in in vitro assays (GANNON and LANE 1987).

To study in more detail the structural and biochemical requirements for this p53-mediated suppression of replication, we constructed a mouse p53 expression library of C → T transition mutants by methoxylamine treatment of + strand DNA in M13 followed by transfer of mutagenised + strand/wild-type — strand heteroduplexes into the SV40 origin-containing vector pBC12CMV (CULLEN 1986). Following transfection into *Escherichia coli* DH5, individual clones were grown up in small scale culture and plasmid DNA assayed for DNA replication by transient

| | Mutation - | | Replication % | Localization | T-binding | Hsp-binding |
| | Amino-Acid residue | Amino-Acid transition | | | | |
| --- | --- | --- | --- | --- | --- | --- |
| Cmv IL2 | - | - | 100 | - | - | - |
| wt p53 | - | - | 6 | n | + | − |
| 4 | 274 | Pro > Ser | 84 | c/n | − | − |
| | 314 | Gln > Ochre | | | | |
| 8 | 274 | Pro > Leu | 78 | c/n | + | + |
| 187 | 1-157 | | 127 | n | + | − |
| 170 | 215 | Pro > Ser | 64 | c/n | + | + |
| 146 | 175 | His > Tyr | 43 | c/n | + | + |
| 115 | 94 | Pro > Leu | 20 | n/c | + | + |

**Fig. 1.** Characterization of mouse p53 point mutants. The coding strand of wild-type mouse p53 in M13 DNA was treated with methoxylamine, resulting in 7-methoxy-dC, to create a library of single-point mutant p53 cDNA. After annealing to the non-coding strand, the double-stranded p53 fragment was transferred into the pBC12CMV expression vector (CULLEN 1986) and the plasmids grown in competent *Escherichia coli* DH5. 7-Methoxy-dC generates C → T transitions in the coding strand of p53 cDNA during plasmid replication in *E. coli*.

Replication experiments were carried out by transfecting these plasmid DNAs containing a functional SV40 origin of replication into SV40-transformed monkey COS cells as described previously (BRAITHWAITE et al. 1987). At 72 h post-transfection plasmid DNA was isolated and digested with *Dpn*I restriction endonuclease; the digestions were used to transform competent DH5 *E. coli*. *Dpn*I-resistant DNA molecules representing the replicated fraction transform *E. coli* efficiently and are thus a measure of plasmid replication. The data concerning subcellular localisation of mutant p53 proteins are based on immunofluorescence studies using monoclonal p53-specific antibodies PAb242, PAb246 (YEWDELL et al. 1986), PAb200-47 (DIPPOLD et al. 1981) and PAb421 (HARLOW et al. 1981), *n*, nucleus; *c*, cytoplasm. T antigen and hsp72/73 binding studies were performed with [$^{35}$S]-methionine-labelled mutant p53 from transfected COS cells as described in Fig 2

expression in COS cells as described elsewhere (BRAITHWAITE et al. 1987; STÜRZ-
BECHER et al. 1988b). Candidate clones which scored as having altered activity in
this assay were grown up in a large scale culture, the plasmid DNA banded
twice in cesium chloride gradients and retested. To eliminate the possibility of gross
cloning artefacts and to obtain preliminary mapping data, the coding sequence of each
candidate clone was subdivided using convenient restriction sites, and each subfragment was transferred into wild-type context by ligation into the appropriate position
in an unmutagenised p53-expressing pBC12CMV construct (CMVmsp53; BRAITH-
WAITE et al. 1987; STÜRZBECHER et al. 1988b). These subfragment clones were then
reassayed repeatedly by transient expression. Figure 1 gives a summary of these
studies for different mutant p53 constructs. The data represent, in each case, the
mean values of at least 35 separate transfections. While wild-type mouse p53
(CMVmsp53) suppresses plasmid replication to about 6% of the non-p53-expressing
control plasmid (CMVori) level, all the plasmids expressing these representative
mutant p53 proteins either inhibit poorly or not at all.

We characterised these mutants in more detail. Figure 2 shows immunoprecipitation
of [$^{35}$S]-methionine-labelled mutant p53 proteins following transient expression in
COS cells. Mutant 8 protein coprecipitates with the T antigen-specific monoclonal

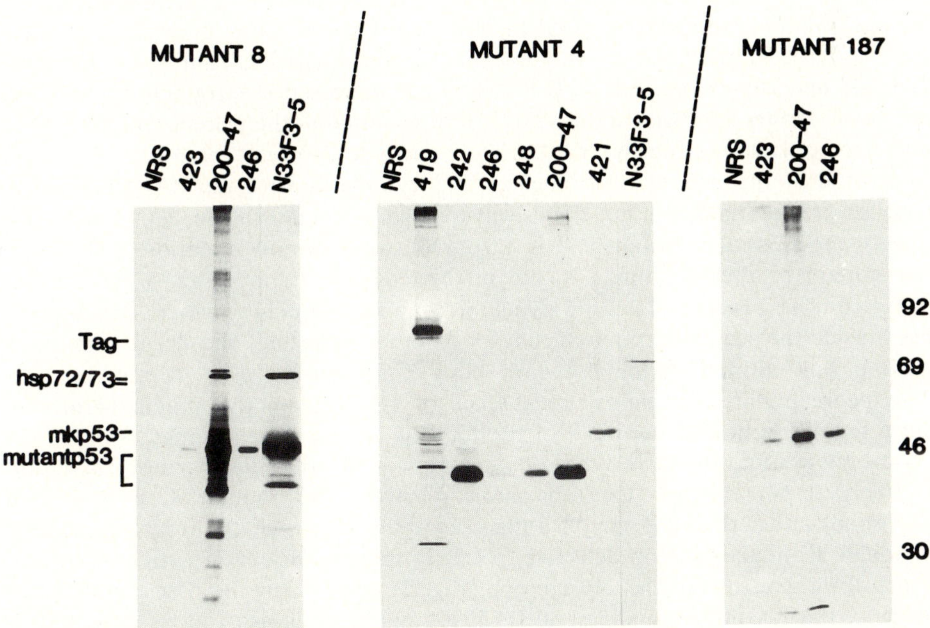


Fig. 2. T antigen and/or hsp 72/73 binding of selected mouse p53 point mutants. [$^{35}$S]-Methionine-
labelled p53 point mutants (see Fig. 1) from transfected COS cells were immunoprecipitated
using different monoclonal antibodies and subjected to SDS-PAGE and autoradiography as described
previously (STÜRZBECHER et al. 1987). *NRS*, normal rabbit serum; *423*, monoclonal T antigen-
specific antibody (HARLOW et al. 1981); *242, 246, 248* (YEWDELL et al. 1986), *200-47* (DIPPOLD
et al. 1981), *419, 421* (HARLOW et al. 1981), monoclonal p53-specific antibodies; *N33F3-5*,
monoclonal hsp72/73-specific antibody (W. J. WELCH, unpublished data); *Tag*, SV40 T antigen;
*hsp72/73*, heat shock proteins 72/73K; *mkp53*, monkey p53

antibody PAb 423 (HARLOW et al. 1981), indicating that this mutant protein retains the ability to complex with T antigen. Additionally, a subset of mutant 8 protein precipitates with mouse p53-specific monoclonal antibody PAb 246, which in previous work has been found to correlate well with T-antigen binding (MILNER and COOK 1986; YEWDELL et al. 1986). However, the bulk of mutant 8 protein fails to express the 246 epitope or to bind T antigen but precipitates with monoclonal PAb 200-47 (DIPPOLD et al. 1981) and in complex with hsp72/73. Indirect immunofluorescence of transfected and fixed cells shows that PAb 200-47 reactive mutant 8 protein distributes throughout both nucleus and cytoplasm, while the PAb 246 reactive subclass is restricted to the nucleus (data not shown). DNA sequence analysis reveals that the point mutation in mutant 8 is, as predicted, a C → T transition, which results in a change from Pro → Leu at amino acid residue 274. Residue 274 lies within the second of the two regions of primary amino acid sequence implicated in stable complex formation with T antigen (JENKINS et al. 1988). An attractive explanation for the replication pheno-type of mutant 8 is, therefore, that this mutant protein is defective in replication suppression simply because the residue 274 lesion is sited within sequences required for T-antigen binding and in some way directly compromises, but does not abolish, such binding.

Mutant 4 is representative of a second distinct class of suppression-defective mutant. Immunoprecipitation of labelled mutant 4 protein (Fig. 2) produces a poly-peptide that is immunoreactive with anti-mouse-p53 monoclonal antibodies PAb242, 246, 248 and 200-47, but not with PAb421, has increased electrophoretic mobility, and binds neither T antigen nor hsp72/73. Indirect immunofluorescence of transfected and fixed cells shows that both PAb200-47 and PAb246 reactive mutant 4 protein is distributed throughout both nucleus and cytoplasm with the PAb246 reactive subclass preferentially, but not exclusively, nuclear (data not shown). DNA sequence analysis reveals that mutant 3 has acquired two separate mutations. The first, like mutant 8, alters residue 274, but in this case results in a Pro → Ser change. The second involves residue 314, giving rise to a Glu → ochre change. Residue 314 lies outside the sequence requirements for T-antigen binding, and deletion mutants lacking this entire C-terminal region bind T antigen in vivo (JENKINS et al.). This ochre mutation explains the absence of PAb421 reactive mutant 4 protein since the 421 epitope C-terminal of residue 314 lies between residues 370 and 378 (WADE-EVANS and JENKINS 1985). It is highly likely that the residue 274 lesion is solely responsible for the replication phenotype of mutant 4, in this case abolishing rather than reducing T-antigen binding. However, the resulting protein, although T-antigen-binding defective, partially retains the conformational require-ments for expression of the 246 epitope (Fig. 2) and is thus the first experimental evidence that T-antigen binding and PAb246 immunoreactivity can be separated.

Mutant 187 is representative of a third distinct class of p53 mutant. Immuno-precipitation (Fig. 2) shows that mutant 187 protein binds to T antigen, expresses the 246 epitope and does not complex with hsp72/73. Indirect immunofluorescence of transfected and fixed cells shows that mutant 187 protein is exclusively nuclear (data not shown). Subfragment mapping data localise the lesion in mutant 187 as being N-terminal of amino acid residue 157, which places it outside the sequence blocks implicated in T-antigen binding (JENKINS et al. 1988). Analysis of mutant 187 p53/T-antigen complex stability by in vitro association and incremental SDS concentration washes (data not shown) shows no decrease in stability. Sucrose gradient analysis of in vivo-labelled mutant 187 p53 shows that this mutant accumulates indistinguishably from wild-type mouse p53 in higher order multimeric forms and in complex with T antigen (data not shown).

We conclude that T-antigen binding, although necessary, is insufficient for p53-mediated suppression of SV40 DNA replication and that a second distinct activity – with sequence requirements in the *N*-terminal region of p53 – is implicated in suppression.

To gain experimental access to the mechanisms underlying suppression of SV40 DNA replication by p53 we reconstructed this activity in an in vitro DNA replication system as described by LI and KELLY (1984) using HeLa cell lysates, immuno-affinity purified T-antigen (SIMMANIS and LANE 1985) and p53 protein. The p53 purification was performed using an immunoaffinity column with covalently linked p53-specific antibody PAb-122. The p53 was eluted from the antibody with excess peptide corresponding to the PAb421/122 epitope as we describe elsewhere (WADE-EVANS and JENKINS 1985). Figure 3 shows different purified p53 proteins and T antigen: dl162p53 (JENKINS et al. 1985) is an internal deletion mutant of mouse p53 which binds to T antigen and inhibits SV40 DNA replication in vivo (BRAITHWAITE et al. 1987; STÜRZBECHER et al. 1988b); dl518p53 (JENKINS et al. 1985) is a non-inhibiting, hsp72/73-binding, mouse p53 mutant (BRAITHWAITE et al. 1987; STÜRZBECHER et al. 1988b).

These purified p53 proteins were tested for their effects on SV40 DNA replication in vitro (Fig. 4). As a control for the specificity of the system, a plasmid carrying a defective SV40 origin of replication was analysed in the presence or absence of T antigen. Whereas SV40 DNA is readily replicated in the presence of T antigen, no replication occurs with the origin minus plasmid. Addition of 100 ng dl162p53

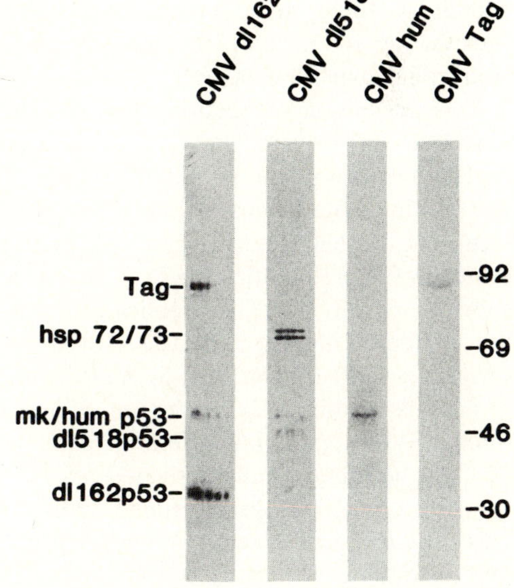


**Fig. 3.** Immunoaffinity purified T antigen and p53 proteins. T antigen was purified from transfected COS cells as described by SIMANIS and LANE (1985). The p53 proteins were purified from transfected COS cells by immunoaffinity chromatography using PAb122 covalently linked to protein A sepharose. The p53 was eluted from the antibody with excess peptide corresponding to the PAb 122 epitope as described before (WADE-EVANS and JENKINS 1985). *Tag*, SV40 T antigen; *hsp 72/73*, heat shock proteins 72/73K; *mk/hum p53*, monkey/human p53; *dl518p53*, *dl162p53*, internal deletion mutants of mouse p53 (JENKINS et al. 1985; STÜRZBECHER et al. 1987)

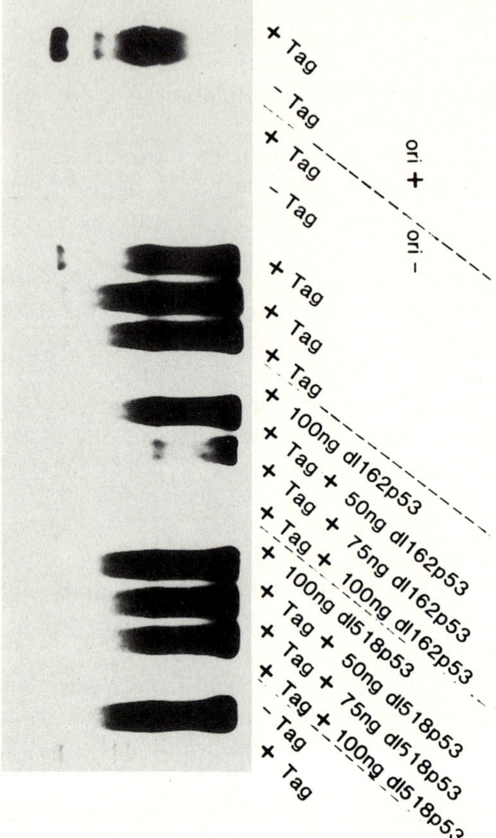



**Fig. 4.** Inhibition of SV40 DNA replication by mouse p53 in vitro. The in vitro DNA replication reaction was performed as described by Li and Kelly (1984) using HeLa cell lysates, immunoaffinity purified T antigen and p53 protein (see Fig. 3) and SV40 DNA. *ori +*, SV40 DNA containing a functional origin of replication; *ori −*, SV40 DNA containing a defective origin of replication (*Sfi*I cut, blunt ended and religated); *Tag*, +450 ng SV40 T antigen; *dl162p53*, internal deletion mutant of mouse p53, competent to bind to T antigen (Jenkins et al. 1985; Stürzbecher et al. 1987); *dl518p53*, internal deletion mutant of mouse p53, incompetent to bind to T antigen (Jenkins et al. 1985; Stürzbecher et al. 1987)

protein to the system almost completely blocks replication. On the other hand, introducing 100 ng dl518p53 into the system has little effect on DNA replication. Equivalent experiments using immunopurified human p53 show that, as in vivo, it does not inhibit replication (data not shown).

A time-course-based analysis of restriction fragments derived from $^{32}$P-labelled replication intermediates indicates that the replication block imposed by mouse p53 in this system must be localised either at or close to the initiation of DNA synthesis, rather than during elongation or decatenation, since there is no evidence for preferential labelling of restriction fragments at or near *ori* (data not shown); or all stages of replication must be blocked to the same extent. We are currently investigating the effects of p53 on the other known biochemical properties of T antigen such as ATPase and helicase activity.

The observation that expression of non-primate but not human p53 proteins can inhibit SV40 replication offers an explanation of the apparent contradiction that p53 can be mutationally activated and scores as a dominant transforming gene in transfection assays, yet is functionally absent in several transformed cell lines. Also, the p53 gene is selectively lost during Friend virus-mediated leukaemogenesis. One intriguing possibility is that activated p53 mutants such as dl162 and dl518

(JENKINS et al. 1985) score in transformation assays simply because they are in some way functionally defective rather than being functionally "enhanced". It is becoming increasingly clear that p53 participates in complexes with a variety of proteins such as T antigen, adenovirus E1b 57K, hsp72/73 and in self oligomers. Any functionally defective p53 proteins which retain the ability to complex with some cellular target could compete with wild-type p53 for these sites, and it may be significant that all the activated p53 mutants analysed are stable and over-expressed relative to wild-type p53. The use of T antigen and SV40 replication as a model system may offer the possibility of identifying the role of p53 in cell proliferation and in malignancy, and the preliminary data we have obtained using this system suggest that p53 as an oncogene may act by perturbing either positively or negatively the recruitment of a subset of cellular DNA replication origins resulting in aberrant S-phase-linked events.

*Acknowledgements.* We thank Jenny Mollan for excellent secretarial services.

# References

Braithwaite AW, Stürzbecher H-W, Addison C, Palmer C, Rudge K, Jenkins JR (1987) Mouse p53 inhibits SV40 origin-dependent DNA replication. Nature 329: 458–460

Crawford L (1983) The 53,000-dalton cellular protein and its role in transformation. Int Rev Exp Pathol 25: 1–50

Cullen BR (1986) *Trans*-activation of human immunodeficiency virus occurs via a bimodal mechanism. Cell 46: 973–982

Dippold WG, Jay G, DeLeo AB, Khoury G, Old LJ (1981) p53 transformation-related protein: detection by monoclonal antibody in mouse and human cells. Proc Natl Acad Sci USA 78: 1695–1699

Eliyahu D, Raz A, Gruss P, Givol D, Oren M (1984) Participation of p53 cellular tumour antigen in transformation of normal embryonic cells. Nature 312: 646–649

Finlay CA, Hinds PW, Tan T-H, Eliyahu D, Oren M, Levine AJ (1988) Activating mutations for transformation by p53 produce a gene product that forms an hsc70-p53 complex with an altered half-life. Mol Cell Biol 8: 531–539

Gannon JV, Lane DP (1987) p53 and DNA polymerase α compete for binding to SV40 T-antigen. Nature 329: 456–458

Harlow E, Crawford LV, Pim DC, Williamson NM (1981) Monoclonal antibodies specific for simian virus 40 tumor antigens. J Virol 39: 861–869

Hinds PW, Finlay CA, Frey AB, Levine AJ (1987) Immunological evidence for the association of p53 with a heat-shock protein, hsc 70, in p53-plus-*ras*-transformed cell lines. Mol Cell Biol 7: 2863–2869

Jenkins JR, Stürzbecher H-W (1988) The p53 oncogene. In: Reddy EP (ed) The oncogene handbook. Elsevier, Amsterdam

Jenkins JR, Rudge K, Currie GA (1984) Cellular immortalization by a cDNA clone encoding the transformation-associated phosphoprotein p53. Nature 312: 651–654

Jenkins JR, Rudge K, Chumakov P, Currie GA (1985) The cellular oncogene p53 can be activated by mutagenesis. Nature 317: 816–818

Jenkins JR, Chumakov P, Addison C, Stürzbecher H-W, Wade-Evans A (1988) Two distinct regions of the murine p53 primary amino acid sequence are implicated in stable complex formation with simian virus 40 T antigen. J Virol 62: 3903–3906

Lane DP, Crawford LV (1979) T-antigen is bound to a host protein in SV40-transformed cells. Nature 278: 261–263

Li JJ, Kelly TJ (1984) Simian virus 40 DNA replication in vitro. Proc Natl Acad Sci USA 81: 6973–6977

Linzer DIH, Levine AJ (1979) Characterization of a 54K dalton cellular SV40 tumor antigen present in SV40 transformed cells and uninfected embryonal carcinoma cells. Cell 17: 43–52

Milner J, Cook A (1986) The cellular tumour antigen p53: evidence for transformation-related, immunological variants pf p53. Virology 154: 21–30

Oren M (1985) The p53 cellular tumor antigen: gene structure, expression and protein properties. Biochim Biophys Acta 823: 67–78

Parada LF, Land H, Weinberg RA, Wolf D, Rotter V (1984) Cooperation between gene encoding p53 tumour antigen and ras in cellular transformation. Nature 312: 649–651

Rotter V, Wolf D (1985) Biological and molecular analysis of p53 cellular-encoded tumor antigen. Adv Cancer Res 43: 113–141

Rovinski B, Benchimol S (1988) Immortalization of rat embryo fibroblasts by the cellular p53 oncogene. Oncogene 2: 445–452

Rovinski B, Munroe D, Peacock J, Mowat M, Bernstein A, Benchimol S (1987) Deletion of 5'-coding sequences of the cellular p53 gene in mouse erythroleukemia: a novel mechanism of oncogene regulation. Mol Cell Biol 7: 847–853

Sarnow P, Ho YS, Williams J, Levine AJ (1982) Adenovirus E1b-58Kd tumor antigen and SV40 large tumor antigen are physically associated with the same 54Kd cellular protein in transformed cells. Cell 28: 387–394

Simanis V, Lane DP (1985) An immunoaffinity purification procedure for SV40 large T antigen. Virology 144: 88–100

Stürzbecher H-W, Chumakov P, Welch WJ, Jenkins JR (1987) Mutant p53 proteins bind hsp 72/73 cellular heat-shock-related proteins in SV40-transformed monkey cells. Oncogene 1: 201–211

Stürzbecher H-W, Addison C, Jenkins JR (1988a) Characterization of mutant p53-hsp72/73 protein-protein complexes by transient expression in monkey COS cells. Mol Cell Biol 8: 3740–3747

Stürzbecher H-W, Braithwaite AW, Addison C, Palmer C, Rudge K, Lynge-Hansen D, Jenkins JR (1988b) p53 inhibits DNA synthesis from the SV40 origin of replication. In: Cancer Cells, vol 6: eukaryotic DNA replication. Cold Spring Harbor Laboratory, Cold Spring Harbor pp. 159–163

Tan T-H, Wallis J, Levine AJ (1986) Identification of the p53 protein domain involved in formation of the simian virus 40 large T-antigen-p53 protein complex. J Virol 59: 574–583

Wade-Evans A, Jenkins JR (1985) Precise epitope mapping of the murine transformation-associated protein, p53. EMBO J 4:699–706

Yewdell JW, Gannon JV, Lane DP (1986) Monoclonal antibody analysis of p53 expression in normal and transfected cells. J Virol 59: 444–452

# Interactions Between SV40 and Cellular Oncogenes in the Transformation of Primary Rat Cells

D. Michalovitz[1], A. Amsterdam[2], and M. Oren[1]

## 1 Introduction

The transforming activity of SV40 large T antigen is due to the ability of this protein to interfere with and modify a variety of cellular regulatory processes (Tooze 1981). Cellular proto-oncogene products have been shown to play key roles in such normal processes (Weinberg 1985; Bishop 1987). It is thus conceivable that SV40-mediated transformation may both affect the functioning of certain proto-oncogenes and, reciprocally, be influenced by the particular proto-oncogene expression patterns and oncogene activation events present in various cells. In order to gain further insight into this aspect of large T-antigen action, we investigated the ability of specific activated cellular oncogenes to play a role in SV40-mediated transformation. The studies described below demonstrate that such cellular proteins may indeed dramatically affect the ability of SV40 to cause transformation. Furthermore, the aberrant expression of cellular oncogenes may greatly influence the biochemical and physiological consequences of large T-antigen action in the transformed cell.

## 2 Methods and Experimental Procedures

### 2.1 Cells, plasmids, and Transfections

Primary rat embryo fibroblasts were prepared from 16-day-old embryos and maintained in 10% DMEM supplemented with 10% fetal calf serum (Eliyahu et al. 1984). Primary granulosa cells were prepared from diethylstilbesterol-treated immature female rats and maintained in DMEM/F12 medium containing 5% fetal calf serum (Ben-Zeev and Amsterdam 1987). The plasmids used were: pSVBam, containing wild-type SV40 DNA cloned in the *Bam*HI site of pBR322 (Michalovitz et al. 1986); pLTRp53cG, directing the expression of mouse p53 (Eliyahu et al. 1985); pLTRcGXK, which is a derivative of pLTRp53cG from which most of the p53 protein coding region has been removed (Kaczmarek et al. 1986); pEJ6.6,

Departments of [1]Chemical Immunology and [2]Hormone Research, The Weizmann Institute of Science, Rehovot 76100, Israel

Current Topics in Microbiology and Immunology, Vol. 144
© Springer-Verlag Berlin · Heidelberg 1989

containing a mutationally activated human *Ha-ras* oncogene (SHIH and WEINBERG 1982); and pF8dl, p536dl (SOMPAYRAC et al. 1983), dl1001, dl1047, dl1055, dl1061, dl1135, dl1151 (PIPAS et al. 1983), and pACTSV (FISCHER-FANTUZZI et al. 1986), all of which carry transformation-defective mutant SV40 DNA variants. Transfections were performed by the calcium phosphate method, as described before (ELIYAHU et al. 1984).

## 2.2 Protein Analysis

Cells were metabolically labeled with [$^{35}$S]methionine and extracted in a buffer containing 50 m$M$ Tris, pH 8.0, 5 m$M$ EDTA, 150 m$M$ NaCl, 0.5% NP40 0.3 mg/ml PMSF, and 1% Aprotinin (Sigma, St. Louis, MO) as described before (MALTZMAN et al. 1981). Aliquots of the extract were subjected to immuno-precipitation with appropriate monoclonal antibodies, and reactive polypeptides were resolved on a 12.5% SDS-polyacrylamide gel as described before (MALTZMAN et al. 1981).

## 2.3 Determination of Progesterone Production

Progesterone concentrations in cells and medium were determined by a radio-immunoassay (KOHEN et al. 1975).

# 3  Results and Discussion

## 3.1 Effect of p53 on SV40-Mediated Transformation

In SV40-transformed cells the viral large T antigen forms a tight specific complex with the cellular p53 protein (LANGE and CRAWFORD 1979; McCORMICK and HARLOW 1980). This is usually correlated with a dramatic stabilization of p53, leading to a significant increase in p53 levels (OREN et al. 1981). Since p53 overproduction can contribute, under some circumstances, to the induction of neoplastic transformation (ELIYAHU et al. 1984; PARADA et al. 1984), it was of interest to find out whether p53-related events may also play a role in this viral transformation process. We therefore explored whether cotransfection with a p53 expression plasmid can enhance the transforming activity of SV40 DNA in the primary rat embryo fibroblast (REF) system.

   The first set of experiments involved the use of wild-type (WT), transformation-competent SV40 DNA. REFs (passage 4 or 5) were transfected with pSVBam (containing strain 776 SV40 DNA cloned into the *Bam*HI site of pBR322), together with pLTRp53cG, a plasmid efficiently directing the expression of mouse p53 (ELIYAHU et al. 1985). It has recently been shown that the p53 protein encoded by this plasmid carries a single amino acid substitution relative to WT

**Table 1.** Transformation of rat embryo fibroblasts by various combinations of plasmids (foci/$10^6$ cells)

| Transfected DNA | Expt I | Expt II | Expt III | Expt IV |
|---|---|---|---|---|
| pLTRp53cG | 0 | 0 | ND[b] | 0 |
| pSVBam + pLTRcGXK | 25 | 15 | ND | 25 |
| pSVBam + pLTRp53cG | 75 | 25 | ND | ND |
| pLTRcGXK | ND | ND | 0 | 0 |
| pF8dl + pLTRp53cG | 16 | 7 | 12 | 3 |
| pF8dl + pLTRcGXK | 0 | 0 | 0 | 0 |
| dl1001 + pLTRp53cG | ND | ND | ND | 5 |
| p536dl + pLTRp53cG | ND | ND | 0 | 0 |

Early passage rat embryo fibroblasts were transfected with the indicated plamids (see below); 5 µg of each DNA were employed except in the case of pSVBam, where 0.5 µg was used. pLTRp53cG encodes full-length p53; pLTRcGXK is similar to the former, but contains a deletion of almost the entire p53 coding sequences; pSVBam contains WT SV40 DNA; pF8dl is a transfor-mation-defective SV40 deletion mutant incapable of binding p53; p536dl is an SV40 mutant with a larger deletion.
[b] Not done.

p53, probably leading to its enhanced transforming activity (FINLAY et al. 1988; ELIYAHU et al. 1988). In a parallel experiment, pLTRp53cG was replaced by pLTRcGXK, which is a derivative of the former from which the majority of p53-encoding sequences have been deleted (KACZMAREK et al. 1986); this plasmid served as a negative control. As seen in Table 1, pSVBam elicited typical SV40-transformed foci irrespective of whether a functional or a deleted p53 plasmid was used for cotransfection. However, the presence of pLTRp53cG consistently led to a higher number of such foci, as observed in multiple, independent experiments. Furthermore, almost all cell lines isolated from such cotransformants did produce very high levels of the transfected mouse p53 (MICHALOVITZ et al. 1986), further supporting the role of this overproduction in the induction of the foci. Hence, it indeed appears that aberrant p53 expression may play a key role in the oncogenic activity of SV40.

The results obtained with WT SV40 were only suggestive, since it can also transform without the addition of a p53 plasmid. To obtain a more definitive answer, use was therefore made of a series of SV40-transformation-defective deletion mutants. These mutants — pF8dl, p536dl, and dl1001 — all fail to transform REF; in addition, they all encode a T antigen which cannot form a detectable complex with p53, and in cells expressing these deleted large T-antigen variants there is no significant increase in cellular p53 levels (SOMPAYRAC et al. 1983). If the transformation defect is, at least partially, due to the inability to affect p53, then it is conceivable that the induction of aberrant p53 expression by independent means may facilitate transformation by these deletion mutants. This concept was tested by cotransfecting each mutant into REF together with either pLTRp53cG or pLTRcGXK. As seen in Table 1 the cointroduction of pLTRp53cG with either pF8dl or dl1001 indeed led to the appearance of morphologically transformed foci; no such foci were obtained with the pLTRcGXK control. Furthermore, the morphology of such cotransfection-derived foci appeared identical to that of REF transformed with WT SV40 (MICHALOVITZ et al. 1986). Thus, it would appear that the p53 plasmid indeed complements the SV40 defect, probably restoring the normal SV40 transformation process. Yet it is noteworthy that the complementation appears rather inefficient when compared with the frequency of focus induction by WT SV40 (Table 1). One trivial explanation could be that since plasmids pF8dl and dl1001 encode extensively deleted versions of SV40 large T antigen, these molecules could lack additional biochemical activities which do not affect p53 expression yet are important for SV40-mediated transformation. This is seemingly in line with the fact that plasmid p536dl, carrying a more extensive deletion in the large T-antigen coding region (SOMPAYRAC et al. 1983), has absolutely no transforming activity even when supplemented with a p53-overexpressing plasmid (Table 1). Alternatively, the biochemical activities induced by the cotransfecting plasmid pLTRp53cG may not be identical to those elicited by p53 in SV40-transformed cells. This latter notion is consistent with the fact that pLTRp53cG causes the overproduction of a mutant p53 (FINLAY et al. 1988; ELIYAHU et al. 1988) whereas the p53 accumulated in SV40-transformed cells is probably of the WT sequence. Furthermore, there may be certain changes in p53 which are specifically dependent on its interaction with an active large T antigen as was suggested from the analysis of p53 phosphorylation sites (SAMAD et al. 1986).

One particular aspect which these experiments allowed us to address was the role of SV40 in the stabilization of p53. It has often been suggested that the stabilization is a direct consequence of the mere formation of p53-large T antigen complexes. Yet recent evidence from DEPPERT and coworkers suggests that this may not be the case (DEPPERT and HAUG 1986; DEPPERT et al. 1987, DEPPERT and STEINMAYER, this volume). In the course of the studies described above, cell lines have been generated which either express both cotransfected SV40 large T antigen and mouse p53 (pSVBam + pLTRp53cG lines) or large T antigen only (pSVBam + pLTRcGXK lines). The latter are typical SV40 transformants, similar to those frequently studied before. In the former cells, however, there is much more p53 (see Fig. 1). Furthermore, the mutant p53 encoded by pLTRp53cG has a reduced affinity for large T antigen when compared with WT p53 (data not shown). This reduced binding to T antigen is correlated with an increased ability to form p53-hsc70 complexes, a property observed for a number of other mutant p53 proteins (TAN et al. 1986; STÜRZBECHER et al. 1987, STÜRZBECHER et al., this volume). Thus, cells cotransfected with pLTRp53cG plus pSVBam are expected to contain a significant amount of non-T-antigen-bound p53. Yet they express a functional, transforming large T antigen and are essentially indistinguishable from regular SV40-transformed cells. It was therefore of interest to ask whether, despite the presence of a fraction of non-T-antigen-bound p53 in the cotransformants, this p53 was still as stable as in standard SV40 transformants. To that end, a pulse chase experiment was performed on S1 cells, transformed by SV40 alone, and on S6 cells, expressing both SV40 large T antigen and the mutant, pLTRp53cG-encoded p53 (Fig. 1).

Several features are evident. First, there is in fact non-T-antigen-bound p53 in S6 cells. Thus, while practically all p53 in S1 cells is brought down by large T antigen-specific antibodies (compare 1-h samples between A and B), in S6 cells the p53-specific antibodies (B) precipitate more p53 that do the large T antigen-specific antibodies (A).

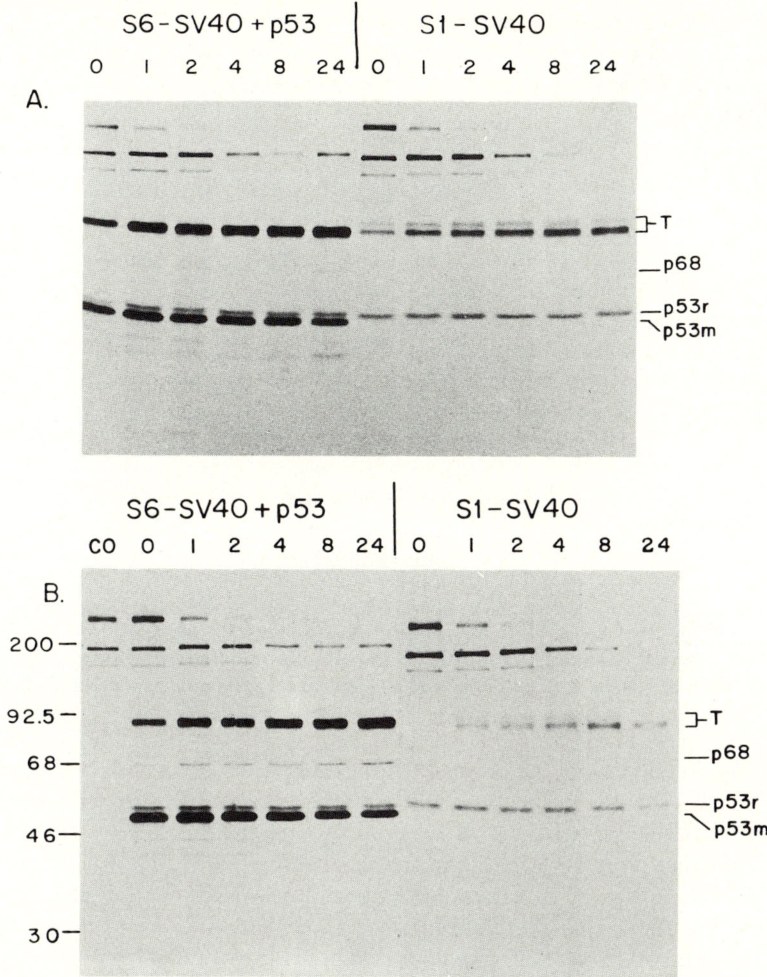



**Fig. 1A, B.** Pulse chase analysis of p53 and large T antigen in SV40-transformed rat embryo fibroblasts. *S1* cells, transformed by pSVBam alone, and *S6* cells, transformed by a combination of pSVBam and pLTRp53cG, were labeled for 45 min with $[^{35}S]$ methionine and subjected to a chase with excess unlabeled methionine for the periods (in hours) indicated on *top*. Extracts containing equal amounts of acid-insoluble radioactivity were immunoprecipitated with either SV40 large T antigen-specific monoclonal PAb419 **(A)** or p53-specific monoclonal PAb421 **(B)**. *T*, large T antigen; *p68*, hsc 70 (Pinhasi-Kimhi et al. 1986); *p53r*, endogenous rat p53; *p53m*, transfected mouse p53

By densitometric scanning (data not shown) it could be shown that after a 1 h-chase about 25% of S6 p53 is not bound to large T antigen. This fraction is probably bound to hsc70 (p68 in Fig. 1), which is more prominent in B than in A. Furthermore, in agreement with previous studies (Oren et al. 1981) the S1 p53 is very stable, and its half-life is well above 24 h (Fig. 2). On the other hand, the p53 in S6, while still quite stable, has a much shorter half-life of about 7 h. Thus, the presence of non-T-antigen-bound p53 does appear to correlate with a more rapid p53 turnover. Furthermore, the decay curve in S6 has a distinct biphasic appearance

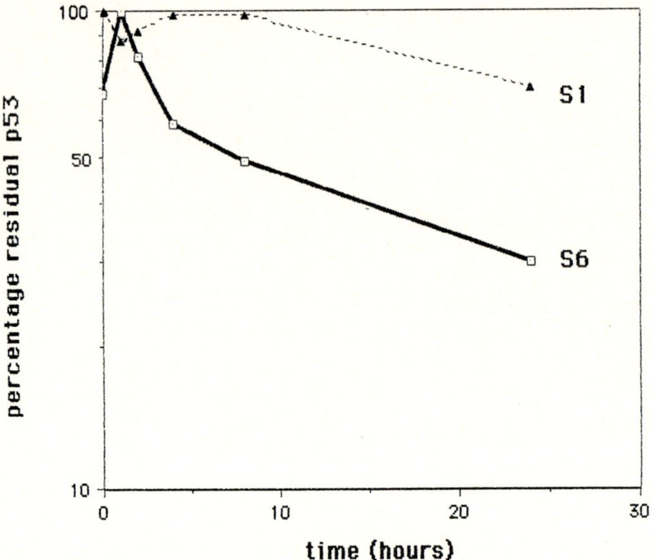



**Fig. 2.** Kinetics of decay of p53 in S1 and S6 cells. The autoradiogram shown in Fig. 1 was quantitated using a Biorad model 620 video densitometer. For the S6 lanes, a shorter autoradiographic exposure was employed in order to increase the accuracy of the measurements. Values are normalized relative to the lane exhibiting the highest p53 levels in each individual line

(Fig. 2), indicating the presence of two subpopulations with different turnover rates. The less stable population has a half-life of about 3 h, very similar to that of p53 complexed with hsc70 in non-SV40-transformed cells (FINLAY et al. 1988). The other population, though, has a half-life of over 20 h. Furthermore, the analysis shown in Fig. 2 suggests that about one–third of the p53 in S6 belongs to the less stable class, a figure in good agreement with the estimate of the fraction of p53 complexed to hsc70 (see above). These facts are highly consistent with the notion that the two stability classes do, indeed, correspond to p53 bound to hsc70 and to large T antigen, respectively.

At face value, these results support the role of a direct p53-large T antigen interaction in the stabilization of the former protein over an indirect, transformation-related change in the cell which eventually slows down p53 degradation. However, we cannot rule out the possibility that the mutation in the transfected p53, which reduces its association with large T antigen, also makes it less susceptible to this indirect stabilization mechanism.

## 3.2 Effect of ras on SV40-Mediated Transformation

In general, p53 is considered a nuclear acting protein, functionally related to the family of "immortalizing" oncogene products (ELIYAHU et al. 1984; PARADA et al. 1984; JENKINS et al. 1984). These proteins are thought to have activities which differ markedly from those of other "cytoplasmic" oncogenes such as the *ras* family

(WEINBERG 1985). It was therefore of interest to see if aberrant *ras* expression could also have an effect on the transforming activity of SV40. To answer that question, various SV40 deletion mutants were tested for their ability to transform REF in the presence of cotransfected activated *ras*. In addition to the previously described plasmids, we used also several other SV40 mutants including pACTSV, which directs the synthesis of a cytoplasmic, rather than a nuclear, SV40 large T antigen (FISCHER-FANTUZZI and VESCO 1985). The results of such cotransformation assays are shown in Table 2. All mutants except p536dl effectively collaborated with activated *ras* in the induction of transformed foci. Cotransformation with *ras* also caused a much more rapid rate of focus induction by WT SV40 (not shown). Unlike in the cotransformation with p53, the *ras* cotransformants had a morphology very different from that of standard SV40 transformants, suggesting that *ras* was affecting cellular processes which are usually not altered by SV40.

The collaboration of pEJ6.6 with pACTSV was particularly unexpected, since both encode cytoplasmic proteins, supposedly incapable of having any direct nuclear activities. To see whether cotransformation was due to the presence of pACTSV revertants with an increased nuclear affinity, we established cell lines from such transformed foci. Analysis of such cells by indirect immunofluorescence failed to reveal any nuclear large T antigen, while this protein was abundant in the cytoplasm (MICHALOVITZ et al. 1987). Thus, although we could not rule out the presence of minute amounts of large T antigen in the nucleus, the results strongly suggest that effective transformation of REF can be achieved by the combined

**Table 2.** Transformation of rat embryo fibroblasts by activated Ha-*ras* and SV40 mutants

| Transfected DNA | Transformation (foci per $10^6$ cells/µg DNA) | |
| --- | --- | --- |
| | Expt I | Expt II |
| p536dl + pEJ6.6 | 0 | 0 |
| pF8dl + pEJ6.6 | 21 | 13 |
| pACTSV + pEJ6.6 | 27 | ND |
| dl1135 + pEJ6.6 | 6 | ND |
| dl1061 + pEJ6.6 | 9 | ND |
| dl1151 + pEJ6.6 | 4 | ND |
| dl1001 + pEJ6.6 | ND | 15 |
| dl1055 + pEJ6.6 | ND | 6 |
| dl1047 + pEJ6.6 | ND | 7 |
| pSVBam | 50 | ND |
| pSVBam + pEJ6.6 | ND | 124 |

$8 \times 10^5$ cells in a 90-mm dish were cotransfected with 4 µg (Expt I) or 5 µg (Expt II) of each plasmid, except pSVBam (0.5 µg). In transfections including pEJ6.6, foci appeared after 9–12 days, whereas with pSVBam alone, foci appeared after 2.5–3 weeks. In parallel, each mutant was cotransfected with pBR322 DNA; no foci were detected in any of these transfections.
ND, Not done.

action of two cytoplasmic proteins. The *ras* cotransformation studies may not be directly related to the natural mode of action of large T antigen, yet they clearly demonstrate that the final outcome of large T-antigen action is greatly dependent on the pattern of expression of cellular regulatory genes, particularly oncogenes.

Analysis of cell lines derived from the *ras* cotransformation studies led to an unexpected finding. When a radiolabeled extract from a pACTSV + *ras* cotransformed line was immunoprecipitated with large T antigen-specific monoclonal antibodies, a new 70K band was observed in addition to the expected large T antigen and p53 (Fig. 3, left). Two-dimensional gel electrophoresis (Fig. 3, right), as well as immunoprecipitation of an extract from heat-shocked cells (data not shown), clearly established that this 70K protein was in fact the rat heat-shock protein cognate (hsc70). Since it has previously been found that hsc70 forms a complex with some types of aberrant p53 (PINHASI-KIMHI et al. 1986; HINDS et al. 1987; STÜRZBECHER et al. 1987), we wondered whether this was the case also in the CTSVR-5 line. A CTSVR-5 extract was therefore subjected to two rounds of precipitation with PAb421, a p53-specific monoclonal antibody (Fig. 4, lanes 1 and 2). The p53-free supernatant (see lane 4) was then reacted with a large T antigen-specific monoclonal. As seen in lane 3, this resulted in the efficient coprecipitation of large T antigen and hsc70, in the obvious absence of any detectable p53. Hence hsc70 must be directly associated with the nonkaryophilic T antigen made in those cells; no such interaction was ever reported for WT large T antigen.

The significance of this hsc70-cytoplasmic T antigen interaction is unclear at the moment. One possibility, however, is that the binding of hsc70 somehow masks the biochemical activity of the misplaced T antigen, probably preventing it from interacting with inappropriate targets in the cytoplasm. In this respect, it is noteworthy that a similar binding to hsc70 is also exhibited by a nontransforming polyomavirus middle T antigen which fails to localize properly in the cell (WALTER et al. 1987). This conclusion is also in line with a suggestion made by PELHAM (1986) concerning the cellular functions of the hsp70 protein family. Nevertheless, we cannot rule out the possibility that the binding of hsc70 actually endows the

**Fig. 3A, B.** Coprecipitation of a nonkaryophilic large T antigen and a 70K polypeptide. Top, ▶ One-dimensional gel analysis. Cells in a 90-mm dish were labelled with 85 μCi [$^{35}$S]-methionine for 4.5 h. Extracts containing equal amounts of trichloroacetic acid-insoluble radioactivity were immunoprecipitated with the following antibodies: *T*, PAb419, anti SV40 T antigen; *T'*, PAb412, anti-SV40 large T antigen; *P*, PAb421, anti-p53; *C*, control hybridoma culture medium. Line *CTSVR-5* was established from an individual focus derived by cotransfection with pEJ6.6 plus pACTSV, encoding a nonkaryophilic large T-antigen mutant. Bottom, Two-dimensional polyacrylamide gel comparison of p70 and hsc70. **A** Total cell extract from Rat-1 cells; **B** total cell extract from stressed Rat-1 cells; **C** immunoprecipitate from CTSVR-5 cells, using PAb419 T antigen-specific antibody. Rat-1 cells were either untreated **(A)** or incubated overnight with 25 μ*M* CdCl$_2$ **(B)**. Cells were labelled with [$^{35}$S]-methionine, and samples were directly processed for electrophoresis. For **C**, CTSVR-5 cells were extracted and immunoprecipitated with pAb419. Immunoprecipitated proteins were subjected to two-dimensional gel electrophoresis. The positions of actin (*a*), vimentin (*v*), the heat-shock protein cognate hsc70 (*70*, panels **A** and **B**), the inducible heat shock protein hsp70 (*68*), and the p70 coprecipitating with the cytoplasmic large T of CTSVR-5 (*70*, panel **C**) are indicated

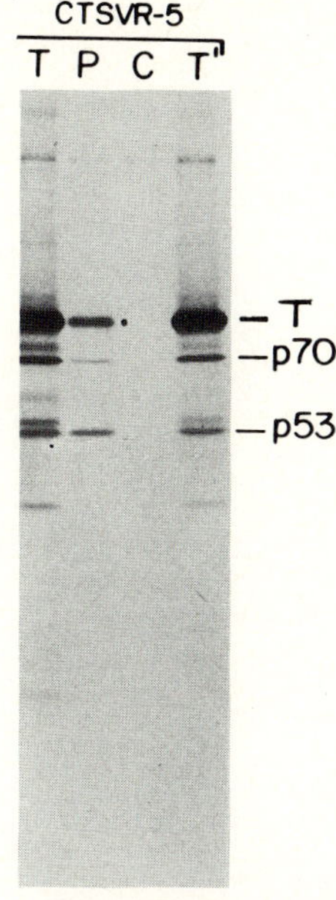
CTSVR-5
T  P  C  T"
—T
—p70
—p53

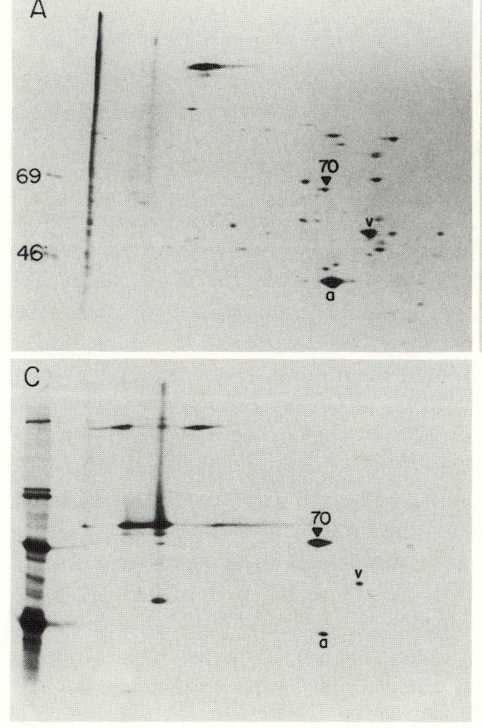
A
69
46
70
v
a
C
70
v
a

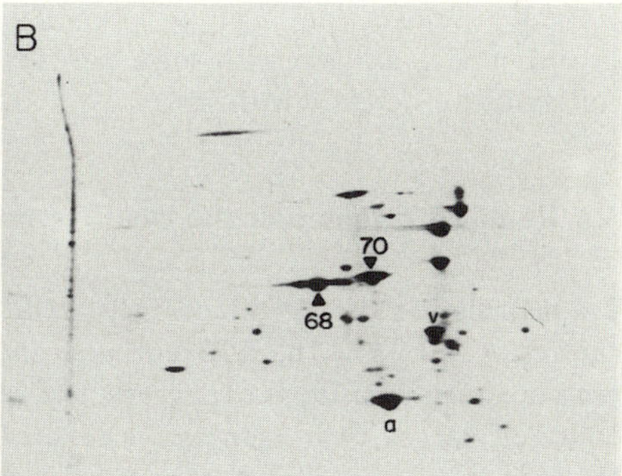
B
70
68
v
a

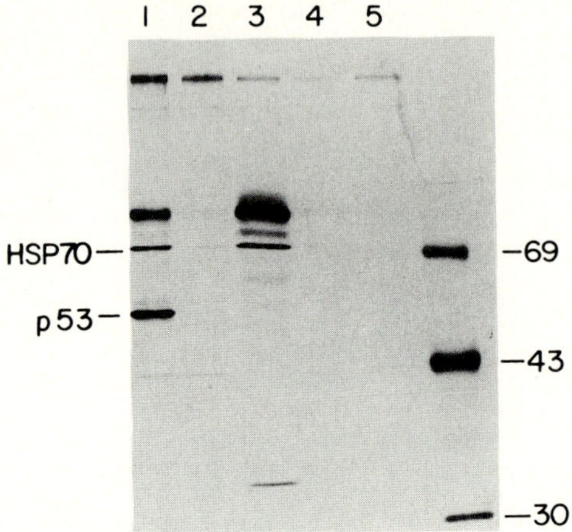



**Fig. 4.** Analysis of hsc70-T antigen complexes in cells expressing a nonkaryophilic large T antigen. CTSVR-5 cells were labelled with [$^{35}$S]methionine, extracted, and subjected to serial immunoprecipitation. First, the extract was cleared twice with PAb421 (lanes *1* and *2*). The resulting supernatant was divided into three equal parts which were reacted with PAb419 (lane *3*), PAb421 (lane *4*), or control hybridoma culture medium (lane *5*). *HSP70* refers to the position of hsc70 (see text)

cytoplasmic T antigen with some new activities, and that these explain its ability to collaborate efficiently with *ras*.

## 3.3 Ras Can Counteract the Dedifferentiation Effect of SV40

Transformation by SV40 of cells exhibiting particular differentiated properties can result in loss of the differentiated phenotype (GARCIA et al. 1986). Experiments presented below, using the rat granulosa cell system, suggest that this dedifferentiation effect of large T antigen may be at least partially counteracted by activated Ha-*ras*.

Granulosa cells exhibit a highly differentiated phenotype. In particular they can be induced to make and secrete progesterone either by treatment with cyclic AMP-dependent gonadotropic hormones or by directly elevating intracellular cAMP levels (KNECHT et al. 1981). The effect of SV40 on these cells was investigated by transfecting primary rat granulosa cells with pSVBam and establishing stably transformed cell lines. When tested for the induction of progesterone either by hormones or by directly increasing cAMP levels, all these lines were totally nonresponsive (Table 3, Gs-8 and Gs-14). However, when similar lines were established by cotransformation of rat granulosa cells with SV40 + *ras*, a different picture emerged. While the resultant lines were still not inducible by hormones, they now exhibited a substantial level of progesterone production when cAMP was directly increased by a variety of agents (Table 3, Grs-21 and Grs-28). These findings strongly suggest that the overexpression of an activated *ras* oncogene can counteract the dedifferentiation effect of SV40 large T antigen. Theoretically, *ras* could influence directly the activity of large T antigen, in which case one

**Table 3.** Progesterone synthesis in primary cultures of granulosa cells (G), in cells transformed by SV40 alone (Gs-8, Gs-14), and in cells cotransformed by SV40 + Ha-*ras* (Grs-21, Grs-28)

| Cell type | G | Gs-8 | Gs-14 | Grs-21 | Grs-28 |
|---|---|---|---|---|---|
| No addition | 0.3 | 0 | 0 | 0 | 0 |
| FSH | 9.0 | 0 | 0 | 0 | 0 |
| 8-Br-cAMP | 14.0 | 0 | 0 | 21.0 | 10.0 |
| Forskolin | 20.0 | 0 | 0 | 13.0 | 10.0 |
| Cholera toxin | 17.0 | 0 | 0 | 10.0 | 9.0 |
| Isoproterenol | 4.8 | 0 | 0 | 4.8 | 2.1 |

Progesterone concentrations in cells and medium were determined by a radio-immunoassay. Data on progesterone levels following incubation with various stimulants are given in $ng/10^6$ cells per 48 h. All determinations were performed twice in duplicate culture plates. (Individual values did not deviate by more than $\pm 7\%$ from the mean). The various stimulants were used at the following concentrations: follicle-stimulating hormone (FSH), 250 ng/ml; 8-bromo-cAMP, $10^{-3}$ $M$; forskolin, $10^{-4}$ $M$; cholera toxin, 100 ng/ml; isoproterenol, $10^{-4}$ $M$.

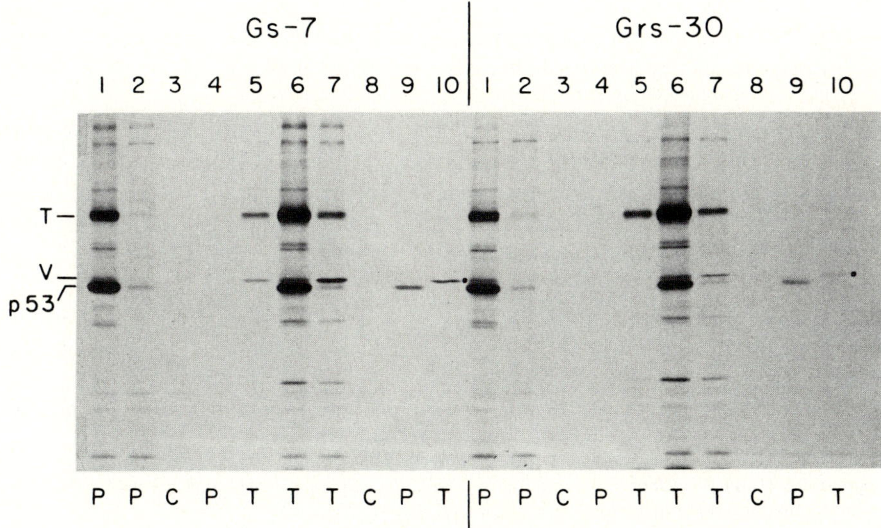


**Fig. 5.** Serial immunoprecipitation of p53 and large T antigen from transformed granulosa cells. Lines *Gs-7* and *Grs-30*, derived by transfection of rat granulosa cells with either SV40 alone or SV40 plus Ha-*ras*, respectively, were labeled with $[^{35}S]$methionine for 3 h. Radioactive extracts of each line containing equal amounts of acid-insoluble radioactivity were first precipitated with p53-specific monoclonal antibody PAb421 (*1*). The supernatant was reprecipitated with the same antibody (*2*). The resultant supernatant was divided into thirds, and each aliquot was reacted with control hybridoma culture medium (*3*), PAb421 (*4*), or the large T antigen-specific monoclonal antibody PAb419 (*5*). Identical extracts were similarly first subjected to two consecutive rounds of precipiation with PAb419 (*6*, *7*), and then the supernatant was divided into thirds and reacted with control medium (*8*), PAb421 (*9*), or PAb419 (*10*). *T*, large T antigen; *V*, a 55K polypeptide which bound nonspecifically (especially with PAb419), identified by two-dimensional gel electrophoresis as vimentin (data not shown)

would expect it to suppress the dedifferentiating effects in many systems. Alternatively, *ras* could play a distinct, specific role in the granulosa system and turn on some differentiated properties in a large T antigen-independent fashion. At present we cannot distinguish between the two alternatives, although there seem to be systems in which *ras* does not reverse the dedifferentiating activity of large T antigen (GARCIA et al. 1986).

In a preliminary attempt to find out whether the presence of the activated *ras* induced gross changes in the behavior of large T antigen, we characterized the interaction between T antigen and p53 in representative cell lines. As can be seen from Fig. 5, the line transformed by *ras* + SV40 (Grs-30) was practically indistinguishable from that transformed by SV40 alone (Gs-7). In both cases, there existed a subpopulation of free large T antigen (lanes 5) as well as of free p53 (lanes 9). Note that each of these lanes represents only one-third of the immunoprecipitation supernatant. Densitometric scanning of lanes 1 plus 2 in comparison with lanes 6 plus 7 revealed that in line Grs-30, 73% of p53 was bound to large T antigen, whereas only 37% of large T antigen was bound to p53; the picture in Gs-7 was very similar. Thus, the apparent reversion of some large T-antigen effects by coexpression of activated *ras* does not appear to operate through perturbation of p53-large T antigen interactions.

All the studies presented above clearly demonstrate that the final consequences of large T antigen action are greatly dependent on the pattern of expression of cellular oncogenes and probably also on other genes in the same cells.

*Acknowledgements.* We wish to thank J. PIPAS, L. SOMPAYRAC, and C. VESCO for the gifts of various mutant SV40 plasmids, and O. PINHASI-KIMHI and G. MEIR for excellent technical assistance. Supported in part by grants from the Minerva Foundation (Munich), the Laub Foundation, and the Leo and Julia Forchheimer Center for Molecular Genetics.

# References

Ben-Zeev A, Amsterdam A (1987) In vitro regulation of granulosa cell differentiation: involvement of cytoskeletal protein expression. J Biol Chem 262: 5366–5376

Bishop JM (1987) The molecular genetics of cancer. Science 235: 305–311

Deppert W, Haug M (1986) Evidence for free and metabolically stable p53 protein in nuclear subfractions of SV40-transformed cells. Mol Cell Biol 6: 2233–2240

Deppert W, Haug M, Steinmayer R (1987) Modulation of p53 protein expression during cellular transformation with SV40. Mol Cell Biol 7: 4453–4463

Eliyahu D, Raz A, Gruss P, Givol D, Oren M (1984) Participation of p53 cellular tumor antigen in transformation of normal embryonic cells. Nature 312: 646–649

Eliyahu D, Michalovitz D, Oren M (1985) Overproduction of p53 antigen makes established cells highly tumorigenic. Nature 316: 158–160

Eliyahu D, Goldfinger N, Pinhasi-Kimhi O, Shaulsky G, Skurnik Y, Arai N, Rotter V, Oren M (1988) Meth A fibrosarcoma cells express two transforming mutant p53 species. (manuscript submitted)

Finlay CA, Hinds PW, Tan TH, Eliyahu D, Oren M, Levine AJ (1988) Activating mutations for transformation by p53 produce a gene product that forms an hsc70-p53 complex with an altered half-life. Mol Cell Biol 8: 531–539

Fischer-Fantuzzi L, Vesco C (1985) Deletion of 43 amino acids in the NH$_2$-terminal half of the large tumor antigen of simian virus 40 results in a non-karyophilic protein capable of transforming established cells. Proc Natl Acad Sci USA 82: 1891–1895

Fischer-Fantuzzi L, Scheidtmann KH, Vesco C (1986) Biochemical properties of a transforming nonkaryophilic T antigen of SV40. Virology 153: 87–95

Garcia I, Sordat B, Rauccio-Farinon E, Dunand M, Kraehenbuhl JP, Diggelman H (1986) Establishment of two rabbit mammary epithelial cell lines with distinct oncogenic potential and differentiated phenotype after microinjection of transforming genes. Mol Cell Biol 6: 1974–1982

Hinds PW, Finlay CA, Frey AB, Levine AJ (1987) Immunological evidence for the association of p53 with a heat-shock protein, hsc 70, in p53-plus-ras-transformed cell lines. Mol Cell Biol 7: 2863–2869

Jenkins JR, Rudge K, Currie GA (1984) Cellular immortalization by a cDNA clone encoding the transformation-associated phosphoprotein p53. Nature 312: 651–654

Kaczmarek L, Oren M, Baserga R (1986) Co-operation between the p53 protein tumor antigen and platelet-poor plasma in the induction of cellular DNA synthesis. Exp Cell Res 162: 268–272

Kohen F, Bauminger S, Lindner HR (1975) In: Cameron EHD, Hillier SG, Griffiths K (ed) Steroid immunoassay, Alpha Omega, Cardiff, pp 11–23

Lane, DP, Crawford LV (1979) T antigen is bound to a host protein in SV-40 transformed cells. Nature 278: 261–263

Maltzman W, Oren M, Levine AJ (1981) The structural relationships between 54,000-molecular-weight cellular tumor antigens detected in viral and nonviral transformed cells. Virology 112: 145–156

McCormick F, Harlow E (1980) Association of a murine 53,000 dalton phosphoprotein with simian virus 40 large-T antigen in transformed cells. J Virol 34: 213–224

Michalovitz D, Eliyahu D, Oren M (1986) Overproduction of protein p53 contributes to simian virus 40-mediated transformation. Mol Cell Biol 6: 3531–3536

Michalovitz D, Fischer-Fantuzzi L, Vesco C, Pipas JM, Oren M (1987) Activated Ha-ras can cooperate with defective simian virus 40 in the transformation of nonestablished rat embryo fibroblasts. J Virol 61: 2648–2654

Oren M, Maltzman W, Levine A (1981) Post-translational regulation of the 54K cellular tumor antigen in normal and transformed cells. Mol Cell Biol 1: 101–110

Parada LF, Land H, Weinberg RA, Wolf D, Rotter V (1984) Cooperation between gene encoding p53 tumor antigen and ras in cellular transformation. Nature 312: 649–651

Pelham H (1986) Speculations on the functions of the major heat shock and glucose-regulated proteins. Cell 46: 959–961

Pinhasi-Kimhi O, Michalovitz D, Ben-Zeev A, Oren M (1986) Specific interaction between the p53 cellular tumor antigen and major heat-shock proteins. Nature 320: 182–185

Pipas JM, Peden KWC, Nathans D (1983) Mutational analysis of Simian Virus 40 T antigen: isolation and characterization of mutants with deletions in the T antigen gene. Mol Cell Biol 3: 203–213

Samad A, Anderson CW, Carroll RB (1986) Mapping of the phosphomonoester and apparent phosphodiester bonds of the oncogene product p53 from SV40-transformed 3T3 cells. Proc Natl Acad Sci USA 83: 897–901

Shih C, Weinberg RA (1982) Isolation of a transforming sequence from a human bladder carcinoma cell line. Cell 29: 161–169

Sompayrac LM, Gurney EG, Danna KJ (1983) Stabilization of the 53,000-dalton nonviral tumor antigen is not required for transformation by simian virus 40. Mol Cell Biol 3: 290–296

Stürzbecher HW, Chumakov P, Welch WJ, Jenkins JR (1987) Mutant p53 proteins bind hsp72/73 cellular heat chock-related proteins in SV40-transformed monkey cells. Oncogene 1: 201–211

Tan TH, Wallis J, Levine AJ (1986) Identification of the p53 protein domain involved in formation of the simian virus 40 large T antigen-p53 protein complex. J Virol 59: 574–583

Tooze J (ed) (1981) DNA tumor viruses. Molecular biology of tumor viruses, pt. 2. Cold Spring Harbor Laboratory, Cold Spring Harbor

Walter G, Carbonne A, Welch WJ (1987) Medium tumor antigen of polyomavirus transformation-defective mutant NG59 is associated with 73-kilodalton heat-shock protein. J Virol 61: 405–410

Weinberg RA (1985) The action of oncogenes in the cytoplasm and nucleus. Science 230: 770–776

# Metabolic Stabilization of p53 in SV40-Transformed Cells Correlates with Expression of the Transformed Phenotype but is Independent from Complex Formation with SV40 Large T Antigen

W. Deppert[1] and T. Steinmayer[2]

## 1 Introduction

One of the most prominent changes in cellular protein expression observed after transformation with SV40 is the drastically enhanced level of the cellular protein p53 (reviewed in Crawford 1983). This protein is involved in the control of cell proliferation and also has a demonstrated oncogenic potential (reviewed in Oren 1985; Rotter and Wolf 1985). This p53 is rapidly turned over in normal cells but becomes metabolically stabilized in SV40-transformed cells. Since p53 in SV40-transformed cells forms a tight complex with the large T antigen, it has long been assumed that in these cells metabolic stabilization of p53 is the result of the physical interaction of these proteins (reviewed in Crawford 1983). However, since SV40-transformed cells, in addition to p53 complexed to large T antigen, also harbor free (i.e., noncomplexed) p53 which is metabolically stable (Deppert and Haug 1986; Deppert et al. 1987), alternative mechanisms for metabolic stabilization of p53 have to be envisioned.

In this paper we address the question whether there is a direct correlation between metabolic stabilization of p53 and expression of the transformed phenotype in SV40-transformed cells. As test system we chose cells derived from the established, normal rat fibroblast line F111 by transformation with the SV40 tsA mutant tsA58 (Pintel et al. 1981). Two sets of transformants were analyzed: (a) N-type transformants, expressing a transformed phenotype at the permissive growth temperature (32 °C) but reverting to the normal phenotype at the nonpermissive growth temperature (39 °C), and (b) A-type transformants, exhibiting a transformed phenotype at both growth temperatures (Pintel et al. 1981). Both types of cells were found to express similar amounts of large T antigen at both growth temperatures and to constitute a matched pair of cells with regard to their cellular phenotypes, growth properties, and the temperature sensitivity of their T antigens at the permissive and at the nonpermissive growth temperatures (Richter and Deppert, manuscript in preparation).

---

[1] Heinrich-Pette-Institut für Experimentelle Virologie und Immunologie an der Universität Hamburg, 2000 Hamburg 20, FRG
[2] Abteilung Biochemie der Universität Ulm, 7900 Ulm, FRG

Current Topics in Microbiology and Immunology, Vol. 144
© Springer-Verlag Berlin · Heidelberg 1989

## 2 Methods and Experimental Procedures

### 2.1 Cells

SV40 wild-type and tsA mutant (tsA58) transformed rat F111 fibroblasts were developed and kindly provided by Dr. Noel Bouck (Chicago). The following cell lines were used and kept in DMEM supplemented with 10% fetal calfserum: FR (wt648), F111 cells transformed with SV40 wild-type virus (strain 648); FR(ts58)A, an SV40 tsA58 mutant virus-transformed N-type transformant of F111 cells; and FR(tsA58)R57, the corresponding SV40 tsA58 mutant virus-transformed A-type transformant of F111 cells (PINTEL et al. 1981). Parental F111 cells grown in DMEM supplemented with 10% fetal calf serum were used as a control.

### 2.2 Labelling of Cells, In Situ Cell Fractionation, Immunoprecipitation

The procedures for cell fractionation (STAUFENBIEL and DEPPERT 1984) and sequential immunoprecipitation of large T antigen and p53 from these extracts (DEPPERT and HAUG 1986; DEPPERT et al. 1987) using the large T antigen-specific monoclonal antibody pAb108 (GURNEY et al. 1986) and the p53-specific monoclonal antibody pAb122 (GURNEY et al. 1980), respectively, have been described in detail in previous publications from this laboratory. A brief description of these procedures is given in the text and the legends to the figures.

## 3 Results

### 3.1 Analysis of Complex Formation and of Metabolic Stabilities

To investigate the relation between metabolic stabilization of p53 and expression of the transformed cellular phenotype, we have compared complex formation and metabolic stabilities of p53 in SV40 wild-type and in SV40 tsA mutant N-type and A-type transformants kept at permissive (32 °C) or at nonpermissive (39 °C) growth temperatures. FR(wt648), FR(ts58)R57, and FR(ts58) A cells were pulse (30 min) and pulse-chase (4-h chase) labelled at 32 °C and at 39 °C, respectively. The labelled cells were then fractionated in situ (STAUFENBIEL and DEPPERT 1984) to yield (i) an NP40 extract, containing the nucleoplasmic subclasses of large T antigen and of p53, (ii) a DNAse/$2\,M$ NaCl extract, containing the chromatin-associated subclasses, and (iii) an Empigen BB extract of the residual cellular structures, containing the subclasses of large T antigen and p53 tightly associated with the nuclear matrix (DEPPERT and HAUG 1986; DEPPERT et al. 1987). This protocol not only allows the preparation of nuclear subclasses of these proteins but also, and more importantly in this regard, the separation of in vivo complexed and noncomplexed subclasses of p53 (DEPPERT and HAUG 1986). The cellular extracts were analyzed by sequential immunoprecipitations first for large T antigen and large T antigen-p53 complexes, using the large T antigen-specific monoclonal antibody pAb108 (GURNEY et al. 1986), and then for free p53, using the p53-specific monoclonal antibody pAb122 (GURNEY et al. 1980).

Figure 1 shows the subnuclear distribution and metabolic stabilities of large T antigen as well as of complexed and free p53. It is evident (Fig. 1A) that,

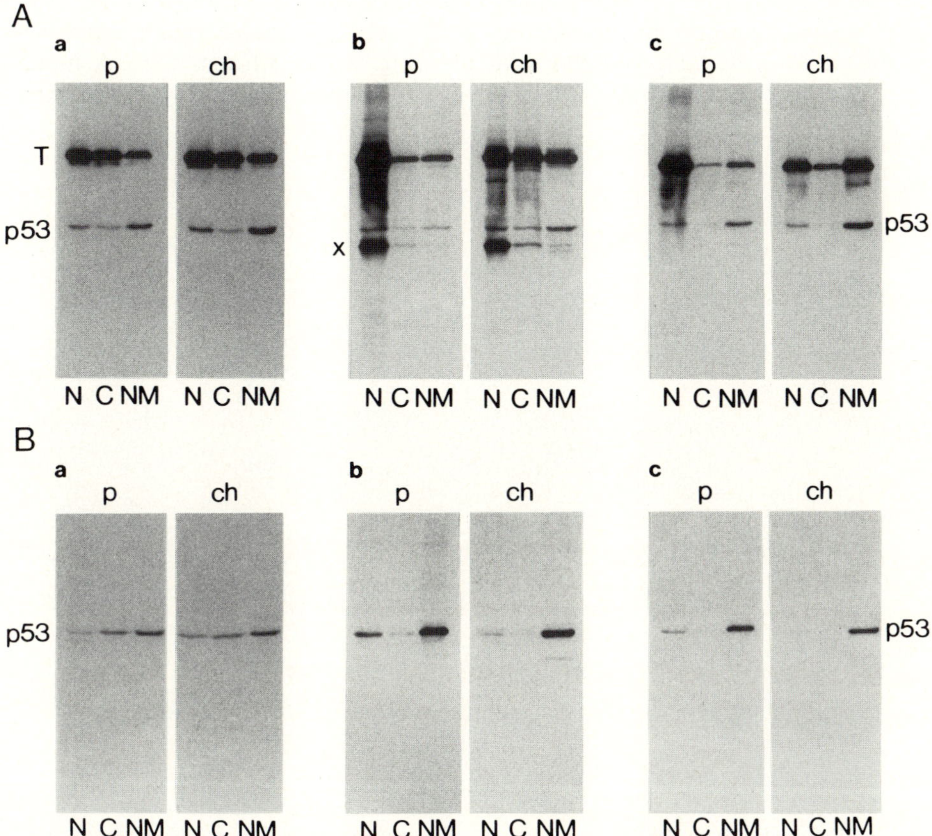

**Fig. 1A, B.** Metabolic stabilities of complexed and free p53 in FR(wt648), in FR(ts58)R57, and in FR(ts58)A cells at 32 °C. A total of $1 \times 10^7$ cells of each group were pulse (30 min) or pulse-chase (4-h chase) labelled with [$^{35}$S]-methionine (50 µCi [$^{35}$S]methionine in 1 ml of labelling medium). Cells were subfractionated into a nucleoplasmic (*N*) a chromatin (*C*) and a nuclear matrix (*NM*) extract as described previously (DEPPERT and HAUG 1986; DEPPERT et al. 1987) and in the text. Nuclear extracts were sequentially immunoprecipitated for large T antigen and large T antigen-p53 complexes with the large T antigen-specific monoclonal antibody pAb108 (GURNEY et al. 1986) (**A**) followed by immunoprecipitation with pAb122 (GURNEY et al. 1980) for free p53 (**B**) using protein A sepharose, as described previously (DEPPERT and HAUG 1986; DEPPERT et al. 1987). *a*, FR(wt648) cells; *b*, FR(ts58)R57 cells; *c*, FR(ts58)A cells; *p*, pulse labelled; *ch*, pulse-chase labelled cells; *x* marks a cellular protein specifically complexing with mutant large T antigen in FR(ts58)R57 cells both in vivo and in vitro (RICHTER and DEPPERT, manuscript in preparation)

at the permissive growth temperature (32 °C), large T antigen in all transformants formed complexes with p53. The p53 in these complexes was metabolically stable. Also the nuclear subclasses of free p53 in these cells were metabolically stable (Fig. 1 B). These results, follow the expected pattern for complex formation and metabolic stability of p53 in SV40-transformed cells.

Figure 2 shows the analyses of the subnuclear distribution and of metabolic stabilities of large T antigen and large T antigen-p53 complexes (Fig. 2A) and of free p53 (Fig. 2B) in cells of all transformants kept and labelled at 39 °C. As already expected from previously published data (MONTENARH et al. 1985), only large T antigen in cells of the SV40 wild-type transformant FR(wt648) is able to form a stable complex with p53 (Fig. 2A,a), whereas the tsA mutant large T antigens in both transformants does not complex p53 to a significant extent

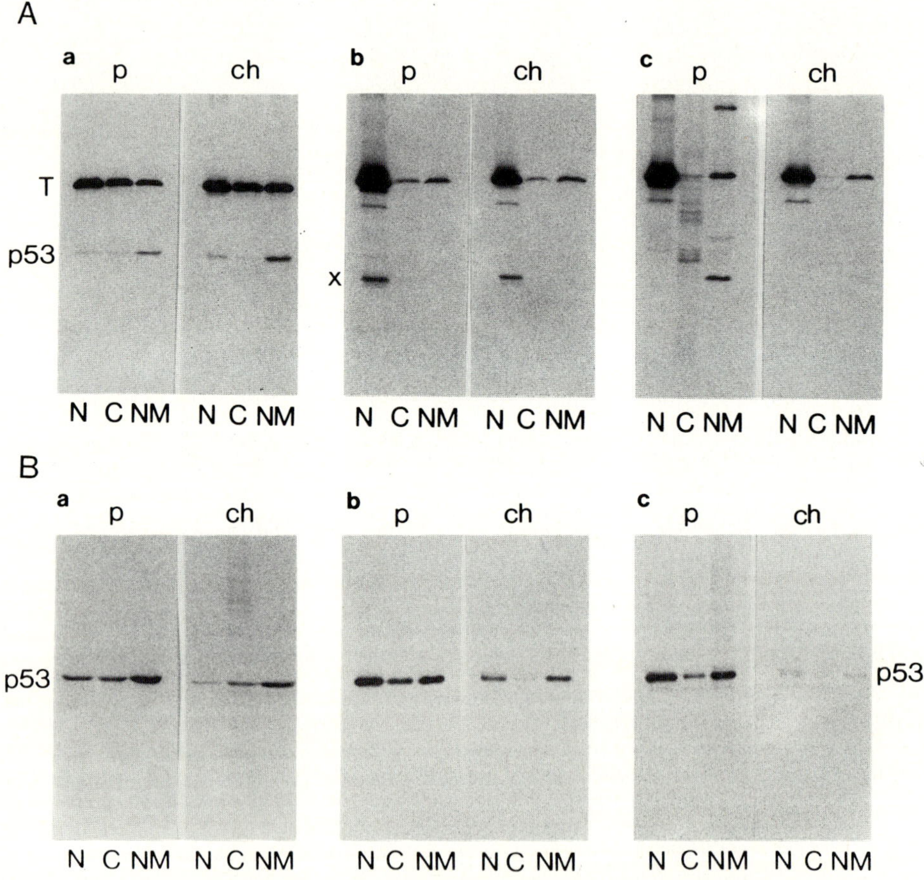

Fig. 2A, B. Complex formation and metabolic stability of p53 in FR(wt648), in FR(ts68)R57, and in FR(ts58)A cells at 39 °C. A total of $1 \times 10^7$ cells of each group, kept at 39 °C for 3 days, were pulse and pulse-chase labelled as described in the legend to Fig. 1. Cells were subfractionated and nuclear extracts analyzed for large T antigen and large T antigen-p53 complexes, as well as for free p53. Designations are as in the legend to Fig. 1

under these conditions (Fig. 2A,b,c). Free p53 was present in all transformants (Fig. 2B). However, this p53 was metabolically stable only in FR(wt648) cells (Fig. 2B,a) and in FR(ts58)R57 cells (Fig. 2B,b) which exhibited a transformed phenotype also at 39 °C. In contrast, in FR(ts58)A cells, which revert to a normal phenotype at 39 °C, p53 was metabolically unstable (Fig. 2B,c).

## 3.2 Analysis of Mutant Large T Antigens for Functional In Vivo Differences

A possible explanation for the results described above might be that cellular transformation in SV40 tsA A-type transformation had become independent from large T-antigen expression. This possibility is difficult to rule out. However, the fact that these cells can be cultured at the nonpermissive temperature for a long period of time without losing large T-antigen expression (RICHTER and DEPPERT, unpublished data) in our opinion strongly argues for a functional role of the mutant T antigen in maintaining the transformed phenotype of these cells even at the nonpermissive growth temperature. If this were the case, one should expect that the mutant large T antigens in SV40 tsA N-type and A-type transformants differ in some functional in vivo aspect(s).

The only functional in vivo difference for SV40 tsA mutant large T antigen compared with wild-type in SV40-transformed cells described so far is its inability to associate with nuclear substructures (in SV40 tsA N-type transformants kept at the nonpermissive growth temperature; HINZPETER and DEPPERT 1987). In line with this observation, the data shown in Fig. 2A for the subnuclear distribution of the mutant large T antigens in cells kept and labelled at 39 °C also demonstrate that the mutant T antigens differ from wild-type in their association with the cellular chromatin and the nuclear matrix. However, closer comparison of the subnuclear distribution of large T antigen in A-type transformant cells (FR(ts58)R57, Fig. 2A,b) with that in N-type transformant cells (FR(ts58)A, Fig. 2A,c) reveals that the association of the mutant large T antigen with nuclear substructures, especially the cellular chromatin, seems to be considerably more stable in cells of the A-type transformants.

## 4 Discussion

It has been suggested for quite some time that the cellular protein p53 plays an important role in cell transformation by SV40 (reviewed in CRAWFORD 1983), since p53 in these cells forms a tight complex with the large T antigen and becomes metabolically stabilized. However, so far no clear correlation has been established between p53 protein stability and SV40 transformation. Furthermore, conflicting data exist as to whether this stabilization is the result of the physical interaction of these proteins (DEPPERT et al. 1987; LINZER et al. 1979).

We previously suggested that metabolic stabilization of p53 is a secondary event in cellular transformation by SV40 which correlates with the expression of the transformed phenotype (DEPPERT et al. 1987). In this paper we further corroborate this hypothesis. Comparative analyses of the phenotypes of a set of matched SV40 (tsA N-type and A-type transformants at the permissive and at the nonpermissive growth temperatures) with the metabolic stability of the p53 expressed under these conditions show that also in this system there is a clear correlation between metabolic stability of p53 and expression of the transformed phenotype. It is important to note that metabolic stabilization of p53 is independent from complex formation with large T antigen, since the mutant large T antigens in cells of both N-type and A-type transformants are unable to complex p53 to any significant extent when the cells are kept at the nonpermissive growth temperature. Thus, metabolic stabilization of free p53 in cells of SV40 A-type transformants kept at 39 °C most likely is not the result of a direct physical interaction of the mutant large T antigen with p53. We, therefore, suggest that complex formation of large T antigen with p53 is not directly related to metabolic stabilization of p53 but primarily reflects a physical property of these proteins. However, this interpretation does not exclude a functional role of the large T antigen-p53 complex in cellular transformation by SV40.

Although our results strongly suggest that a so far unknown cellular mechanism mediates the metabolic stabilization of p53 during cellular transformation with SV40, they are still compatible with the assumption that metabolic stabilization of p53 in SV40-transformed cells is dependent on the expression of a functional large T antigen, i.e., that this cellular mechanism is controlled by large T antigen. At first glance this seems difficult to reconcile with our data demonstrating that p53 is metabolically stable in cells of the A-type transformant FR(ts58)R57 expressing a mutant T antigen both at permissive and at nonpermissive growth temperatures (RICHTER and DEPPERT, manuscript in preparation). Instead, this finding rather seems to argue for a large T antigen-independent mechanism. However, the analysis of the subcellular distribution of the mutant large T antigens in cells of both types of transformants suggests the exciting possibility that, at the nonpermissive temperature, these T antigens might functionally differ in their ability to associate stably with nuclear substructures in vivo, despite being indistinguishable genetically, as well as in biochemical in vitro properties (RICHTER and DEPPERT, manuscript in preparation). Therefore, further analyses of this system should be useful for identifying the functional role of large T antigen in the process of metabolic stabilization of p53 during SV40-mediated cell transformation.

*Acknowledgements.* This study was supported by grants De 212/6-1 and Fa 138/3-1 from the Deutsche Forschungsgemeinschaft, and by the Fonds der Chemischen Industrie. The Heinrich-Pette-Institut is financially supported by Freie und Hansestadt Hamburg and Bundesministerium für Jugend, Familie, Frauen, und Gesundheit.

# References

Crawford LV (1983) The 53,000-dalton cellular protein and its role in transformation. Int Rev Exp Pathol 25: 1–50

Deppert W, Haug M (1986) Evidence for free and metabolically stable p53 in nuclear subfractions of simian virus 40-transformed cells. Mol Cell Biol 6: 2233–2240

Deppert W, Haug M, Steinmayer T (1987) Modulation of p53 protein expression during cellular transformation with simian virus 40. Mol Cell Biol 7: 4453–4463

Gurney EG, Harrison RO, Fenno J (1980) Monoclonal antibodies against simian virus 40 T antigens: evidence for distinct subclasses and for similarities among nonviral T antigens. J Virol 34: 752–763

Gurney EG, Tamowsky S, Deppert W (1986) Antigenic binding sites of monoclonal antibodies specific for simian virus 40 large T antigen. J Virol 57: 1168–1172

Hinzpeter M, Deppert W (1987) Analysis of biological and biochemical parameters for chromatin and nuclear matrix association of SV40 large T antigen in transformed cells. Oncogene 1: 119–129

Linzer DIH, Maltzman W, Levine AJ (1979) The SV40 A gene is required for the production of a 54,000 MW cellular tumor antigen. Virology 98: 308–318

Montenarh M, Kohler M, Henning R (1985) Structural prerequisites of simian virus 40 large T antigen for the maintenance of cell transformation. EMBO J 4: 2941–3947

Oren M (1985) The p53 cellular tumor antigen: gene structure, expression and protein properties. Biochim Biophys Acta 823: 67–78

Pintel D, Bouck N, Di Mayorca G (1981) Separation of lytic and transforming functions of the simian virus 40 A region: two mutants which are temperature sensitive for lytic functions have opposite effects on transformation. J Virol 38: 518–528

Rotter V, Wolf D (1985) Biological and molecular analysis of p53 cellular-encoded tumor antigen. Adv Cancer Res 43: 1041–1047

Staufenbiel M, Deppert W (1984) Preparation of nuclear matrices from cultured cells: subfractionation of nuclei in situ. J Cell Biol 98: 1886–1894

# SV40 Large T Antigen Induces a Protein Kinase Responsible for Phosphorylation of the Cellular Protein p53

K. H. SCHEIDTMANN

The overall phosphorylation state of p53 from normal versus transformed cells was compared by labeling cells with [$^{35}$S]methionine or [$^{32}$P]phosphate and isolating p53 from cell extracts by immunoprecipitation and subsequent SDS-polyacrylamide gel electrophoresis. In transformed cells both T antigen and p53 are heavily phosphorylated (Fig. 1, lane a). Phosphorylation of p53 occurs mainly at serine and to a minor extent at threonine residues (not shown). In contrast, p53 from normal cells is hardly phosphorylated (not shown). The rate of synthesis of p53 appears to be similar, whereas the metabolic stability differs considerably in both cell lines, the half-lives being less than 30 min in normal cells and more than 5 h in transformed cells. In the transformed cells, p53 seems to be converted to a second form with a slightly higher electrophoretic mobility (Fig. 1, lane c). This conversion is complete in about 5 h with a half-time of about 2 h (data not shown). Interestingly, the $^{32}$P-labeled

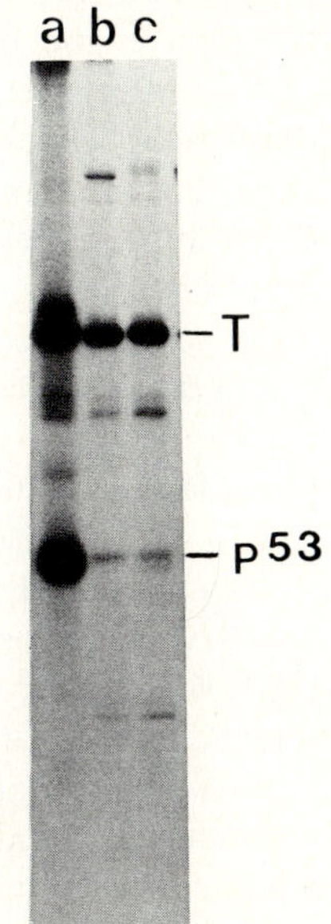



**Fig. 1.** Phosphorylation and conversion of p53. The SV40-transformed rat cell line D29 (KALDERON and SMITH 1984) was labeled with $^{32}$P for 2 h (*a*), or with [$^{35}$S]methionine for 30 min (*b*), and chased for 1 h (*c*). Cells were extracted with isotonic buffer containing 0.5% Nonidet-P40, and the T antigen-p53 complex was isolated by immunoprecipitation with polyclonal T antigen-specific serum and *Staphylococcus aureus* as described previously (SCHEIDTMANN 1986). The samples were run on a 10% SDS-polyacrylamide gel and fluorographed

Institut für Immunbiologie der Universität Freiburg, Stefan-Meier-Str. 8, 7800 Freiburg, FRG

Current Topics in Microbiology and Immunology, Vol. 144
© Springer-Verlag Berlin · Heidelberg 1989

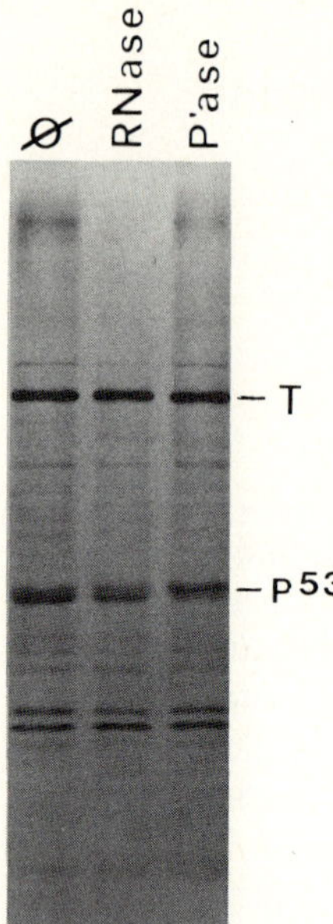


**Fig. 2.** Effect of dephosphorylation on p53. The D29 cells were labeled with [$^{35}$S]methionine for 30 min and chased for 2 h. The T antigen-p53 complex was isolated as in Fig. 1 and either left untreated (Ø), or treated with RNase A (10 µg/ml) or with alkaline phosphatase (from calf intestine, Boehringer, Mannheim; 10 units in 50 µl) at 30 °C for 30 min; as protease inhibitors, 1 m$M$ phenyl methyl sulfonylfluoride and 50 µ$M$ leupeptin were added. Analysis was performed on a 10% polyacrylamide gel. Reconversion of the lower to the upper p53 band was even more complete after treatment with acid phosphatase

protein seems to comigrate with the second form (Fig. 1), suggesting that this conversion is due to the higher extent of phosphorylation in transformed cells. This assumption was ascertained in three ways. i) Dephosphorlytion of p53 in vitro lead to reconversion of form 2 to form 1 (Fig. 2, lane c). ii) Treatment with RNase had no effect (lane b). iii) Analyses of the Phosphopeptides revealed characteristic differences between the two forms (not shown). On the other hand, the methionine- or proline-containing peptides, which in the murine p53 are indicative for the N- or C-terminus, respectively (JENKINS et al. 1984), are identical, thus excluding proteolytic processing.

To address the question whether increased phosphorylation of p53 is due to the action of SV40 T antigen or is instead a consequence of the transformed phenotype, normal rat F111 cells were infected with SV40, and p53 was analyzed 1 or 2 days postinfection. After 2 days, a quantitative and qualitative increase in phosphorylation of p53 was observed and concomitantly its stabilization (not shown). This response occurred in monkey and mouse cells as well. Using the temperature-sensitive mutant tsA58, which is replication and transformation defective at the restrictive temperature (39,5 °C), enhanced phosphorylation and conversion were only observed at the permissive temperature (32,5 °C) (Fig. 3). These findings suggest that SV40 T antigen induces or activates a protein kinase which phosphorylates p53. Moreover, the coincidence of enhanced phosphorylation and stabilization of p53 might indicate a possible relationship between these phenomena.

To see whether induction of protein kinase(s) and enhanced phosphorylation of p53 might play a role in SV40-mediated transformation, we analyzed p53 from a set of rat cell lines which had been transformed with different transformation-defective (td) mutants. These mutants induced transformed foci in vitro at a lower frequency and after a longer latency period than WT (KALDERON and SMITH 1984;

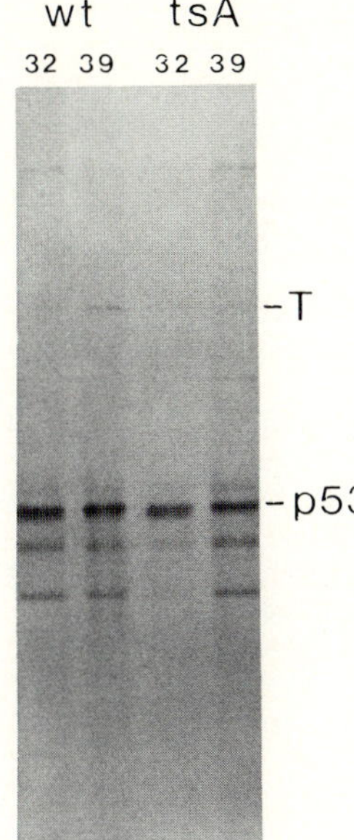


**Fig. 3.** Effect of SV40 infection on p53. Normal rat F111 cells were infected with SV40 wild-type (*wt*) (VA 45–54) or with the temperature-sensitive mutant tsA58. Cells were kept at 32.5 °C for 3 days or at 39.5 °C for 2 days and labeled with [$^{35}$S]methionine at the appropriate temperature for 3 or 2 h, respectively. The p53 was then isolated by immunoprecipitation with the p53-specific monoclonal antibody PAb122 (Gurney et al. 1980) and protein A-sepharose and analyzed as in Fig. 1. Conversion of p53 to the second form is observed upon wild-type infection at both temperatures but after tsA infection, only at the permissive temperature

Schneider and Fanning 1988). Interestingly, p53 from td-transformants was less phosphorylated, conversion of form 1 to form 2 was very inefficient, and the phosphopeptide pattern showed characteristic differences compared with p53 from WT-transformed cells. These data suggest that T-antigen mutants with reduced transforming capacity are impaired in kinase induction.

The data described here show that infection and transformation of cells with SV40 lead to enhanced phosphorylation and stabilization of p53. Whether increased phosphorylation is responsible for the metabolic stabilization remains to be clarified. In rat cells, enhanced phosphorlytion of p53 generates a second form with altered electrophoretic mobility. It is conceivable that phosphorylation of p53 modulates its functions and potentiates its transforming activity. It is possible that in normal cells this modulation takes place only at a certain stage of the cell cycle. This is presently under investigation. In mouse lymphocytes, immunologically distinct forms of p53 have been observed in resting versus growing cells (Milner 1985) which seem to be generated by alternative splicing (Araj et al. 1986). Perhaps mouse and rat cells use different mechanisms to generate functionally distinct forms. Enhanced phosphorylation of p53 appears to be related to SV40-induced transformation, since it is impaired in cells transformed by td-mutants. I propose that T antigen induces (or activates) a protein kinase and that this kinase is involved in SV40-mediated transformation by modulating the activities of p53, and possibly other substrates, through phosphorylation.

*Acknowledgements.* The excellent technical assistance of Angelika Haber and Marion Buck is gratefully acknowledged. This investigation was supported by the Deutsche Forschungsgemeinschaft through SFB 31 and by Baden-Württemberg through Schwerpunkt 09.

# References

Arai N, Nomura D, Yokota K, Wolf D, Brill E, Shohat O, Rotter V (1986) Immunologically distinct p53 molecules generated by alternative splicing. Mol Cell Biol 6: 3232–3239

Gurney EG, Harrison RO, Femo J (1980) Monoclonal antibodies against simian virus 40 T antigens: evidence for distinct subclasses of large T antigen and for similarities among nonviral T antigens. J Virol 34: 752–763

Jenkins JR, Rudge K, Redmond S, Wade-Evans A (1984) Cloning and expression analysis of full length mouse cDNA sequences encoding the transformation associated protein p53. Nucleic Acids Res 12: 5609–5626

Kalderon D, Smith AE (1984) In vitro mutagenesis of a putative DNA binding domain of SV40 large T. Virology 139: 109–137

Milner J (1985) Different forms of p53 detected by monoclonal antibodies in non-dividing and dividing lymphocytes. Nature 310: 143–145

Scheidtmann KH (1986) Phosphorylation of simian virus 40 large T antigen: cytoplasmic and nuclear phosphorylation sites differ in their metabolic stability. Virology 150: 85–95

Schneider J, Fanning E (1988) Mutations in the phosphorylation sites of simian virus 40 (SV40) T antigen alter its origin DNA-binding specificity for sites I or II and affect SV40 DNA replication activity. J Virol 62: 1598–1605

# Evidence for Threshold Effects in Transformation of Pancreatic β Cells by SV40 T Antigen in Transgenic Mice

S. Efrat[1] and D. Hanahan[2]

## 1 Introduction

Transgenic mice harboring hybrid insulin-T antigen genes manifest heritable patterns of tumor formation (Hanahan 1985; Efrat and Hanahan 1987; Adams et al. 1987; Teitelman et al. 1988). In all mice inheriting the hybrid oncogene, denoted RIP1-Tag, its cell-specific expression in β cells in all the pancreatic islets of Langerhans causes β-cell transformation and proliferation, which results in islet hyperplasia. This is followed by development of only a few islets into solid, vascularized tumors and consequent premature death. It thus appears that T antigen is necessary but not sufficient for tumor formation and that additional events may be required.

SV40 T antigen is capable of both adapting primary cells to continuous growth in culture and transforming cultured cells into a tumorigenic condition. Most known oncogenes are proficient in one of these activities but not both. It is possible that part of the activities of large T antigen are mediated through its association with p53, a cellular phosphoprotein (Crawford et al. 1980; Jay et al. 1980) present at high levels in a variety of transformed cells (for reviews, see: Crawford 1983; Oren 1985; Rotter and Wolf 1985). In normal cells p53 has a short half-life (approximately 20 min in fibroblasts) and is present at very low levels. In contrast in cells transformed by SV40 or adenovirus, its half-life is dramatically increased, to about 24 h (Oren et al. 1981). This is achieved by formation of a stable, non-covalent complex between p53 and the oncogene product. The murine p53 can immortalize primary rodent cells when expressed at high levels (Jenkins et al. 1984) and can cooperate with *ras* in transformation of these cells in a similar manner to *myc* or E1A (Eliyahu et al. 1984; Jenkins et al. 1984; Parada et al. 1984). These results suggest an important role for p53 in the regulation of cell growth. We investigated the possibility that expression of the endogenous p53 gene may have a role in the progression from islets to tumors in insulin-T antigen transgenic mice (Efrat et al. 1987). Immunohistochemical analysis of mice of various ages shows a complete correspondence between the targeted expression of T antigen and the appearance of the endogenous murine p53 protein in β cells. A pronounced increase in endogenous p53 levels was observed in every cell expressing T antigen.

Cold Spring Harbor Laboratory, Cold Spring Harbor, New York 11724, United States of America

[1] Present address: Department of Molecular Pharmacology, Albert Einstein College of Medicine, 1300 Morris Park Ave., Bronx, New York, NY 10461, USA
[2] Present address: Department of Biochemistry and Biophysics, University of California, San Francisco, Ca 94143, USA

Similar data were obtained with mice expressing T antigen in the α cells of the endocrine pancreas, under control of the rat preproglucagon regulatory region (EFRAT et al., unpublished results). These results suggest that differences in p53 cannot simply account for the observed progression of normal islets to hyperplastic islets and to infrequent solid tumors.

A phenotype similar to that of the RIP1-Tag lineages develops in mice harboring a hybrid gene, in which the rat insulin regulatory region is placed in reverse orientation with respect to the T antigen coding sequence (denoted RIR1-Tag) (HANAHAN 1985). A promoter element acting on the opposite strand and in the opposite direction from the insulin coding region has been identified, which directs expression of T antigen specifically to β cells (EFRAT and HANAHAN 1987). Deletions constructed in the insulin region of the hybrid gene indicate that the enhancer elements are necessary for reverse promoter activity in transgenic mice. The levels of large T antigen in β cells of mice harboring these hybrid genes are markedly reduced when compared with RIR1-Tag mice. The p53 levels in β cells are also proportionally lower. Mice from these lineages do not develop tumors, suggesting that certain threshold levels of T antigen or p53 in the β cells are necessary for tumorigenesis.

## 2 Materials and Methods

### 2.1 Constructs

The RIR2-Tag insert was generated by cutting the RIR1-Tag plasmid (HANAHAN 1985) with *Bam*HI and *Bgl*I. RIR3-Tag was constructed from RIR1-Tag by digestion with *Stu*I and religation.

### 2.2 Transgenic Mice

The generation of transgenic mice has been described previously (HANAHAN 1985). RIR3-Tag was injected as a linearized plasmid. RIR2-Tag lacks vector sequences. Two founder mice were generated with each construct, with transgene copy number ranging between 1 and approximately 10 per haploid genome. Only one founder harboring each construct has transmitted the gene and given rise to a lineage. RIR1-Tag is described elsewhere (EFRAT and HANAHAN 1987). In the study presented here we have used lineage 2 of RIR1-Tag.

### 2.3 Immunohistochemistry

Fresh-frozen pancreas sections were stained as detailed (EFRAT and HANAHAN 1987) with rabbit Tag-specific diluted serum 1 : 5000 or with rabbit p53-specific serum diluted 1 : 1000. The bound antibody was visualized with horseradish peroxidase-conjugated goatrabbit-specific IgG.

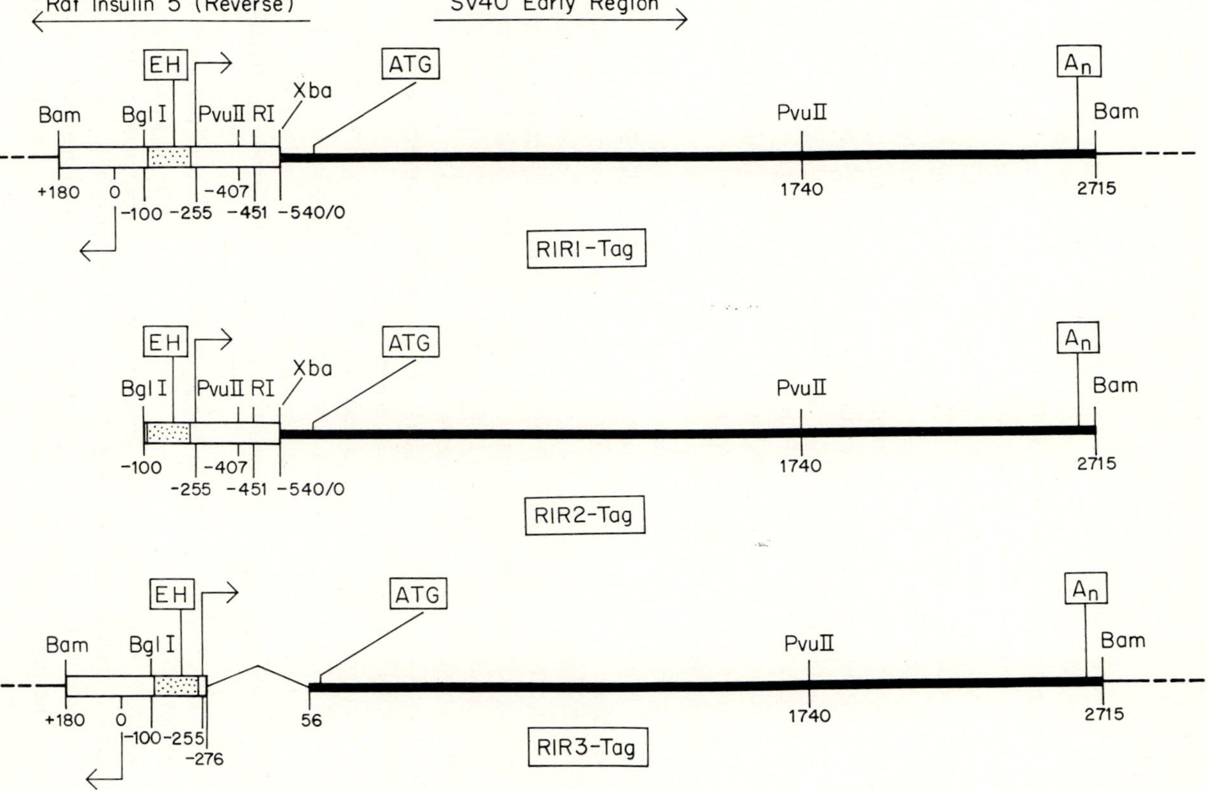


**Fig. 1.** The RIR-Tag constructs. Insulin gene regions are shown in *light bars* [the *dotted part* marks the approximate boundaries of the enhancer (*EH*) region]. *Arrows* indicate transcription initiation sites from the insulin promoter and the reverse promoter. The T-antigen coding region is represented by *dark bars*. *ATG*, translation initiation site; $A_n$, polyadenylation site. The *dashed line* indicates vector sequences

## 3 Results

Rodent genomes contain two nonallelic and highly homologous preproinsulin genes. Cell transfection analyses have identified four elements in the rat insulin I gene regulatory region which are necessary for transcription from the insulin promoter (KARLSSON et al. 1987). In the rat insulin II gene, which is the one used in our studies, these elements can be located (by sequence homology, as presented by SOARES et al. 1985) as follows: a promoter element at —22 to —33, and three enhancer elements at —93 to —100 (E1, —223 to —231 (E2), and —293 to —300 (E3). To determine which of these elements is required for reverse promoter activity in transgenic mice, we have constructed two deletions in the insulin regulatory region (Fig. 1). In RIR2-Tag a portion of the insulin gene coding region, the insulin promoter, and one of the enhancer elements (E1) have been deleted from the original RIR1-Tag hybrid gene. The RIR3-Tag construct lacks only the E3 enhancer element. In both constructs, T-antigen transcription is effected by the reverse promoter.

Immunohistochemical analysis of pancreas sections from young adult mice harboring the RIR2-Tag and RIR3-Tag hybrid genes detects considerably less large T antigen protein in the nuclei of β cells when compared with RIR1-Tag mice (Fig. 2). The deletions do not seem to affect the specificity of transgene expression in β cells, as far as other cell types in the pancreas are concerned. No expression of large T antigen has been detected in a series of other tissues analyzed by immunohistochemistry (including kidney, liver, and small intestine). Analysis of total pancreas RNA is underway, to determine the extent of reduction in T-antigen transcripts in β cells, as well as a survey of other tissues for inappropriate expression of the transgene. However, these qualitative results indicate that the reverse promoter requires the same enhancer elements used by the insulin promoter.

The levels of endogenous p53 are also reduced in the nuclei of β cells from RIR2-Tag and RIR3-Tag mice, in comparison with RIR1-Tag mice (Fig. 2). The decrease appears to be proportional to the reduction in large T antigen, which is consistent with the notion that the large T protein stabilizes p53 by forming a complex with it.

The phenotype of RIR2-Tag and RIR3-Tag mice differs dramatically from that of RIR1-Tag or RIR1-Tag mice. In young adult mice the islets of Langerhans, all of which express low levels of T antigen, appear histologically normal. Older mice (around 7–9 months of age) show moderate hyperplasia of β cells in some of the islets. This is manifested in increased islet size and in a higher number of β cells per islet (increased nuclei to cytoplasm ratio). With one exception, no tumors have been found in mice from these lineages, some of which are more than 2 years old, and none of them has died prematurely. The exception was a RIR3-Tag mouse (out of some 50 mice) which died at the age of 20 months, bearing two small insulinomas. The levels of large T antigen in the tumor cells were much higher than in the islet β cells of this mouse and similar to that of RIR1-Tag islets. It is possible that a rare event of transgene rearrangement or amplification increased the expression of T antigen in some β cells, resulting in the development of tumors. In addition, blood glucose levels are normal, even in old animals. No abnormalities have been observed in any other tissue in these mice. In contrast,

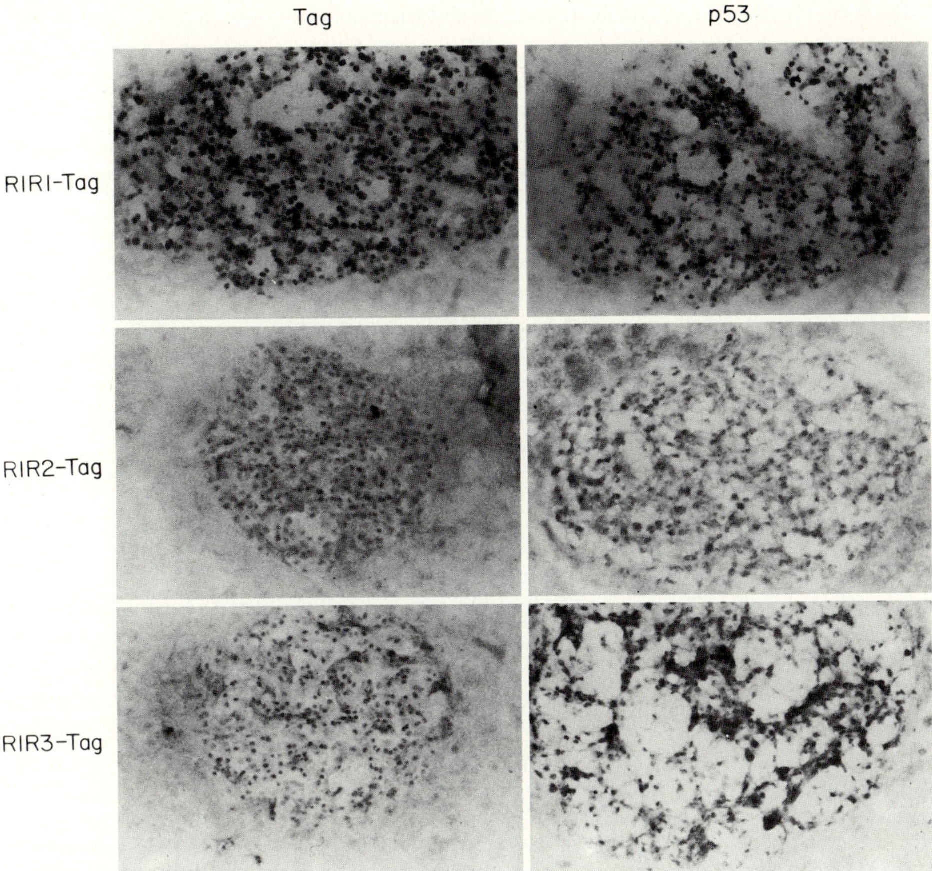


**Fig. 2.** Immunohistochemical analysis of large T and p53 proteins in islets of RIR-Tag mice. Thin sections of the pancreas were stained with antigen-specific or p53-specific sera, which were visualized by horseradish peroxidase-conjugated second antibody. The ages of the mice shown were 5 months (RIR1-Tag), 4 months (RIR2-Tag), and 8 months (RIR3-Tag). Magnification is ×230

RIR1-Tag mice develop several β-cell tumors each and die by 5 months of age. These results suggest that the low levels of T antigen, and perhaps of p53, are not sufficient for formation of β-cell tumors.

## 4 Discussion

Expression of the SV40 T-antigen gene under the control of truncated insulin gene enhancer regions and a reverse promoter element results in very low levels of large T protein in the nuclei of β cells in transgenic mice. In both RIR2-Tag and RIR3-Tag lineages the levels of large T antigen are several-fold lower compared with RIR1-Tag mice, in which expression of T antigen is regulated by the intact insulin gene

5′-flanking region (but again in reverse orientation). The levels of the endogeneous p53 protein in β cells are also proportionally reduced but are nevertheless detectable and clearly elevated over those seen in normal β cells. This fits the observation that in β-cell tumors the majority of p53 is found in complex with the large T antigen (EFRAT et al. 1987). If this is also the case in β cells in preneoplastic islets, then the availability of less large T antigen to stabilize p53 would result in expression of lower levels of p53 in the cells.

RIR2-Tag and RIR3-Tag mice do not develop β-cell tumors, such as are observed in RIP1-Tag or RIR1-Tag mice. Although the precise onset of expression of the transgene has not been determined yet in the RIR2-Tag and RIR3-Tag lineages, the analysis of young adult mice (3–4 months old) shows expression of large T antigen in virtually all β cells in all of the islets. The longest period between onset of expression and appearance of tumors observed in other mice harboring insulin-promoted T antigen genes is 6 months (HANAHAN 1985; ADAMS et al. 1987). Therefore, the lack of tumors in RIR2-Tag and RIR3-Tag mice cannot be explained by late onset of oncogene expression. These results suggest that the low levels of large T antigen, and perhaps of p53, are below a threshold required for tumorigenesis. It is unclear at present whether these levels are sufficient to induce β-cell transformation but are insufficient to promote progression of islets to solid tumors. The ability of these subthreshold levels of T antigen to induce abnormal β-cell proliferation can be assessed by measuring the β-cell mitotic index and by the presence of markers of dividing β cells (see TEITELMAN et al. 1988). Moreover, the fact that SV40 T antigen has oncogene activities similar to both the *myc* family of nuclear oncogenes and the *ras* family of cell-surface/cytoplasmic oncogenes might imply that only one of these activities is sub-threshold, which could be addressed in genetic complementation experiments using other hybrid insulin oncogenes.

*Acknowledgements.* We thank D. Lane for p53 antiserum, J. Duffy for artwork, and D. Green for photography. S.E. is supported by the Cancer Research Institute/ David Jacobs Memorial Fellowship. This work was funded by Monsanto Company.

# References

Adams T, Alpert S, Hanahan D (1987) Non-tolerance and autoanitibodies to a transgenic self antigen expressed in pancreatic β cells. Nature 325: 223–228

Crawford LV (1983) The 53,000-dalton cellular protein and its role in transformation. Int Rev Exp Pathol 25: 1–50

Crawford LV, Lane DP, Denhardt DT, Harlow EE, Nicklin PM, Osborn K, Pim DC (1980) Characterization of the complex between SV40 large T antigen and the 53K host protein in transformed mouse cells. Cold Spring Harbor Symp Quant Biol 44: 179–187

Efrat S, Hanahan D (1987): Bidirectional activity of the rat insulin II 5′-flanking region in transgenic mice. Mol Cell Biol 7: 192–198

Efrat S, Baekkeshov S, Lane D, Hanahan D (1987) Coordinate expression of the endogenous p53 gene in β cells of transgenic mice expressing hybrid insulin-SV40 T antigen genes. EMBO J 6: 2699—2704

Eliyahu D, Raz A, Gruss P, Givol D, Oren M (1984) Participation of p53 cellular antigen in transformation of normal embryonic cells. Nature 312: 646–649

Hanahan D (1985) Heritable formation of pancreatic β cell tumors in transgenic mice expressing recombinant insulin-simian virus 40 oncogenes. Nature 315: 115–122

Jay G, DeLeo AB, Appella E, Dubois GC, Law LW, Khoury G, Old LJ (1980) A common transformation-related protein in murine sarcomas and leukemias. Cold Spring Harbor Symp Quant Biol 44: 659–664

Jenkins JR, Rudge K, Currie GA (1984) Cellular immortalization by a cDNA clone encoding the transformation-associated phosphoprotein p53. Nature 312: 651–654

Karlsson O, Edlund T, Barnett-Moss J, Rutter WJ, Walker MD (1987) A mutational analysis of the insulin gene transcription control region: expression in beta cells is dependent on two related sequences within the enhancer. Proc Natl Acad Sci USA 84: 8819–8823

Oren M (1985) The p53 cellular tumor antigen: gene structure, expression and protein properties. Biochim Biophys Acta 823: 67–78

Oren M, Maltzman W, Levine AJ (1981) Post-translation regulation of the 54K cellular tumor antigen in normal and transformed cells. Mol Cell Biol 1: 101–110

Parada LF, Land H, Weinberg RA, Wolf D, Rotter V (1984) Cooperation between gene encoding p53 tumor antigen and *ras* in cellular transformation. Nature 312: 649–651

Rotter V, Wolf D (1985) Biological and molecular analysis of p53 cellular-encoded tumor antigen. Adv Cancer Res 43: 113–141

Soares MB, Schon E, Henderson A, Karathanasis SK, Cate R, Zeitlin S, Chirgwin J, Efstradiatis A (1985) RNA-mediated gene duplication: the rat preproinsulin I gene is a functional retroposon. Mol Cell Biol 5: 2090–2103

Teitelman G, Alpert S, Hanahan D (1988) Proliferation, senescence, and neoplastic progression of β cells in hyperplastic pancreatic islets. Cell 52: 97–105

# The Biological Activity of Early SV40 Antisense RNA and DNA Molecules

M. Graessmann and A. Graessmann

## 1 Introduction

It is of basic and applied scientific interest to develop reliable techniques which allow the down regulation of single genes either in isolated cells or in cells of an intact organism. Such an approach should give us a more detailed insight into the particular role of single genes for growth and cell differentiation and offer us the possibility to cure cells in cases in which overproduction of a specific protein is the reason for the cellular malfunction. One promising method to decrease selectively the expression rate of either endogeneous or transgenes at the post-transcriptional level is the use of antisense RNA and DNA molecules.

Izant and Weintraub (1984) first demonstrated that antisense molecules can be effective inhibitors of gene expression in mammalian tissue culture cells. A strong reduction of thymidine kinase (TK) gene activity in TK-minus mouse cells was demonstrated following coinjection of the HSV-TK gene together with a high copy number of large antisense RNA and DNA molecules. In subsequent experiments it was shown that short antisense RNA molecules, complementary either to the 5′ or 3′ portion of the mRNA, were even more effective inhibitors than the full size antisense molecules (for review, see Green et al. 1986).

So far, it is not known how antisense molecules operate as inhibitors of gene expression in eukaryotic cells. Whatever the mechanism is (e.g., destabilization of the mRNA, inhibition of transport of the mRNA from the nucleus to the cytoplasm, interference at the level of translation), it is tempting to speculate that the inhibition requires hybrid formation between the antisense and the mRNA molecules.

The purpose of our study is to characterize the extent and the efficiency of inhibition of SV40 gene expression by microinjection of antisense RNA and DNA molecules together with SV40 DNA into tissue culture cells. The main emphasis of this work is to demonstrate a possible correlation between the secondary structure of the target sequence and the activity of antisense molecules.

Institut für Molekularbiologie und Biochemie, Freie Universität Berlin, Arnimallee 22, 1000 Berlin 33, FRG

## 2 Results and Discussion

In earlier investigations we demonstrated that in vitro-synthesized SV40 cRNA was efficiently processed into mature mRNA and translated into wild-type (WT) T antigen (GRAESSMANN and GRAESSMANN 1982; GRAESSMANN et al. 1983). To find out whether full-size, early SV40 antisense RNA molecules can inhibit T-antigen synthesis, the early SV40 coding region (*BgI*I-*Bam*H1 DNA fragment) was cloned in sense (pTysT7) and antisense (pTyasT7) orientation under the transcriptional control of the prokaryotic T7 promoter as illustrated in Fig. 1a, b. For in vitro transcription the purified pTysT7 and pTyasT7 plasmids were linearized with *Sal*I restriction enzyme and capping of the run-off transcripts was obtained by addition of diguanosyl-5′-triphosphate to the reaction mixture. We first analyzed the biological activity of the in vitro-synthesized sense RNA following microinjection into TC7 cells. We found that 24 h after injection about 60–70% of the recipient cells were T-antigen positive. To answer the question whether T-antigen synthesis can be blocked by the full length antisense RNA, the sense RNA was mixed with different concentrations

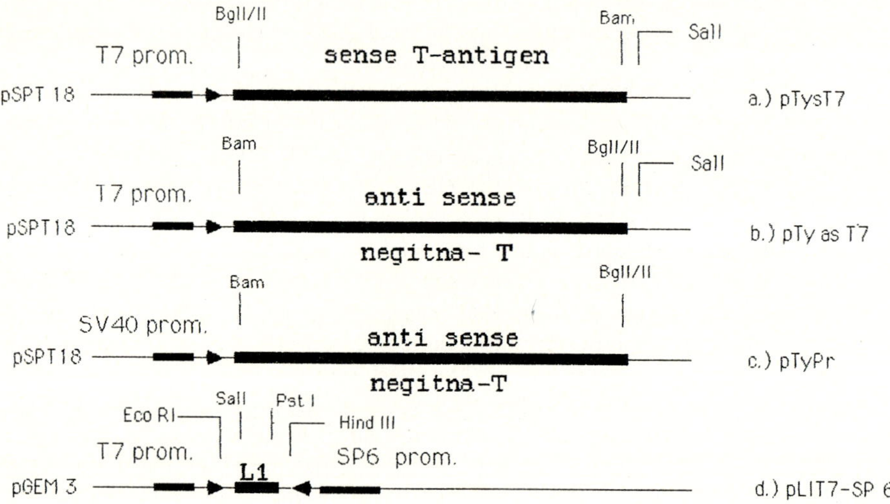


**Fig. 1a–d.** Schematic representation of sense and antisense constructs used. **a** In the pTysT7 construct the early SV40 coding region (*BgI*I-*Bam*HI fragment, where the *BgI*I site was converted into a *BgI*II site, a gift of Y. Gluzman, Cold Spring Harbor Laboratory) was inserted into the *Bam*HI site of the pSPT18 plasmid DNA under the control of the T7 promoter in the sense orientation. **b** The pTyasPr plasmid contains the same early SV40 fragment as **a**, but in the antisense orientation to the T7 promoter. Both plasmids **a** and **b** were linearized with *Sal*I endonuclease for in vitro transcription experiments. **c** The pTyPR plasmid contains, inserted into the *Sal*I position of the multiple cloning site of the pSPT18 DNA, the early SV40 promoter/enhancer fragment *Hpa*II/*BgI*I (both ends modified by *Sal*I linkers; GRAESSMANN et al. 1984), and in the *Bam*H1 site the early SV40 coding region (*BgI*I converted in *BgI*II-*Bam*H1 fragment) in the antisense orientation to the SV40 promoter. **d** The pLIT7-SP6 plasmid contains the L1 DNA (see Fig. 3a) inserted between the multiple cloning sites *Sal*I and *Pst*I of the pGEM3 DNA, so that the T7 promoter generates the antisense L1 RNA of the *Hind*III endonuclease-linearized plasmid DNA

of antisense RNA and injected into the nuclei of TC7 cells. However, inhibition of T-antigen synthesis was not demonstrable. Also at a 50-fold excess of the antisense RNA the number of T-antigen-positive cells and the intensity of the intranuclear T-antigen-specific fluorescence were similar to those of control cells injected with only the sense RNA. We also coinjected the antisense RNA with 2–4 SV40 DNA molecules and fixed and stained the cells for T antigen at different times (hours) after injection. T-antigen synthesis always occurred with the same efficiency as in control cells injected with only SV40 DNA.

To determine whether continuous in vivo synthesis of the full length antisense RNA can block SV40 gene expression, the early SV40 coding region was cloned in the antisense orientation under the transcriptional control of the early SV40 promotor-enhancer part as illustrated in Fig. 1c. The pTyPr DNA was then mixed with SV40 DNA I and microinjected into TC7 cells, but this antisense construct failed even at a 1000-fold excess to inhibit SV40 gene expression (T- and V-antigen; Table 1). From these experiments we conclude that large early SV40 antisense RNA molecules cannot block viral gene expression.

Since it has been shown (IZANT and WEINTRAUB 1984) that large HSV-TK antisense RNA molecules inhibit TK expression, we asked why our full length, in vitro- and in vivo-synthesized, SV40 antisense RNA did not prevent T-antigen synthesis. Our initial assumption was that the secondary structure of the RNA molecules must be of importance for the antisense activity, especially if the inhibition requires a hybrid formation between the sense and antisense RNA. Therefore, computer modeling experiments were performed to obtain an estimate of the

**Table 1.** Extent of inhibition of SV40 gene expression by various DNA molecules

| DNA molecules injected per cell | | Percentage of SV40-positive cells | |
|---|---|---|---|
| pPL1 | + SV40 | T antigen | V antigen |
| 0 | 2– 4 | 140 | 135 |
| 2– 4 | 2– 4 | 62 | 35 |
| 20– 40 | 2– 4 | 40 | 15 |
| 200– 400 | 2– 4 | 14 | 8 |
| 2000–4000 | 2– 4 | 0.5 | 0 |
| 20– 40 | 20– 40 | 60 | 45 |
| 200– 400 | 20– 40 | 46 | 32 |
| 200– 400 | 200–400 | 72 | 66 |
| pTyPr | SV40 | | |
| 2000–4000 | +2– 4 | 140 | 135 |
| pPL1 | + PV DNA | % of PV-T antigen | |
| 2000–4000 | 2– 4 | 125 | |

The data given represent the mean value of three independent injection experiments with 100 injected cells each. The number of injected cells were taken as the baseline of 100%, and higher values represent cell division. Cells were fixed and stained for T and V antigen 24 h after injection.

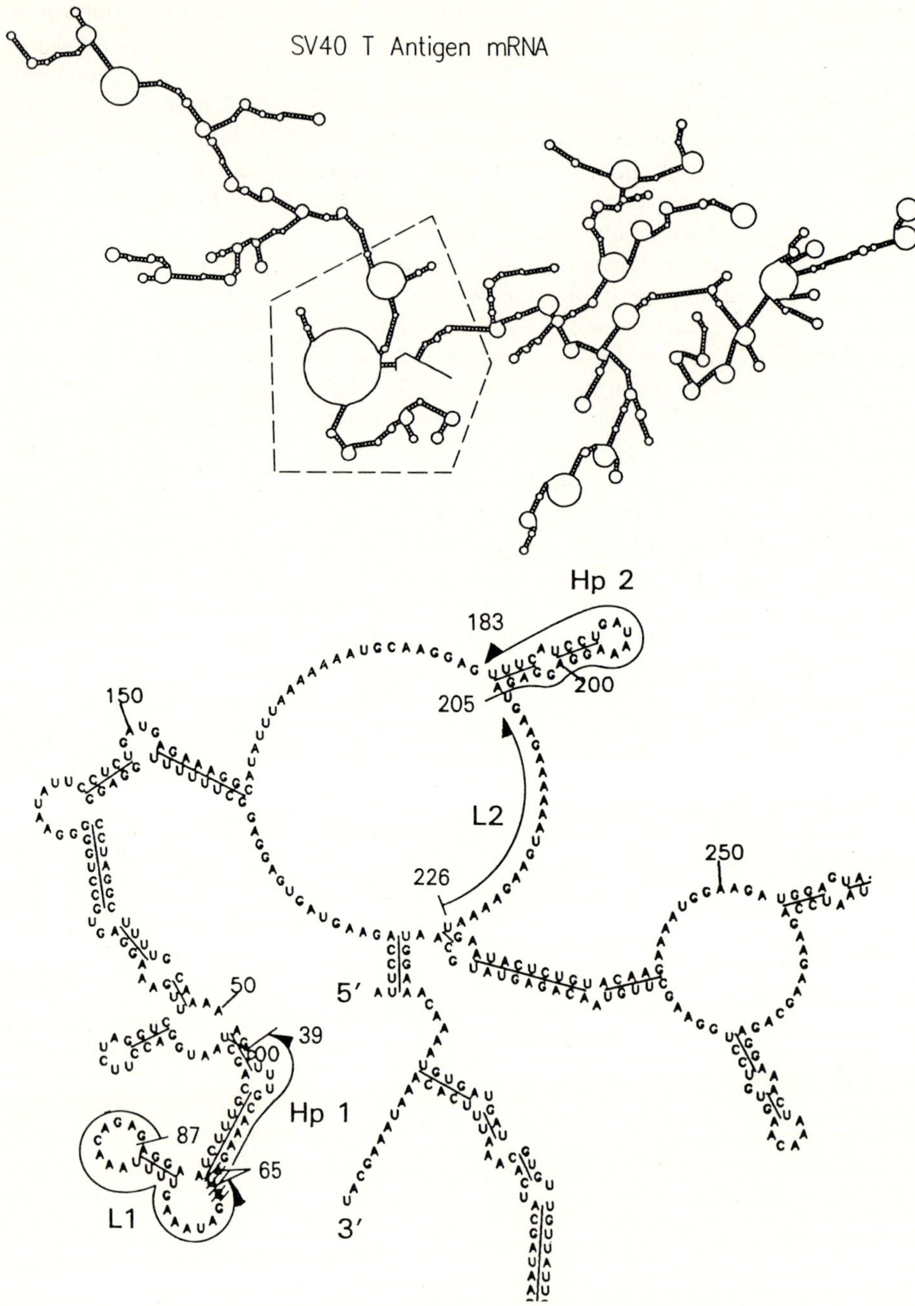


**Fig. 2.** Computer folding model of the early SV40 19S mRNA. The blow up of the 19S mRNA shows the localization of the sequences *L1*, *Hp1*, *L2*, and *Hp2* in the reverse orientation, indicated by the *arrows*. (Computer folding analysis was kindly done by Dr. G. Michaels at the National Institutes of Health, Bethesda)

possible secondary structure of the early SV40 sense and the corresponding anti-sense RNA. Based on the nucleotide sequence (2280 nucleotides) the computer predicted that the mRNA can form a highly complex secondary structure with dominant intramolecular double-stranded segments (Fig. 2). The calculated free energy ($\Delta$G) for the mRNA at 37 °C was $-386$ kcal/mol. This highly negative free energy value suggests that the predicted secondary structure in the absence of other macromolecules is very stable, thus excluding a spontaneous RNA unwinding and hybrid formation with the likewise complex antisense RNA ($\Delta$G $-315$ kcal/mol). A critical part of these computer modeling experiments is the fact that they cannot consider all possible interactions of the RNA with other cellular macromolecules which could have a significant influence on its configuration. However, the computer model should be a reasonable prediction of the RNA structure at least until microinjection.

In contrast to large RNA molecules, short RNA segments are predicted to have a less stable secondary structure that would be easily disrupted in an ATP-independent fashion. This implies that short RNA molecules will hybridize more easily with their corresponding target mRNA molecules. Therefore we selected four different target segments each of 22–24 nucleotides from the early SV40 mRNA sequence. The first segment chosen was L1 (bases 65–87) localized in a partial loop-hairpin structure at the 5′-translated region of the early 19S mRNA (Fig. 2). The triplet sequence 63–65 of the mRNA is the translation initiation codon AUG. This 22-oligonucleotide sequence was synthesized in vitro as a DNA template (Fig. 3a) and cloned in the antisense orientation under the control of either the T7 promotor as pL1T7-SP6 DNA (see Fig. 1d) or the early SV40 promotor/enhancer segment as pPL1 DNA (Fig. 3b). To test the biological activity of the L1 RNA we transcribed the pLT7-SP6 in vitro with the T7 RNA polymerase and coinjected the L1 antisense RNA together with SV40 DNA into the nuclei of TC7 cells. These experiments revealed that the short L1 antisense RNA strongly inhibited SV40 gene expression. During the first 20 h after injection both the number of T-antigen-positive cells and the intensity of the intranuclear T-antigen-specific fluorescence was significantly reduced at all test points·when compared with cells injected with only SV40 DNA.

A prolonged inhibition of early and late SV40 gene expression was obtained after coinjection of the pPL1 DNA (Fig. 3) with SV40 DNA I. As summarized in Table 1, inhibition of T-antigen and V-antigen synthesis was already demonstrable after injection of only 2–4 pPL1 DNA molecules. Under these conditions only 62% of the injected cells synthesized T antigen, and 35% of the recipient cells supported late SV40 gene expression (V-antigen) 24 h after injection. With increasing concentrations of the pPL1 DNA in the injection mixture, T- and V-antigen synthesis were more strongly inhibited. With a 1000-fold excess of the pPL1 DNA, less than 1% of the recipient cells allowed SV40 gene expression. This inhibitory effect continued for over 48 h after microinjection. Cells which were fixed and stained 92 h after injection exhibited an increased number of T- and V-antigen-positive cells, some of them outside the area where injection occurred. This indicates that some of injected SV40 DNA molecules escaped the inhibitory effect of the L1 antisense RNA and infected some of the surrounding cells. As a further step we asked whether the efficiency of inhibition correlates with the relative position of the antisense RNA to the initiation codon AUG and with the predicted secondary structure of the target

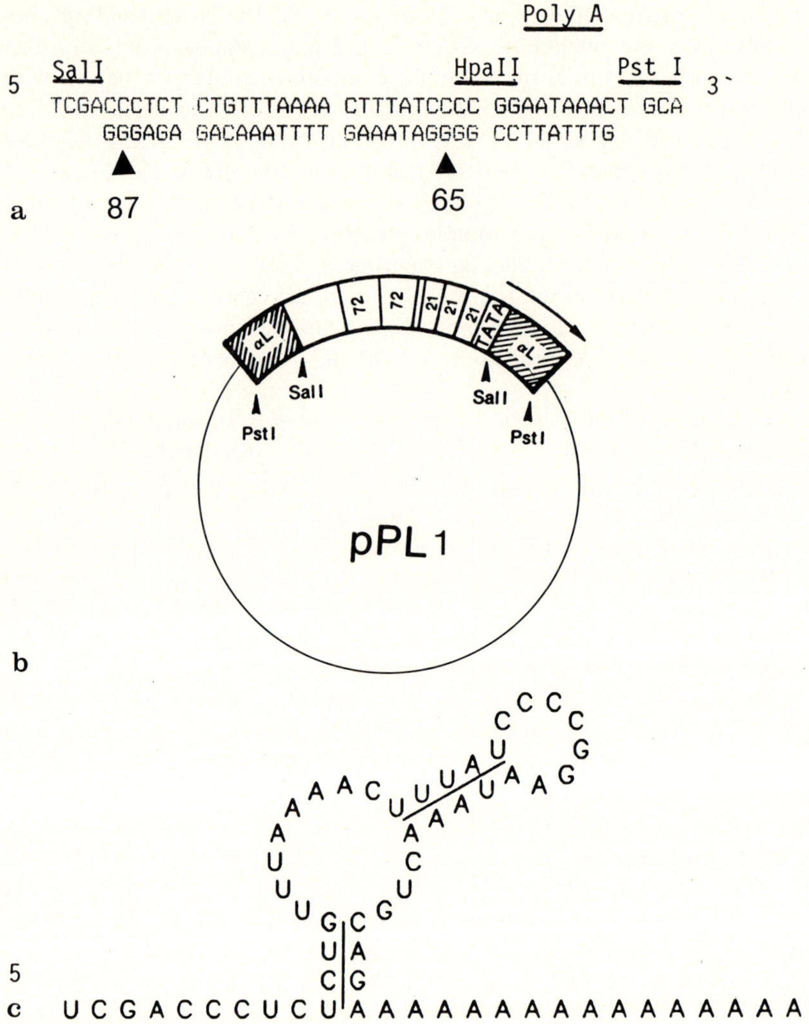
Poly A
SalI
HpaII
Pst I
5
3`
TCGACCCTCT CTGTTTAAAA CTTTATCCCC GGAATAAACT GCA
GGGAGA GACAAATTTT GAAATAGGGG CCTTATTTG
a
87
65
72
72
21
21
21
TATA
αL
αL
SalI
SalI
PstI
PstI
pPL1
b
5
c
UCGACCCUCU AAAAAAAAAAAAAAAAA

**Fig. 3a–c.** The pPL1 DNA construct. **a** The in vitro-synthesized, double-stranded L1 DNA with the *Sal*I and *Pst*I linkers at the respective ends of the oligomere and the position of the polyadenylation (*Poly A*) signal. The *arrows* indicate the nucleotides number 87 and 65 of the early 19S mRNA. The antisense DNA strand is the *lower* one. **b** The two *hatched boxes* represent the L1-DNA inserted into the multiple cloning site (*Pst*I site) of the pGEM DNA and the *open box*, the early SV40 promoter-enhancer part (*Hpa*II-*Bgl*I fragment modified with *Sal*I linkers at both ends) with the two 72- and three 21-base pair repeats ligated to the *Sal*I sites of two *L*1-DNA segments. The *arrow* indicates the transcription orientation. The *solid line* denotes the pGEM3 DNA. **c** The computer-predicted secondary structure based on the nucleotide sequence of the L1-RNA

mRNA. For this reason we analyzed the biological activity of the antisense DNAs pPHp1, pPL2, and pPHp2. These molecules were constructed in the same way as described for the pPL1 DNA. As shown in Fig. 2, the sequence 39–65 (Hp1) is a 24-nucleotide segment localized at the untranslated 5′-end region of the SV40 mRNA. The L2 segment (nucleotide sequence 205–226) is part of the largest predicted loop region of the early mRNA, and the Hp2 segment (nucleotides 183–205) forms a hairpin structure within this loop. The inhibitory activity of the pPHp1, pPL2, and pPHp2 DNAs was tested after coinjection with SV40 DNA molecules into TC7 cells. We found that these antisense contructs were equaly strong inhibitors of SV40 gene expression as the pPL1 DNA. These experiments have shown that in contrast to the large antisense RNA molecules the constructs with the short antisense information (22–24 oligonucleotides) are strong inhibitors of early and late SV40 gene expression. This indicates that the accessibility of the small antisense RNA molecules to their corresponding short target regions at the mRNA is probably better than that of the large RNA molecules. Unexpected, however, was the observation that the efficiency of inhibition did not correlate with the computer-predicted loop or hairpin structures of the mRNA.

*Acknowledgments.* We thank E. Guhl, J. Fuhrhop, and C. Ziechmann for excellent technical assistance. The work was supported by the Deutsche Forschungs-gemeinschaft and the Verband der Chemischen Industrie.

# References

Graessmann M, Graessmann A (1982) Simian virus 40 cRNA is processed into functional mRNA in microinjected monkey cells. EMBO J 1: 1081–1088

Graessmann M, Graessmann A, Westphal H (1983) Microinjected simian virus cRNA is spliced as evidenced by electron microscopy. J Virol 48: 296–299

Graessmann M, Guhl E, Bumke-Vogt C, Graessmann A (1984) The second large simian virus T-antigen exon contains the information for maximal cell transformation J Mol Biol 180: 111–129

Green PJ, Pines O, Inouye M (1986) The role of antisense RNA in gene regulation. Ann Rev Biochem 55: 569–597

Izant JG, Weintraub H (1984) Inhibition of thymidine kinase gene expression by anti-sense RNA: A molecular approach to genetic analysis Cell 36: 1007–1015

# Part II. Polyoma Virus Middle T Antigen

# Introduction

More than 30 years ago an endemic mouse virus was discovered that induced tumors in newborn animals. Tumors developed in many different organs of the infected mouse, hence the name polyomavirus. The virus also transforms cultured rodent embryo cells, which give rise to tumors when injected into susceptible animals (TOOZE 1981).

The genome of the polyomavirus is a circular, double-stranded DNA of 5297 base pairs, organized in vivo as a minichromosome just like simian virus (SV40) DNA. The genes of both these viruses are organized into two regions, an early region coding for the regulatory T antigens, and a late region coding for structural viral proteins. However, polyomavirus differs from SV40 in the organization of its origin structure, of its promoter-enhancer elements, and in gene expression. But, most importantly for the present discussion, polyomavirus codes for three early proteins, the large, middle, and small tumor antigens, of approximate molecular weight of 90K, 55K, and 23K, respectively. The function of the small polyomavirus tumor antigen may be comparable to that of SV40 small tumor antigen, an auxillary protein that increases the efficiency of transformation but is dispensable for an infection process under laboratory conditions.

Polyomavirus large T antigen is a nuclear phosphoprotein. Like its SV40 counterpart it is able to bind specifically to DNA, it exhibits ATPase activity, and it is absolutely required for viral DNA replication. But unlike the SV40 protein, polyomavirus large T antigen is not able to transform infected cells to tumorigenicity. The transformation function of polyomavirus is located on a separate protein, middle T antigen, which is sufficient to transform established cell culture cells and to maintain the transformed phenotype of the cells. However, middle T antigen, an extranuclear protein, cannot induce the conversion of primary cells to "immortalized" or established cell lines. For this process, a function of the polyomavirus large T antigen is required. Thus, polyomavirus large and middle T antigens must cooperate to achieve the full transformation of primary cells. Note that SV40 large T antigen carries an immortalization and a transformation function on separated domains of one polypeptide chain, whereas polyomavirus needs two proteins to achieve a similar result. An important contribution to the continuing research on the transformation activity of polyomavirus is the discovery that middle T antigen associates with cellular proteins, most notably to the cellular proto-oncogene product $pp60^{src}$, whose protein kinase activity is affected by bound middle T antigen.

SMITH and his associates (CHENG et al.) describe below the properties of the $pp60^{c-src}$-middle T complex, and COURTNEIDGE summarizes the present knowledge about the interaction of middle T antigen with cellular proteins and its possible relation to the polyomavirus-induced transformation process.

This part ends with a contribution by STRAUSS and coworkers who describe a peculiar and interesting property of polyomavirus large T antigen, namely, its ability to increase the mutation rate of infected cells. They propose that the mutagenic activity of large T antigen may be somehow connected to its immortalization function.

Current Topics in Microbiology and Immunology, Vol. 144
© Springer-Verlag Berlin · Heidelberg 1989

# References


Fried M, Prives C (1986) The biology of simian virus 40 and polyoma virus. Cancer Cells 4: 1–16

Tooze J (ed) (1981) DNA tumor viruses. Molecular biology of tumor viruses, pt 2. Cold Spring Harbor Laboratory, Cold Spring Habor

# Mechanism of Activation of Complexed pp60$^{c\text{-}src}$ by the Middle T Antigen of Polyomavirus


S. H. Cheng[1], R. Harvey[1], H. Piwnica-Worms[2], P. C. Espino[1], T. M. Roberts[2], and A. E. Smith[1]


## 1 Introduction

Ever since polyomavirus was found to be capable of transforming the growth of cells in tissue culture in a manner that at least resembled its ability to form tumors in animals, most research on the virus has attempted to elucidate the mechanism whereby this occurs. As our knowledge has increased and our molecular analysis of the virus has become increasingly detailed, the exact question asked by particular investigators has changed. However, it is striking that the central questions today are similar to those of 25 years ago. What are the precise molecular triggers fired by the intervention of the viral genome in cells? What is the resulting series of steps that ultimately leads to increased cellular proliferation and consequent uncontrolled growth?

Early genetic analysis implied that polyomavirus DNA codes for so-called early proteins, or tumor antigens that are involved in transformation (for reviews, see Griffin and Dilworth 1983; Smith and Ely 1983; Fried and Prives 1986). Knowledge of the genetic organization of the virus complicated the issue in that the early proteins were found to consist of at least three molecular species, encoded by overlapping DNA sequences. It was not until mRNAs for each of the early proteins were cloned separately that each could be tested for their role in transformation (for review, see Cuzin 1984). These results showed that the large T antigen appears to be responsible for cellular immortalization or establishment, and the middle T antigen, for cellular transformation. The role of the small T antigen is less clear, but several roles in controlling the quality of transformation and tumorigenesis have been postulated.

The quest for an explanation of polyomavirus transformation next switched to an analysis of the biochemical properties associated with the middle T antigen. The deduced amino acid sequence revealed little other than a hydrophobic C-terminal sequence, which together with biochemical fractionation data suggested that the protein was membrane associated (Carmichael et al. 1982). A tyrosine-specific protein kinase was detected in immunoprecipitates containing middle T antigen (Eckhart et al. 1979; Schaffhausen and Benjamin 1979; Smith et al. 1979), but it proved


[1] Laboratory of Cellular Regulation, Integrated Genetics Inc., One Mountain Road, Framingham, MA 01701, USA

[2] Dana-Farber Cancer Institute and Department of Pathology, Harvard Medical School, Boston, MA 02115, USA



Current Topics in Microbiology and Immunology, Vol. 144
© Springer-Verlag Berlin · Heidelberg 1989

impossible to show that middle T antigen had intrinsic kinase activity as had previously been shown for $pp60^{v\text{-}src}$, the transforming protein of Rous sarcoma virus and prototype tyrosine kinase (COLLETT and ERIKSON 1978).

This dilemma was resolved when it was found that middle T antigen forms a complex with the cellular homologue of $pp60^{v\text{-}src}$ called $pp60^{c\text{-}src}$ (COURTNEIDGE and SMITH 1983). The latter protein had tyrosine kinase activity but was itself nontransforming. The presence of a complex between middle T antigen and $pp60^{c\text{-}src}$ in vitro did not establish that this complex was important in transformation or that it existed in vivo. However, results of a thorough mutagenic study of middle T antigen showed an extremely good correlation between the ability of middle T-antigen variants to complex with $pp60^{c\text{-}src}$ and their ability to transform cells (COURTNEIDGE and SMITH 1984; CHENG et al. 1986). In addition, such experiments also indicated the middle T antigen requires membrane association to be active (CARMICHAEL et al. 1982; MARKLAND et al. 1986). The molecular analysis of middle T antigen also gave some indication of its domain structure and suggested, for example, that the $pp60^{c\text{-}src}$ binding site is present within the N-terminal half of the molecule (CHENG et al. 1986; MARKLAND and SMITH 1987a).

These results implied that by binding to $pp60^{c\text{-}src}$, middle T antigen somehow activated the proto-oncogene product thus enabling it to deregulate cellular growth and thereby generate the transformed phenotype. Indeed, early studies of the tyrosine kinase activity of $pp60^{c\text{-}src}$ in complex indicated a five- to tenfold increase in enzymatic activity (BOLEN et al. 1984; COURTNEIDGE 1985). The experiments described here attempt to understand the molecular mechanism of activation of $pp60^{c\text{-}src}$ by generating specific mutations within the protein-coding sequence. These were designed to mimic the effect of middle T-antigen binding.

## 2 Methods and Experimental Procedures

### 2.1 Mutagenic Procedures

All the mutations were introduced into the plasmid, pGC11 (Cm), which contains the cDNA clone for avian $pp60^{c\text{-}src}$, essentially as described by PIWNICA-WORMS et al. (1987) and CHENG et al. (1988). The details of the procedure for site-directed mutageneis using single and mixed oligonucleotides have been published elsewhere (CHENG et al. 1986, 1988). The chimeric HY c/v and HY v/c plasmids were constructed as reported by CHENG et al. (1988).

### 2.2 Generation of Cell Lines

Genes encoding the c-*src* mutants were cloned into a murine retroviral vector, pLJ, as described by PIWNICA-WORMS et al. (1987). Following transfection into $\psi$-2 packaging cells, recombinant viral stocks were harvested and used to introduce the genes encoding the variant forms of $pp60^{c\text{-}src}$ into NIH 3T3 cells. Stable NIH 3T3 cell lines expressing the various mutant pp60$^{c-src}$ proteins were isolated by clonal expansion of G418-resistant colonies.

## 2.3 In Vitro Kinase Assays

Immune-complex kinase assays and quantitative kinase assays using enolase were conducted as described previously (COURTNEIDGE and SMITH 1983; CHENG et al. 1986, 1988). Conditions for labelling with [$^{35}$S]methionine or $^{32}$P$_i$, preparation of cell lysates, immunoprecipitation, and sodium dodecyl sulphate-polyacrylamide gel electrophoresis have all been described elsewhere (SMITH et al. 1979; COURTNEIDGE 1985).

## 2.4 Two-Dimensional Tryptic Phosphopeptide Mapping

Samples were prepared for phosphopeptide mapping using a modification of the procedure described by SMITH et al. (1978) and analyzed as detailed by HUNTER and SEFTON (1980).

## 2.5 Transcription and Translation In Vitro

The pp60$^{c-src}$ variants containing premature termination signals and the chimeric plasmids were cloned into the SP65 transcription vector (Promega Biotec.) Following linearization with *Sal*I, the vectors were transcribed with SP6 RNA polymerase essentially as recommended by the manufacturer. RNA transcripts were purified and translated in vitro in a rabbit reticulocyte lysate in the presence of [$^{35}$S]methionine.

# 3 Results

## 3.1 Lack of phosphorylation at Tyrosine 527 activates pp60$^{c-src}$

The fraction of pp60$^{c-src}$ molecules present in complex with middle T antigen displays an altered pattern of tyrosine phosphorylation when compared with its unassociated counterpart (COURTNEIDGE 1985; YONEMOTO et al. 1985; CARTWRIGHT et al. 1986). Further analysis of the pattern of phosphorylation of pp60$^{c-src}$ suggests that the level of tyrosine modification in the C-terminal region of the protein is inversely related to tyrosine kinase activity (COURTNEIDGE 1985). Since pp60$^{c-src}$ differs from pp60$^{v-src}$ in only a few amino acids at the C-terminus (TAKEYA and HANAFUSA 1983), we argue that the tyrosine residue bestowing the unique activational properties of pp60$^{c-src}$ must be either tyrosine 519 or 527. We therefore created mutants in which these residues were changed to phenylalanine, an amino acid very similar to tyrosine but incapable of undergoing phosphorylation. At the same time, we also created a similar mutation at tyrosine 416, a residue phosphorylated in the in vitro kinase reaction. We combined the three mutations together in all possible different combinations.

The mutated DNA sequence was ligated into retrovirus vectors and introduced into NIH 3T3 cells. Labelling experiments (Fig. 1A) show that approximately equivalent amounts of pp60$^{c\text{-}src}$ could be immunoprecipitated from the cells. However, when the amount of kinase activity was quantitated using enolase as an

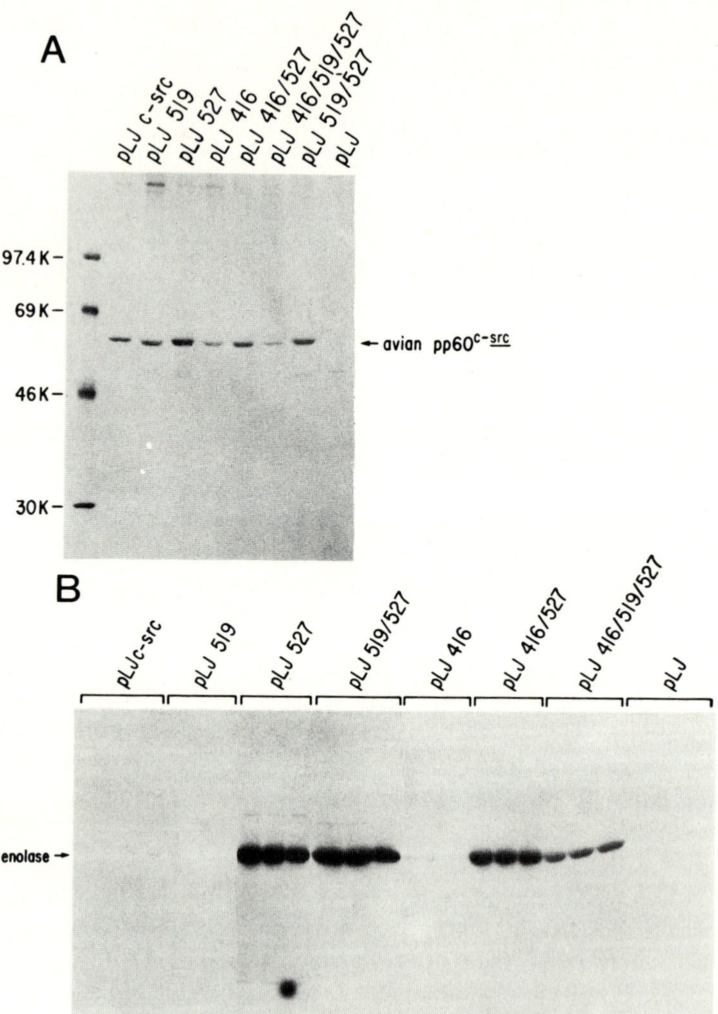



**Fig. 1A, B.** Tyrosine kinase activity of mutant pp60$^{c\text{-}src}$. Cell lines expressing the mutant avian pp60$^{c\text{-}src}$ were labeled with [$^{35}$S]methionine. Cell lysates were prepared, normalized for protein content, and then immunoprecipitated with EC10 serum. Immunoprecipitates obtained were either analyzed directly by SDS-polyacrylamide gel electrophoresis (**A**) or incubated with enolase and [$\gamma^{32}$P]ATP (**B**)

exogenous substrate, all the variants with the tyrosine 527 mutation were found to have elevated enzymatic activity (Fig. 1B). Examination of the properties of cells containing the mutated sequences show a parallel correlation between a transformed phenotype and mutations at 527. Thus, the normally nontransforming pp60$^{c-src}$ was activated to give rise in some cases to transformed colonies of cells (approximately 60% of the G418-resistant colonies displayed foci formation), and this correlated with increased tyrosine kinase activity of the pp60$^{c-src}$ in these cells. A further, unexpected result was that this activation of both transforming and kinase activities was reversed by mutation of tyrosine 416 to phenylalanine. Hence, whilst the mutants 527 and 519/527 exhibited a transformed phenotype, the mutants 416/527 and 416/519/527, although displaying partially elevated kinase activity, were nevertheless nontransforming.

## 3.2 Tyrosine Phosphorylation at 416 Is Involved in Activation

Analysis of the various tyrosine mutants implies that structural features in the C-terminal region of pp60$^{c-src}$, such as the absence of a phosphate group at residue 527, activate the kinase activity of the protein. This, in turn, activates the transforming ability of the pp60$^{c-src}$ via an event involving tyrosine 416. To test whether this event involves phosphorylation of tyrosine 416, we fingerprinted $^{32}$P-labelled pp60$^{c-src}$ isolated from cells containing the variant forms of the protein. The results shown in Fig. 2 of fingerprints of a number of variants allow identification of at least three phosphopeptides. Consistent with the hypothesis being tested, the 527 mutant clearly shows elevated phosphorylation on the peptide containing tyrosine 416.

Extensive mutagenesis of the region surrounding tyrosine 527 shows that none of the other mutations or deletions tested in the region surrounding the residue significantly activated the protein (CHENG et al. 1988). This suggests that the modification involved is phosphorylation and that the enzymes catalysing the reaction have little specificity for the region adjacent to tyrosine 527.

These results, therefore, lead to a model proposing that phosphorylation at tyrosine 527 acts normally to keep pp60$^{c-src}$ in an inactivated, quiescent state. Removal of phosphate or mutation to prevent its addition activates the protein, and concomitant with that is phosphorylation at tyrosine 416.

## 3.3 The Middle T-Antigen Binding Site

Middle T antigen does not bind to pp60$^{v-src}$, implying that the unique C-terminal region of pp60$^{c-src}$ is important in complex formation (CHENG et al. 1988). To investigate this further, we produced a series of deletion mutants lacking short sequences at the C-terminus (503T, 518T, 525T, and 527T), and we created hybrid c-*srv*/v-*src* molecules (HY c/v and HY v/c). Figure 3A shows the proteins encoded by the mutants and demonstrates that the molecular weights of the truncated mutants are consistent with the size of the deletions. Figure 3B shows in vitro kinase assays of immunoprecipitates from extracts of cells expressing the variant pp60$^{c-src}$ molecules which have been either mock-infected or super-infected with polyomavirus. The

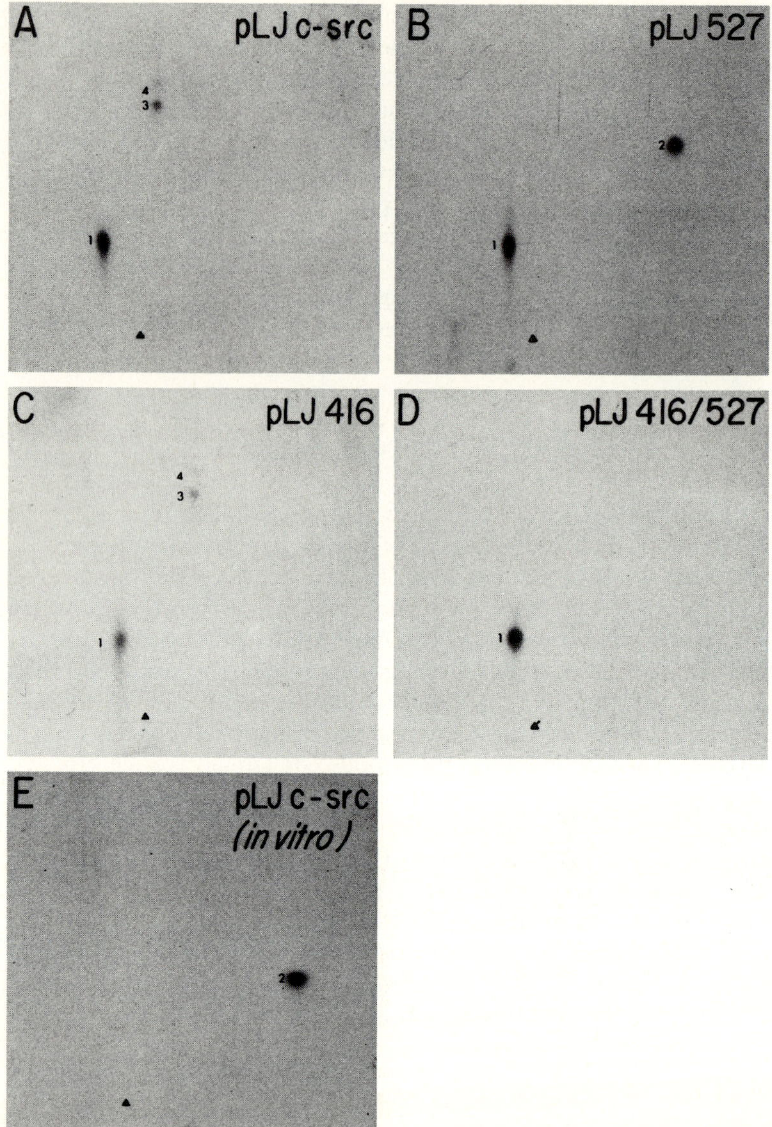



**Fig. 2A–E.** Tryptic phosphopeptide maps of pp60[c-src] mutants. [32]P-labeled pp60[c-src] were purified and analyzed as described in Sect. 2. The origins are marked with *arrowheads*. Spot numbering: *1*, peptides containing Ser 17; *2*, peptides containing Tyr 416; *3* and *4*, peptides containing Tyr 527

monoclonal antibody used (EC10) is specific for the introduced avian pp60[c-src] (PARSONS et al. 1984) and is able to coprecipitate and phosphorylate middle T antigen from a number of cell lines. Complexes are not formed, however, with variants lacking residues 518–533 or in hybrids containing the pp60[v-src] C-terminus. Taken together, these results indicate that the sequences between 518–527 of pp60[c-src]

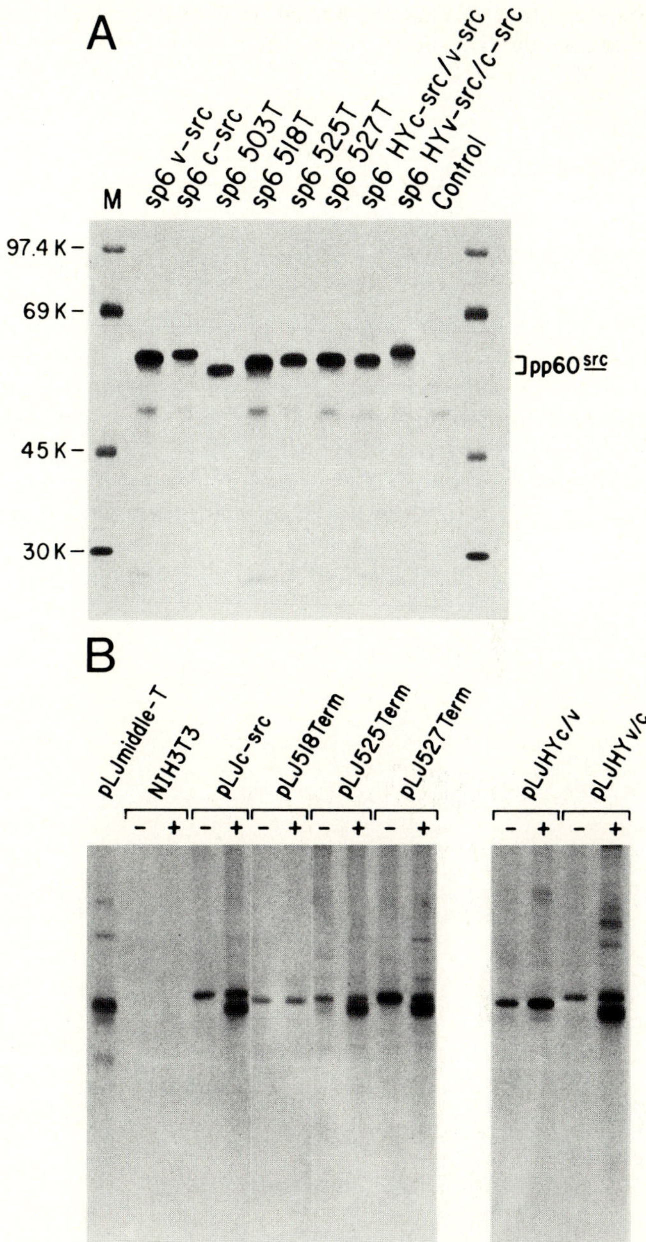


**Fig. 3A, B.** Association of truncated pp60$^{c\text{-}src}$ mutants with middle T antigen. **A** Fluorograph of a gel showing the molecular weights of the truncated and chimeric mutants following in vitro transcription and translation in a rabbit reticulocyte lysate in the presence of [$^{35}$S]methionine. **B** Cells expressing these mutant proteins were mock infected (−) or infected with polyomavirus (+) and then analyzed as described by CHENG et al. (1988)

are important in complex formation, and they imply that middle T antigen may bind directly to sequences in this region.

## 4 Discussion

### 4.1 Model for Interaction Between Middle T Antigen and pp60$^{c\text{-}src}$

Genetic and biochemical analysis of both the middle T antigen and pp60$^{c\text{-}src}$ have facilitated the assignment of various functional domains to the proteins (Fig. 4). Both middle T antigen and pp60$^{c\text{-}src}$ are associated with membranes. The former is anchored by its C-terminus and the latter by its N-terminal myristic acid. The binding site for pp60$^{c\text{-}src}$ on the middle T antigen is in its N-terminal end. The catalytic domain on pp60$^{c\text{-}src}$ is in its N-terminal domain, and its middle T-antigen binding site is towards the C-terminus (Fig. 4). Together these data suggest a model for the interaction between the two proteins involving association between their membrane distal domains, that is the C-terminus of pp60$^{c\text{-}src}$ and the N-terminus

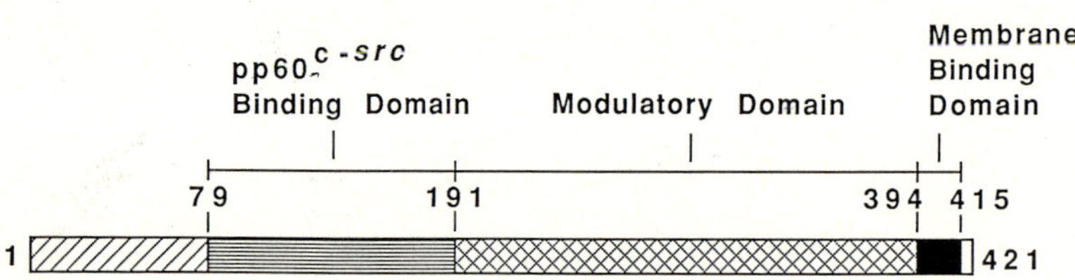


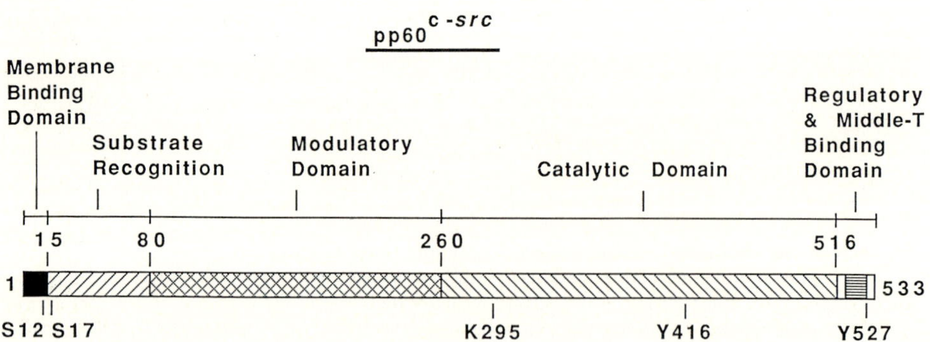


Fig. 4. Functional domains on middle-T antigen and pp60$^{c\text{-}src}$. The numbers refer to amino acid numbers in the respective proteins

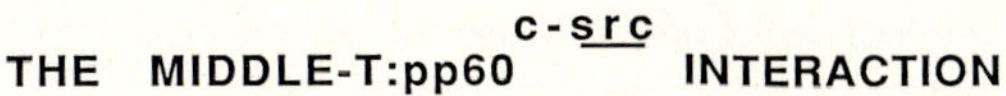


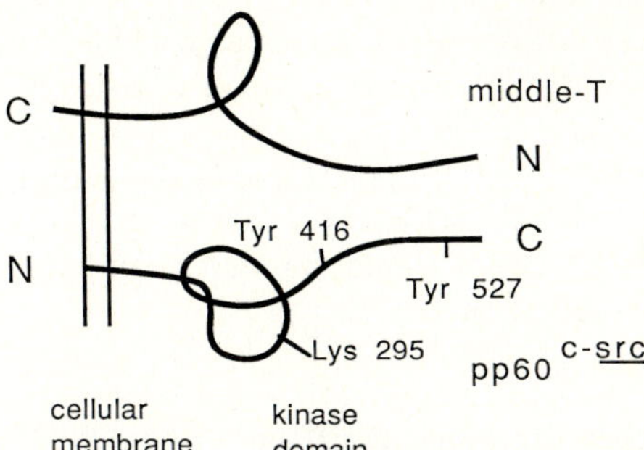


**Fig. 5.** Schematic representation of possible interaction between middle-T, pp60$^{c\text{-}src}$ and cellular membranes. Landmark amino acids in pp60$^{c\text{-}src}$ are indicated

of middle T antigen (Fig. 5). This suggestion is consistent with all the mapping data of which we are aware, particularly the results implying that middle T antigen binds to pp60$^{c\text{-}src}$ at sequences adjacent to tyrosine 527. This suggests a simple model for activation of the proto-oncogene product involving direct steric hindrance by middle T antigen of the kinases that normally add phosphate to tyrosine 527 to maintain the molecule in its quiescent state. As such, the middle T antigen may represent an aberrant version of the normal cellular regulator of pp60$^{c\text{-}src}$.

## 4.2 Composition of the Complex

Early experiments to characterize the complex between pp60$^{c\text{-}src}$ and middle T antigen indicated that it sedimented at a rate corresponding to a molecular weight of 200000 (WALTER et al. 1982; COURTNEIDGE and SMITH 1983). Since the combined molecular weights of middle T antigen and pp60$^{c\text{-}src}$ are approximately 100000, there is considerable uncertainty as to the components present in the complex. What little data are available argues for a one:one relationship between them. This possibly means that two molecules of each are present, but other combinations cannot be ruled out.

Another possibility is that other proteins also interact with the complex. Candidates that have been suggested include phosphatidylinositol kinase (WHITMAN et al. 1985; COURTNEIDGE and HEBER 1987; KAPLAN et al. 1987) and unidentified molecules reported to have molecular weights ranging between 30K and 85K. At present, it is difficult to judge the importance of the possible contribution of these proteins to the activity of the complex. However, a correlation between transforming activity and associated phosphatidylinositol kinase has been reported (KAPLAN et al. 1986, 1987; COURTNEIDGE and HEBER 1987). This could indicate a second type of signal transduction within the middle T antigen-transformed cell.

## 4.3 Modes of Action of the Middle T Antigen in Polyomavirus-Transformed Cells

The experiments reported so far have examined only some of the potential regulatory sites on pp60$^{c\text{-}src}$ and their involvement in regulation by middle T antigen. Cells containing pp60$^{c\text{-}src}$ altered at position 527 although clearly different from their parental cells, may not be as fully transformed as cells containing polyomavirus (KMIECIK and SHALLOWAY 1987; PIWNICA-WORMS et al. 1987; CARTWRIGHT et al. 1987). This raises questions as to whether middle T antigen does more than simply activate pp60$^{c\text{-}src}$ via interaction at tyrosine 527.

Several scenarios are possible. Middle T antigen could do other things to pp60$^{c\text{-}src}$; it could interact with any of a number of other cellular tyrosine kinases similar to pp60$^{c\text{-}src}$, or it could interact with other proteins which play a key role in cellular regulation.

## 4.4 Other Biochemical Effects of Middle T Antigen on Complexed pp60$^{c\text{-}src}$

So far, only limited studies have been reported identifying potential interactions between middle T antigen and other functional domains on pp60$^{c\text{-}src}$. There are several other known or potential phosphorylation sites in the molecule such as Ser 12 and Ser 17 and on as yet unmapped N-terminal tyrosines and threonines (GOULD et al. 1985; YONEMOTO et al. 1985; PATSCHINSKY et al. 1986; CHAKALAPARAMPIL and SHALLOWAY 1988). The kinase domain is localized, and the site of interaction with membranes is well characterized. Further mutagenic studies are required to measure the influence of these on the biological activity of pp60$^{c\text{-}src}$ and on complex function. In addition, appropriate deletions and chimeras need to be constructed to identify the substrate binding site on the molecule and any potential changes in their activity upon complex formation.

Equally, there remain major questions as to the domain structure of middle T antigen itself. Potential phosphorylation sites on tyrosine, serine, and threonine have been mapped, but so far their role in the activity of the protein is controversial (for review, see MARKLAND and SMITH 1987b).

## 4.5 Association of Middle T Antigen with Other Members of the *src* Family of Tyrosine Kinases

Since there is considerable homology within the tyrosine kinase family, particularly at the C-terminus, it is not unreasonable to think that middle T antigen binds to more than one member of this family of enzymes. Indeed, KORNBLUTH et al. (1987) reported that the c-*yes* gene product forms such a complex. Assuming other proteins such as the c-*fyn* and c-*fgr* gene products also interact with middle T antigen, several interesting questions arise. Is middle T antigen in complex with several proto-oncogene products in one large multi-molecular complex? If so, what are the substrates of the activated complexed versions of the proto-oncogene products? Alternatively, does middle T antigen form a separate complex with each of the different tyrosine kinases and does each complex control a different signalling pathway which ultimately leads to, or contributes to, cellular proliferation?

These final questions, in one sense, highlight the molecular detail to which our understanding of the transformation process has progressed and the sophistication of the questions currently asked. On the other hand, the quest to identify those intermediates that lead from the initial signal controlled by the transforming protein (or for that matter, by almost any growth factor or hormone) and which ultimately result in increased DNA synthesis and cellular proliferation seems as inconclusive as ever. We appear to have progressed just one molecule down that route from middle T antigen to pp60$^{c-src}$. If the pathway involves tens of different molecules and takes hours in time, there would appear to be much still to learn.

*Acknowledgements.* We thank Mary Lewis for typing the manuscript. This work was supported in part by industrial postdoctoral fellowships granted by Integrated Genetics and by Public Health Service grant R01 CA43186-02 from the National Cancer Institute to S.H.C., R.H., P.C.E., and A.E.S.; and by CA07634 to H.P.W. and CA43803 to T.M.R.

# References


Bolen JB, Thiele CJ, Israel MA, Yonemoto W, Lipsich LA, Brugge JB (1984) Enhancement of cellular *src* gene product associated tyrosyl kinase activity following polyoma virus infection and transformation. Cell 38: 767–777

Carmichael GG, Schaffhausen BS, Dorsky DI, Oliver DB, Benjamin TL (1982) The carboxy terminus of polyoma middle-T antigen is required for attachment to membranes, associated protein kinase activities, and cell transformation. Proc Natl Acad Sci USA 79: 3579–3583

Cartwright CA, Kaplan PL, Cooper JA, Hunter T, Eckhart W (1986) Altered sites of tyrosine phosphorylation in pp60$^{c-src}$ associated with polyomavirus middle tumor antigen. Mol Cell Biol 6: 1562–1570

Cartwright CA, Eckhart W, Simon S, Kaplan PL (1987) Cell transformation by pp60$^{c-src}$ mutated in the carboxy terminal regulatory domain. Cell 49: 83–91

Chackalaparampil I, Shalloway D (1988) Altered phosphorylation and activation of pp60$^{c-src}$ during fibroblast mitosis. Cell 52: 801–810

Cheng SH, Markland W, Markham AF, Smith AE (1986) Mutations around the NG59 lesion indicate an active association of polyoma virus middle-T antigen with pp60$^{c-src}$ is required for cell transformation. EMBO J 5: 325–334

Cheng SH, Piwnica-Worms H, Harvey RW, Roberts TM, Smith AE (1988) The carboxy terminus of pp60$^{c-src}$ is a regulatory domain and is involved in complex formation with the middle-T antigen of polyomavirus. Mol Cell Biol 8: 1736–1747

Collett MS, Erikson RL (1978) Protein kinase activity associated with the avian sarcoma virus *src* gene products. Proc Natl Acad Sci USA 75: 2021–2024

Courtneidge SA (1985) Activation of the pp60$^{c-src}$ kinase by middle-T antigen binding or by dephosphorylation. EMBO J 4: 1471–1477

Courtneidge SA, Heber A (1987) An 81kd protein complexed with middle-T antigen and pp60$^{c-src}$: a possible phosphatidylinositol kinase. Cell 50: 1031–1037

Courtneidge SA, Smith AE (1983) Polyoma virus transforming protein associates with the product of the c-*src* cellular gene. Nature 303: 435–439

Courtneidge SA, Smith AE (1984) The complex of polyoma virus middle-T antigen and pp60$^{c-src}$. EMBO J 3: 585–591

Cuzin F (1984) The polyomavirus oncogenes: coordinated functions of three distinct proteins in the transformation of rodent cells in culture. Biochim Biophys Acta 781: 193–204

Eckhart W, Hutchinson MA, Hunter T (1979) An activity phosphorylating tyrosine in polyoma T-antigen immunoprecipitates. Cell 18: 925–933

Fried M, Prives C (1986) The biology of simian virus 40 and polyomavirus. Cancer Cells 4: 1–16

Gould KL, Woodgett JR, Cooper JA, Buss JE, Shalloway D, Hunter T (1985) Protein kinase C phosphorylates pp60$^{c-src}$ at a novel site. Cell 42: 849–857

Griffin BE, Dilworth SM (1983) Polyomavirus: an overview of its unique properties. Adv Cancer Res 39: 183–268

Hunter T, Sefton BM (1980) Transforming gene product of Rous sarcoma virus phosphorylates tyrosine. Proc Natl Acad Sci USA 77: 1311–1315

Kaplan DR, Whitman M, Schaffhausen B, Raptis L, Garcea RL, Pallas D, Roberts TM, Cantley L (1986) Phosphatidylinositol metabolism and polyoma-mediated transformation. Proc Natl Acad Sci USA 83: 3624–3628

Kaplan DR, Whitman M, Schaffhausen B, Pallas DC, White M, Cantley L, Roberts TM (1987) Common elements in growth factor stimulation and oncogenic transformation: 85kd phosphoprotein and phosphatidylinositol kinase activity. Cell 50: 1021–1029

Kmiecik TE, Shalloway D (1987) Activation and suppression of pp60$^{c-src}$ transforming ability by mutation of its primary sites of tyrosine phosphorylation. Cell 49: 65–73

Kornbluth S, Sudol M, Hanafusa H (1987) Association of polyomavirus middle-T antigen with c-*yes* protein. Nature 325: 171–173

Markland W, Smith AE (1987a) Mapping of the amino-terminal half of polyomavirus middle-T antigen indicates that this region is the binding domain for pp60$^{c-src}$. J Virol 61: 285–292

Markland W, Smith AE (1987b) Mutants of polyomavirus middle-T antigen. Biochim Biophys Acta 907: 299–321

Markland W, Cheng SH, Oostra BA, Smith AE (1986) In vitro mutagenesis of the putative membrane-binding domain of polyomavirus middle-T antigen. J Virol 59: 82–89

Parsons SJ, McCarley DJ, Ely CM, Benjamin DC, Parsons JT (1984) Monoclonal antibodies to Rous sarcoma virus pp60$^{c-src}$ react with enzymatically active cellular pp60$^{c-src}$ of avian and mammalian origin. J Virol 51: 272–282

Patschinsky T, Hunter T, Sefton BM (1986) Phosphorylation of the transforming protein of Rous sarcoma virus: direct demonstration of phosphorylation of serine 17 and identification of an additional site of tyrosine phosphorylation in p60$^{v-src}$ of Praque sarcoma virus. J Virol 59: 73–81

Piwnica-Worms H, Saunders KB, Roberts TM, Smith AE, Cheng SH (1987) Tyrosine phosphorylation regulates the biochemical and biological properties of pp60$^{c-src}$. Cell 49: 75–82

Schaffhausen BS, Benjamin TL (1979) Phosphorylation of polyoma T-antigens. Cell 18: 935–946

Smith AE, Ely BK (1983) The biochemical basis of transformation by polyoma virus. Adv Viral Oncol 3: 3–30

Smith AE, Smith R, Paucha E (1978) Extraction and fingerprint analysis of simian virus 40 large and small T-antigens. J Virol 28: 140–153

Smith AE, Smith R, Griffin B, Fried M (1979) Protein kinase activity associated with polyoma virus middle-T antigen in vitro. Cell 18: 915–924

Takeya T, Hanafusa H (1983) Structure and sequence of the cellular gene homologous to the RSV *src* gene and the mechanism for generating the transforming virus. Cell 32: 881–890

Walter G, Hutchinson MA, Hunter T, Eckhart W (1982) Purification of polyoma virus medium-sized tumor antigen by immunoaffinity chromatography. Proc Natl Acad Sci USA 79: 4025–4029

Whitman M, Kaplan DR, Schaffhausen B, Cantley L, Roberts TM (1985) Association of phosphatidylinositol kinase activity with polyoma middle-T competent for transformation. Nature 315: 239–242

Yonemoto W, Yarvis-Morar M, Brugge JS, Bolen JB, Israel M (1985) Tyrosine phosphorylation within the amino-terminal domain of pp60$^{c-src}$ molecules associated with polyoma virus middle-sized tumor antigen. Proc Natl Acad Sci USA 82: 4568–4572

# Further Characterisation of the Complex Containing Middle T Antigen and pp60

S. A. COURTNEIDGE

## 1 Introduction

Polyomavirus is a DNA tumor virus which has the capacity to cause tumours in susceptible hosts, as well as converting fibroblasts to the transformed phenotype in vitro. This transformation ability resides in the early region of the genome, which encodes three proteins, called the large, middle and small T antigens (TOOZE 1981). There is overwhelming evidence that only the middle T antigen is required for the transformation of established cells (reviewed in SMITH and ELY 1983). Transformation by polyomavirus has therefore proved to be an interesting model system for dissecting the multitude of effects a single protein can exert on a cell.

Middle T antigen is a protein of 421 amino acids, phosphorylated predominantly on serine, but also to a lesser extent on threonine and tyrosine residues in vivo (SEGAWA and ITO 1982). It is associated with membranes (ITO 1979; ZHU et al. 1984; DILWORTH et al. 1986) via a stretch of hydrophobic amino acids near the C-terminus (NOVAK and GRIFFIN 1981; TEMPLETON and ECKHART 1981; CARMICHAEL et al. 1982; MARKLAND et al. 1986; CHENG et al., this volume). The protein appears to have no posttranslational modifications apart from phosphorylation, and is not exposed on the cell surface (with the possible exception of the six C-terminal amino acids). Middle T antigen is heterodisperse on sucrose density gradients, with the majority of the protein sedimenting with a molecular weight of greater than 100K, suggesting the existence of multimers or heterogeneous complexes (SCHAFFHAUSEN and BENJAMIN 1981; WALTER et al. 1982).

In recent years analysis of transformation by middle T antigen has concentrated on its interactions with host proteins (reviewed in COURTNEIDGE 1986). One type of complex which middle T antigen can form involves protein tyrosine kinases. The first reported and best studied of these is the complex between middle T antigen and pp60$^{c-src}$ (COURTNEIDGE and SMITH 1983, 1984). Middle T antigen can also complex with two other tyrosine kinases: the products of the *yes* (KORNBLUTH et al. 1987) and *fyn* (R. M. KYPTA, A. HEMMING and S. A. COURTNEIDGE, 1988) genes, as well as with proteins which do not appear to have intrinsic tyrosine kinase activity. The most abundant of these latter complexes contains a 61K cellular protein of unknown function (GRUSSENMEYER et al. 1985, 1987). The contribution of each of these complexes to transformation is not yet clearly understood,

European Molecular Biology Laboratory, Postfach 10.2209, Meyerhofstrasse 1, 6900 Heidelberg, FRG

Current Topics in Microbiology and Immunology, Vol. 144
© Springer-Verlag Berlin · Heidelberg 1989

although mutant analysis suggests that the pp60$^{c\text{-}src}$-middle T antigen complex is necessary (COURTNEIDGE and SMITH 1984). It is indeed probable that all complexes are required, and that each specifies one particular part of the range of properties in which transformed cells differ from their normal counterparts.

The study of middle T antigen does not just tell us about transformation. Rather, middle T antigen is proving to be a valuable tool with which to study the normal cell proteins with which it interacts. For example, following on from the demonstration that middle T antigen can increase the kinase activity of pp60$^{c\text{-}src}$ (BOLEN et al. 1984) there came the description of how this was done, namely by the masking of a regulatory phosphorylation site (COURTNEIDGE 1985; CARTWRIGHT et al. 1986). Thus, studying middle T antigen gave us some insights into the normal regulation of pp60$^{c\text{-}src}$. One of the most interesting areas now is to define substrates for pp60$^{c\text{-}src}$, and we are at present concentrating on those proteins which become phosphorylated on tyrosine as a result of the presence of middle T antigen. One good candidate for such a pp60$^{c\text{-}src}$ substrate is an 81K protein which is present in the complex with middle T antigen and pp60$^{c\text{-}src}$. This protein, which appears to have the properties of a phosphatidylinositol (PI) kinase, will be described here.

## 2 Methods and Experimental Procedures

### 2.1 Cells

NIH 3T3 cells and their derivatives containing middle T antigen were as described in COURTNEIDGE and HEBER (1987).

### 2.2 Antibodies

The pp60$^{c\text{-}src}$ was immune precipitated using monoclonal antibody 327 (LIPSICH et al. 1983) purchased from Oncogene Sciences, Mineola NY. Middle T antigen was immunoprecipitated using antibodies specific for the C-terminus of the protein (anti-mt.c; COURTNEIDGE and SMITH 1984).

### 2.3 Methods

Our methods for the labelling of cells, immunoprecipitation of proteins, tyrosine kinase assays, PI kinase assays, SDS-polyacrylamide gel electrophoresis, etc. have all been described before (COURTNEIDGE and SMITH 1983, 1984; COURTNEIDGE and HEBER 1987).

Phosphatase experiments were based on the protocol described in COOPER and KING (1986). Potato acid phosphatase was purchased from Boehringer Mannheim as a suspension in ammonium sulphate. The enzyme was sedimented by centrifugation, resuspended to 1 mg/ml in 20 m$M$ HEPES, pH 7.5, 0.5 m$M$ MgCl$_2$, 0.5 m$M$ dithioerythritol (DTT), 50% glycerol, and stored at $-70\ °C$. To use, immune complexes bound to *Staphylococcus aureus* (Pansorbin, Calbiochem) were resuspended to 2.5% (w/v) in 20 m$M$ Hepes, pH 7.5, 1 m$M$ DTT, 1% aprotinin, 20 μ$M$ leupeptin, and either 2 μg bovine serum albumin (BSA) or 2 μg potato acid phosphatase added. After mixing, the reactions were incubated at 30 °C for 10 min, washed three times and then assayed. Under these conditions, approximately 50%–75% of the phosphate was removed from test proteins (data not shown).

# 3 Results and Discussion

It is by now well established that when immune complexes prepared using antibodies either to middle T antigen or to pp60$^{c\text{-}src}$ and lysates of polyomavirus-infected or -transformed cells are incubated in the presence of ATP and divalent cation, middle T antigen becomes phosphorylated on its tyrosine residues (for review, see COURTNEIDGE 1986). Under some conditions (depending particularly on the antibody used to make the immune complexes), the associated pp60$^{c\text{-}src}$ can also be detected by autophosphorylation. Although initially only these two proteins were identified, previous descriptions of the complex had estimated the molecular weight to be 200–220K (WALTER et al. 1982; COURTNEIDGE and SMITH 1983; COURTNEIDGE et al. 1984), indicating that either the complex consists of multimers of the two known components or that other protein(s) are also present. With the use of better reagents (e.g. monoclonal antibodies), more sensitive kinase assays and longer auto-radiographic exposure times, it became clear that the latter interpretation was correct (COURTNEIDGE and HEBER 1987). Another protein, of approximate molecular mass 81K, could also be detected in kinase assays, an example of which is shown in Fig. 1. When pp60$^{c\text{-}src}$ was immune precipitated from non-transformed mouse fibroblasts and assayed for kinase activity, pp60$^{c\text{-}src}$ itself became phosphorylated (lane 2). No specifically co-precipitating proteins could be detected using this assay, even after prolonged autoradiographic exposure. However, when pp60$^{c\text{-}src}$ immune complexes were derived from polyoma-virus-transformed cells, other co-precipitating proteins were detected (lane 4). The most abundant of these was middle T antigen, but the presence of an 81K protein was also noted. All three proteins were labelled on tyrosine residues in the kinase assay (data not shown). The same three proteins were seen when middle T antigen-specific immune precipitates were assayed for kinase activity (data not shown). Note that the p81 was not precipitated from normal cells with either pp60$^{c\text{-}src}$-specific or middle T anti-gen-specific antibodies. Several different antibodies (peptide, monoclonal, and poly-clonal) have been used directed against both middle T antigen and pp60$^{c\text{-}src}$, and p81 is always detected in the complex. In addition, this complex is not restricted to in vitro-transformed fibroblasts; in both endothelial cells (WILLIAMS et al. 1988) and cortical adrenal cells (R. L. WILLIAMS, E. F. WAGNER and S. A. COURTNEIDGE, manuscript in preparation) transformed with middle T antigen in vivo, a co-precipita-ting p81 protein could also be detected.

These observations, together with the fact that all three proteins co-migrate on sucrose gradients (COURTNEIDGE and HEBER 1987), suggest that they are all present

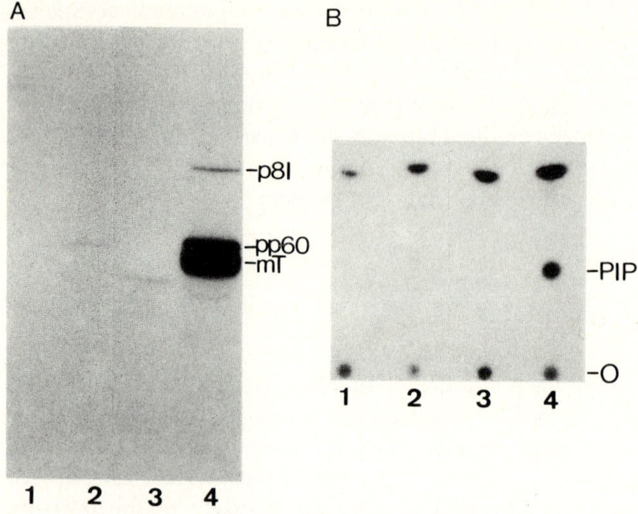



**Fig. 1A, B.** Kinase activities associated with pp60^(c-src) in normal and polyomavirus-transformed cells. **A** Tyrosine kinase assay. Indicated are the positions of middle T antigen (*mT*), pp60^(c-src) (*pp60*) and the associated 81K protein (*p81*). **B** Phosphatidylinositol kinase assay. Indicated are the origin (*O*) and the position of phosphorylated phosphatidylinositol (*PIP*). Immune complexes were prepared using rabbit mouse-specific antibodies alone (lanes *1* and *3*) or a monoclonal antibody specific for pp60^(c-src) and rabbit mouse-specific antibodies (lanes *2* and *4*), and lysates of either non-transformed cells (lanes *1* and *2*) or polyomavirus-transformed cells (lanes *3* and *4*)

in the same complex. However, it is not known at this point whether all the pp60^(c-src) molecules complexed to middle T antigen also have associated p81, or whether this only represents a subset of the middle T antigen-pp60^(c-src) complexes. From the kinase assays the stoichiometry of p81 binding would appear to be very low, but analyses of metabolically labelled cells seem to contradict this (COURTNEIDGE and HEBER 1987, and unpublished observations). I am currently undertaking a purification of the complex in order to address this issue.

In order to determine the possible importance of the p81 protein, a survey was carried out for its presence in complexes between pp60^(c-src) and various mutant middle T antigens. In general the presence of p81 in middle T-antigen immuno-precipitates parallels the presence of pp60^(c-src); all transforming middle T antigens co-immune precipitate with both pp60^(c-src) and p81, whereas those middle T antigen mutants which fail to bind to pp60^(c-src) also fail to associate with p81. This suggests that in order to transform, the complex must contain all three proteins. However, I have previously reported that some mutant middle T antigens (those belonging to class II) were able to bind to pp60^(c-src) yet failed to transform (COURTNEIDGE and SMITH 1984). When I examined the middle T antigen-pp60^(c-src) complex more thoroughly in these cases, I found two different phenotypes. In the first case, represented by mutants such as dl1015 and MG13, all three proteins were present (COURTNEIDGE and HEBER 1987). In the second case, comprising mutants dl23 (COURTNEIDGE and HEBER 1987) and 82JF3 (S. A. COURTNEIDGE and M. FRIED, manuscript in preparation), no associated p81 protein could be detected. The existence of this last class of mutants suggests that middle T antigen is able to bind directly to pp60$^{c\text{-}src}$. Likewise, since under certain circumstances a p81 protein can be detected associated with pp60$^{v\text{-}src}$ (see below), I believe that p81 also binds to pp60$^{c\text{-}src}$, rather than to the middle T antigen.

If p81 binding is indeed critical to transformation by polyomavirus, then this could explain why those class II mutants which do not have associated p81 fail to transform. However, we still lack an explanation as to why the other mutants of class II are non-transforming. Perhaps the answer will come when the ability of these mutants to associate with other host proteins is examined in more detail.

In searching for a function for p81, I recalled the description by WHITMAN et al. (1985) of a phosphatidylinositol (PI) kinase activity present in middle T-antigen immunoprecipitates. To test whether p81 could be involved in such an activity, the immunoprecipitates were screened for the presence of PI kinase activity. These results are shown in Fig. 1 B. When immune complexes are derived from non-transformed cells, no lipid kinase activity is detected. However, when the immune complexes are derived from polyomavirus-transformed cells, they are found to contain an activity which could phosphorylate PI. Although the predominant phosphorylated product of PI found in vivo is PI-4-P, WHITMAN et al. (1988) have recently reported that the PI kinase associated with middle T antigen phosphorylates PI at the 3-position of the inositol ring. In keeping with their data I have found that the product of the enzyme activity measured does not exactly co-migrate with authentic PI-4-P (data not shown), suggesting that I too am looking at this unusual PI kinase. In order to determine whether there is a relationship between the presence of p81 in immune complexes and the detection of PI kinase activity, immune precipitates of the various middle T mutants mentioned above were analysed. A complete correlation between the presence of p81 and PI kinase activity could be shown, indicating that p81 is probably a component of a PI kinase. Using a different approach, KAPLAN et al. (1987) have demonstrated the simultaneous appearance of an 85K phosphotyrosine-containing protein and PI kinase activity shortly after PDGF treatment of quiescent cells. It seems highly likely that the p81 described here and the 85K protein are similar or identical, which suggests that stimulation of this PI kinase is not restricted to transformation but may also be a part of the response of cells to normal growth stimuli.

PI kinase activity has been reported to be associated with (SUGIMOTO et al. 1984), although not a property of (MACDONALD et al. 1985; SUGANO and HANAFUSA 1985; SUGIMOTO and ERIKSON 1985) pp60$^{v\text{-}src}$, and the transforming Phe 527 pp60$^{c\text{-}src}$ mutant (PIWNICA-WORMS et al. 1987) also has associated PI kinase activity (unpublished observations). In both these cases the transformed cells contain the p81 protein, which is immunoprecipitable with a phosphotyrosine-specific antibody (data not shown). I believe that in these cells p81 is a substrate of the *src* proteins, although it does not exist in a stable complex with them. In the polyoma-virus-transformed cells, however, the presence of the middle T antigen appears to stabilise the binding of p81 to pp60$^{c\text{-}src}$. In all cases I propose that tyrosine phosphorylation of p81 by pp60$^{c\text{-}src}$ increases the associated PI kinase activity. What evidence do we have for this?

Some of the preliminary observations are shown in Fig. 2, where to analyse whether PI kinase activity is affected by phosphorylation the effect of dephosphorylation on

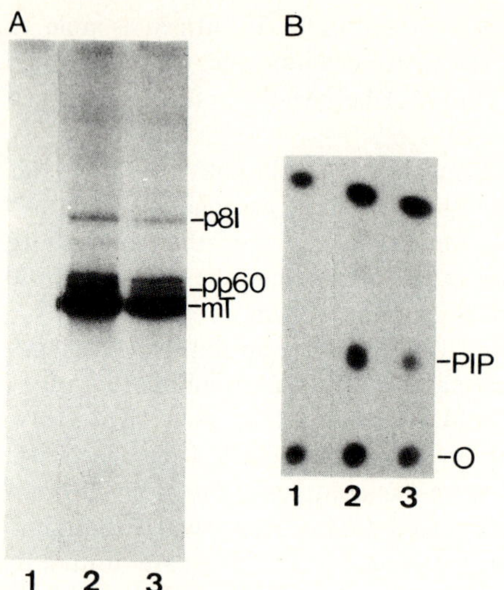



**Fig. 2A, B.** Tyrosine kinase and phosphatidylinositol kinase activities following dephosphorylation of immune complexes. **A** Tyrosine kinase assay. For an explanation of labelling, see legend to Fig. 1. **B** Phosphatidylinositol kinase assay. For an explanation of labelling, see legend to Fig. 1. Immune complexes were prepared from lysates of middle T antigen-transformed cells and mt.c-specific antibodies (lanes 2 and 3) or mt.c-specific antibodies pre-blocked with specific peptide (lane 1), treated with either BSA (lanes 1 and 2) or potato acid phosphatase (lane 3), washed and assayed for kinase activities

PI kinase activity in vitro was determined. I employed the enzyme potato acid phosphatase, which under the conditions used removed approximately 50%–75% of the phosphate from substrate proteins and showed no preference for serine or tyrosine residues. When immune complexes from polyomavirus-transformed cells were first dephosphorylated and then tyrosine kinase activity was measured, a decrease of approximately 50% was seen compared with control values (Fig. 2A). This is in accord with published findings that mutation of the major autophosphorylation site of pp60$^{c\text{-}src}$ reduces its kinase activity (PIWNICA-WORMS et al. 1987; KMIECIK and SHALLOWAY 1987, CHENG et al., this volume). As shown in Fig. 2B, dephosphorylation prior to kinase assay also reduced the PI kinase activity subsequently measured, in this case to approximately 30% of control values.

I do not believe that the effects seen were due to non-specific inactivation of the enzymes. For example, under the same conditions, the tyrosine kinase activity of pp60$^{c\text{-}src}$ from normal cells was increased (data not shown), as one would expect since pp60$^{c\text{-}src}$ is negatively regulated by tyrosine phosphorylation (COURTNEIDGE 1985; COOPER and KING 1986). These results are therefore consistent with the hypothesis that the PI kinase is positively regulated by phosphorylation. However, since potato acid phosphatase (PAP) dephosphorylates phosphoserine, phosphothreonine and phosphotyrosine residues, these data do not tell us whether tyrosine phosphorylation alone can control the enzyme activity. As a first attempt to investigate this the PI kinase activity was determined after lysing cells in the presence or absence of the phosphotyrosine phosphatase inhibitor, sodium orthovanadate, and the same results were obtained i.e. loss of phosphotyrosine from the molecule resulted in a significant loss of PI kinase activity (data not shown). I conclude from these studies that the PI kinase activity associated with pp60$^{c\text{-}src}$ in a polyomavirus-transformed cell may be regulated by tyrosine phosphorylation.

The working model is that p81 is in some way associated with a PI kinase. The PI kinase, or its regulator, becomes phosphorylated on tyrosine residues by pp60$^{c\text{-}src}$ and therefore activated. Several questions then arise. Firstly, is p81 itself the PI kinase? Secondly, how much of this PI kinase activity is complexed to pp60$^{c-src}$? Thirdly, can one measure an increase in PI kinase activity in polyomavirus-transformed cells? Fourthly, what function, if any, does PI-3-P play in vivo? The first three questions are difficult to answer at present because the PI kinase can only be measured in immune complexes. I have therefore begun a purification of middle T antigen to assess in greater detail associated proteins and enzyme activities. As to the fourth question, I have started to address this by looking at PI turnover in polyomavirus-transformed cells. No changes were detected in $(1,4,5)IP_3$ production compared with normal cells; this pathway appeared to be neither constitutively activated nor turning over more rapidly (E. T. ULUG and S. A. COURTNEIDGE, unpublished observations). I thus doubt that PI-3-P contributes to the known inositol signalling pathway. However, there are many inositol-containing compounds in cells, and these will have to be studied in more detail in order to elucidate the role of increased PI-3-P in transformation and perhaps normal growth.

*Acknowledgements.* I thank my colleagues Robert Kypta and Emin Ulug for useful discussions.

# References


Bolen JB, Thiele CJ, Israel MA, Yonemoto W, Lipsich LA, Brugge JS (1984) Enhancement of cellular *src* gene product associated tyrosyl kinase activity following polyoma virus infection and transformation. Cell 38: 767–777

Carmichael GG, Schaffhausen BS, Dorsky DI, Oliver DB, Benjamin TL (1982) Carboxy terminus of polyoma middle-sized tumor antigen is required for attachment to membranes, associated protein kinase activities, and cell transformation. Proc Natl Acad Sci USA 79: 3579–3583

Cartwright CA, Kaplan PL, Cooper JA, Hunter T, Eckhart W (1986) Altered sites of tyrosine phosphorylation in pp60$^{c-src}$ associated with polyoma virus middle tumor antigen. Mol Cell Biol 6: 1562–1570

Cooper JA, King CS (1986) Dephosphorylation or antibody binding to the carboxy terminus stimulates pp60$^{c-src}$. Mol Cell Biol 6: 4467–4477

Courtneidge SA (1985) Activation of pp60$^{c-src}$ kinase by middle T antigen binding or by dephosphorylation. EMBO J 4: 1471–1477

Courtneidge SA (1986) Transformation by polyoma virus middle T antigen. Cancer Surv 5: 173–182

Courtneidge SA, Heber A (1987) An 81 kd protein complexed with middle T antigen and pp60$^{c-src}$: a possible phosphatidylinositol kinase. Cell 50: 1031–1037

Courtneidge SA, Oostra B, Smith AE (1984) Tyrosine phosphorylation and polyoma virus middle T the c-*src* gene. Nature 303: 435–439

Courtneidge SA, Smith AE (1984) The complex of polyoma virus middle T antigen and pp60$^{c-src}$. EMBO J 3: 585–591

Coutneidge SA, Oostra B, Smith AE (1984) Tyrosine phsosphorylation and polyoma virus middle T protein. Cancer Cells 2: 123–131

Dilworth SM, Hansson H-A, Darnfors C, Bjursell G, Streuli CH, Griffin BE (1986) Subcellular localisation of the middle and large T-antigens of polyoma virus. EMBO J 5: 491–499

Grussenmeyer T, Scheidtmann KH, Hutchinson MA, Eckart W, Walter G (1985) Complexes of polyoma virus medium T antigen and cellular proteins. Proc Natl Acad Sci USA 82: 7952–7954

Grussenmeyer T, Carbone-Wiley A, Scheidtmann KH, Walter G (1987) Interaction between polyoma virus medium T antigen and three cellular proteins of 88, 61, and 37 kilodalton. J Virol 61: 3902–3909

Ito Y (1979) Polyoma virus-specific 55K protein isolated from the plasma membrane of productively infected mouse cells is virus-coded and important for cell transformation. Virology 98: 261–266

Kaplan DR, Whitman M, Schaffhausen B, Pallas DC, White M, Cantley L, Roberts TM (1987) Common elements in growth factor stimulation and oncogenic transformation: 85 kd phosphoprotein and phosphatidylinositol kinase activity. Cell 50: 1021–1029

Kmiecik TE, Shalloway D (1987) Activation and suppression of pp60$^{c\text{-}src}$ transforming ability by mutation of its primary sites of tyrosine phosphorylation. Cell 49: 65–73

Kornbluth S, Sudol M, Hanafusa H (1987) Association of the polyomavirus middle T antigen with c-*yes* protein. Nature 325: 171–173

Kypta RM, Hemming A and Courtneidge SA (1988) Identification and characterisation of p59$^{tyr}$ (a *src*-like protein tyrosine kinase) in normal and polyoma virus transformed cells. EMBO J 7: 3837–3844

Lipsich LA, Lewis AJ, Brugge JS (1983) Isolation of monoclonal antibodies that recognize the transforming proteins of avian sarcoma viruses. J Virol 48: 352–360

MacDonald ML, Kuenzel EA, Glomset JA, Krebs Ed (1985) Evidence from two transformed cell lines that the phosphorylations of peptide tyrosine and phosphatidylinositol are catalyzed by different proteins. Proc. Natl Acad Sci USA 82: 3993–3997

Markland W, Cheng SH, Oostra BA, Smith AE (1986) In vitro mutagenesis of the putative membrane binding domain of polyoma virus middle T antigen. J. Virol 59: 82–89

Novak U, Griffin BE (1981) Requirement for the C-terminal region of middle T antigen in cellular transformation by polyoma virus. Nucleic Acids Res 9: 2055–2073

Piwnica-Worms H, Saunders KB, Roberts TM, Smith AE, Cheng SH (1987) Tyrosine phosphorylation regulates the biochemical and biological properties of pp60$^{c\text{-}src}$. Cell 49: 75–82

Schaffhausen BS, Benjamin TL (1981) Protein kinase activity associated with polyoma virus middle T antigen. Protein Phosphorylation 8: 1281–1298

Segawa K, Ito Y (1982) Differential subcellular localization of in vivo-phosphorylated and non-phosphorylated middle-sized tumor antigen of polyoma virus and its relationship to middle-sized tumor antigen phosphorylating activity in vitro. Proc Natl Acad Sci USA 79: 6812–6816

Smith AE, Ely BK (1983) The biochemical basis of transformation by polyoma virus. Adv Viral Oncol 3: 3–30

Sugano S, Hanafusa H (1985) Phosphatidylinositol kinase activity in virus-transformed and non-transformed cells. Mol Cell Biol 5: 2399–2404

Sugimoto Y, Erikson RL (1985) Phosphatidylinositol kinase activities in normal and Rous sarcoma virus-transformed cells. Mol Cell Biol 5: 3194–3198

Sugimoto Y, Whitman M, Cantley LC, Erikson RL (1984) Evidence that the Rous sarcoma virus transforming gene product phosphorylates phosphatidylinositol and diacylglycerol. Proc. Natl Acad Sci USA 81: 2117–2121

Templeton D, Eckhart W (1981) Mutation causing premature termination of the polyoma virus medium T antigen blocks cell transformation. J Virol 41: 1014–1024

Tooze J (1981) DNA tumor viruses. Molecular biology of tumor viruses pt 2. Cold Spring Harbor Laboratory, Cold Spring Harbor

Walter G, Hutchinson MA, Hunter T, Eckhart W (1982) Purification of polyoma virus middle T tumor antigen by immunoaffinity chromatography. Proc Natl Acad Sci USA 82: 4568–4572

Whitman M, Kaplan DR, Schaffhausen B, Cantley L, Roberts TM (1985) Association of phosphatidylinositol kinase activity with polyoma middle T competent for transformation. Nature 315: 239–242

Whitman M, Downes CP, Keeler M, Keller T, Cantley L (1988) Type I phosphatidylinositol kinase makes a novel phospholipid, phosphatidylinositol-3-phosphate. Nature 332: 644–646

Williams RL, Courtneidge SA, Wagner EF (1988) Embryonic lethalities and endothelial tumors in chimeric mice expressing polyoma virus middle T oncogene. Cell 52: 121–131

Zhu Z, Veldman GM, Cowie A, Carr A, Schaffhausen B, Kamen R (1984) Construction and functional characterisation of polyoma virus genomes separately encoding the three early proteins. J Virol 51: 170–180

# The Mutagenic and Immortalizing Potential of Polyoma virus Large T Antigen

M. Strauss[1], L. Lübbe[1], U. Kiessling[1], M. Platzer[1], and B. E. Griffin[2]

## 1 Introduction

Our laboratory has shown that SV40 is mutagenic in infected or transfected cells (Theile et al. 1976, 1980; Lübbe et al. 1982). This activity could be assigned to the function of large T antigen (Theile et al. 1980, 1987). It would be of great interest to know if there is any link between the mutagenic activity and one of the other known functions of this antigen, which include stimulation of replication and gene expression (Kingston et al. 1986). Using T-antigen mutants, we tried to separate individual domains of the SV40 large T antigen (Strauss et al. 1987; McVey, Strauss and Gluzman, unpublished data). Preliminary data of a deletion analysis suggest that immortalization is in some way related to mutagenesis; DNA binding of large T antigen appears to be involved in both processes (Strauss, unpublished data).

In order to obtain a more definitive answer regarding the role of mutagenesis and its correlation to other viral functions we switched to the polyomavirus model.

In the following discussion we demonstrate the mutagenic activity of polyomavirus and discuss its likely correlation with the immortalizing activity of large T antigen. Our data show that low level T-antigen expression results in reversible immortalization.

## 2 Mutagenic Activity

Cloned complete polyoma viral DNA, as well as individual T-antigen genes, were tested for their mutagenic activity in a transfection-transient expression assay. The results are summarized in Table 1.

The data convincingly show that the cloned polyomavirus DNA is mutagenic in this assay. However, the individually cloned T antigens do not exert any effect. This might be due to the lack of a complementing function provided by one of the other T antigens. Indeed, mixed transfections clearly demonstrate a cooperation of large and small T antigens, whereas middle T antigen has no effect. In this

[1] Akademie der Wissenschaften der DDR, Zentralinstitut für Molekularbiologie, Berlin-Buch, 1115 GDR

[2] Royal Postgraduate Medical School, Hammersmith Hospital, Du Cane Road, London W12 ONN, United Kingdom

Current Topics in Microbiology and Immunology, Vol. 144
© Springer-Verlag Berlin · Heidelberg 1989

**Table 1.** Frequency of 6-thioguanine resistant mutants induced by polyoma virus recombinants

| Plasmid | Functional component | Mutation frequency | |
|---|---|---|---|
| | | Exp. 1 | Exp. 2 |
| pAT153 | none | $1.0 \times 10^{-5}$ | $0.8 \times 10^{-5}$ |
| pA2 | complete polyoma | $12 \times 10^{-5}$ | $8 \times 10^{-5}$ |
| pPyLT1 | large T antigen (LT) | $1.0 \times 10^{-5}$ | $2.0 \times 10^{-5}$ |
| pMTE | middle T antigen (mT) | $0.8 \times 10^{-5}$ | $1.0 \times 10^{-5}$ |
| pSTE | small T antigen (sT) | $0.8 \times 10^{-5}$ | $1.2 \times 10^{-5}$ |
| pPyLT1 + pMTE | LT + mT | $1.4 \times 10^{-5}$ | n.d. |
| pPyLT1 + pSTE | LT + sT | $16 \times 10^{-5}$ | n.d. |
| pPyLT1 + pMTE + pSTE | LT + mT + sT | $16 \times 10^{-5}$ | n.d. |
| pEA2 | sT, mT, 40 % LT | $12 \times 10^{-5}$ | $4 \times 10^{-5}$ |
| pEd18 | sT, mT, 40 % LT | $4 \times 10^{-5}$ | $(6 \times 10^{-5})$ |

*Mutagenesis assay*: Mouse L cells $5 \times 10^5$ were transfected in 60 mm dishes with 6 µg of either one plasmid or mixtures of equal amounts of 2 or 3 plasmids, left with precipitates overnight, subcultured, and selected with 2 µg/ml of 6-thioguanine after 6 days. Colonies were counted after 2 weeks. Each figure is a mean value of four plates. Values of $4 \times 10^{-5}$ and above are statistically significant. The mutation frequency was calculated:

$$MF = \frac{\text{mean number of colonies}}{\text{number of cells} \times \text{plating efficiency}}$$

Maps of plasmids are given elsewhere (STRAUSS et al., in preparation). The pE recombinants contain the *BamHI/Eco*RI fragment of polyomavirus.
n.d., no data.

respect it is interesting to note that small T antigen has recently been shown to be located to a large extent in the nucleus (NODA et al. 1986); large T antigen is also nuclear (TOOZE 1981; DILWORTH et al. 1986).

As small T antigen is almost identical with the *N*-terminal part of large T antigen, it may either enhance functions exerted by this domain just by providing additional functional protein or actually by complementing a missing function. A partial answer to this question may come from the transfections of truncated early regions (Table 1). The early region of the wild-type virus containing the complete small and middle T antigen genes, but only 40 % of the *N*-terminus of the large T gene, is fully mutagenic whereas, in one experiment, the same region of the deletion mutant d18 (GRIFFIN and MADDOCK 1979) was found to be partially impaired in its mutagenic function (Table 1). The deletion in this mutant only affects large T antigen and middle T antigen. All data suggest that large T antigen is an essential component of mutagenesis, although the mechanism of action remains elusive. Final evidence for this assumption is provided by transfections with recombinants in which T-antigen coding sequences are under control of regulated foreign promoters (Table 2). Whereas expression of large T antigen from the MMTV promoter does not result in a significant mutagenic activity, its activity is high when the expression is driven by the mouse metallothionein promoter (Table 2). It should be noted that the level of gene expression is more than one order of magnitude lower in the former case, even in the induced state (data not shown).

**Table 2.** Mutagenic activity of induced T antigens

| Plasmid | Promoter | Induction | Mutation frequency | |
|---|---|---|---|---|
| | | | Exp. 1 | Exp. 2 |
| pMTV—MTE | MMTV | — | $1.4 \times 10^{-5}$ | $1.0 \times 10^{-5}$ |
| pMTV—MTE | MMTV | + | $2.2 \times 10^{-5}$ | $1.6 \times 10^{-5}$ |
| pMTV—LT2 | MMTV | — | $1.2 \times 10^{-5}$ | $0.8 \times 10^{-5}$ |
| pMTV—LT2 | MMTV | + | $4.2 \times 10^{-5}$ | $1.2 \times 10^{-5}$ |
| pMT—LT2 | mouse metallothionein | — | $3.5 \times 10^{-5}$ | $2.0 \times 10^{-5}$ |
| pMT—LT2 | mouse metallothionein | + | $16 \times 10^{-5}$ | $12 \times 10^{-5}$ |

Mutagenesis transfection assay as in Table 1. Middle or large T-antigen coding sequences starting at the *Bst*XI site (nucleotide 165) have been linked to the MMTV or the mouse metallothionein promoter (STRAUSS et al., in preparation). Induction was done with either $10^{-6}$ *M* dexamethasone or 1 µ*M* cadmium sulphate and 0.1 m*M* zinc sulphate.

Thus, we can conclude that the large T antigen of polyomavirus is mutagenic when expressed at a high level.

The mutagenic role of large T antigen raises several questions which have to be answered by further experimentation: (1) what type of mutations are produced in large T antigen-induced mutants; (2) by what mechanism does T antigen induce mutations; (3) is there a link to one of the other activities of large T antigen?

Our former studies with SV40 suggest that point mutations are predominantly found in T antigen-induced mutants (THEILE and STRAUSS 1977; STRAUSS et al. 1981). It has recently been shown that polyoma virus large T antigen increases the probability of sister chromatid exchanges, which involve recombination and rearrangements (CERNI et al. 1986). We have developed a gene rescue protocol which should allow us to answer this question (STRAUSS et al. 1985). Concerning the mechanism of mutation we can only speculate at the moment. However, results to be discussed later suggest a role for increased replication.

The most tempting conclusion regarding the third question is the existence of a correlation between the mutagenic activity and an immortalization function. In order to investigate the role of large T antigen in these processes in more detail we have established a modified immortalization assay.

# 3 Immortalization of Human Fibroblasts

Polyomavirus large T antigen is known to immortalize primary rodent cells with high efficiency (RASSOULZADEGAN et al. 1983).

The restriction to rodent cells as hosts is presumably due to the narrow host range of the polyomavirus early promoter (HERBOMEL et al. 1984). We were interested to investigate polyomavirus large T antigen as an immortalizing agent in human fibroblasts. To this end we linked the coding sequence of large T antigen to the MMTV promoter and used this construction to transfect early passaged human embryonic lung (HEL) fibroblasts (STRAUSS et al. 1988). A protocol devised for selecting immortalized mouse fibroblasts (TEVETHIA 1984) has been adopted which basically requires transfection at such low cell density that outgrowth of primary cells to colonies is impossible.

Figure 1 shows a typical result of an immortalization assay with HEL fibroblasts using the plasmid pMTV-LT2. Control transfections with either the cloning vector or with a plasmid containing only the MMTV promoter did not give rise to any colonies. Dexamethasone induction leads to an increase in the number of colonies by five- to tenfold. A total of 40 induced and noninduced colonies were isolated and were checked for immortalization properties, e.g., continuous passaging, growth in the presence of low serum concentrations (1 %) and plating efficiency at low density. All of the noninduced colonies turned out to be immortal but only 10 % of the colonies isolated in the presence of the inducer were actually immortalized on the basis of these criteria (STRAUSS et al. 1988). The majority of colonies have presumably been stimulated for growth by a transient high level of expression of large T antigen.

## 3.1 Immortalization Can Be Reversible

Four of the immortal clones were analyzed in more detail (Table 3). Two of them have been kept in culture for more than 120 generations (HEL8-1, 8-3). In all

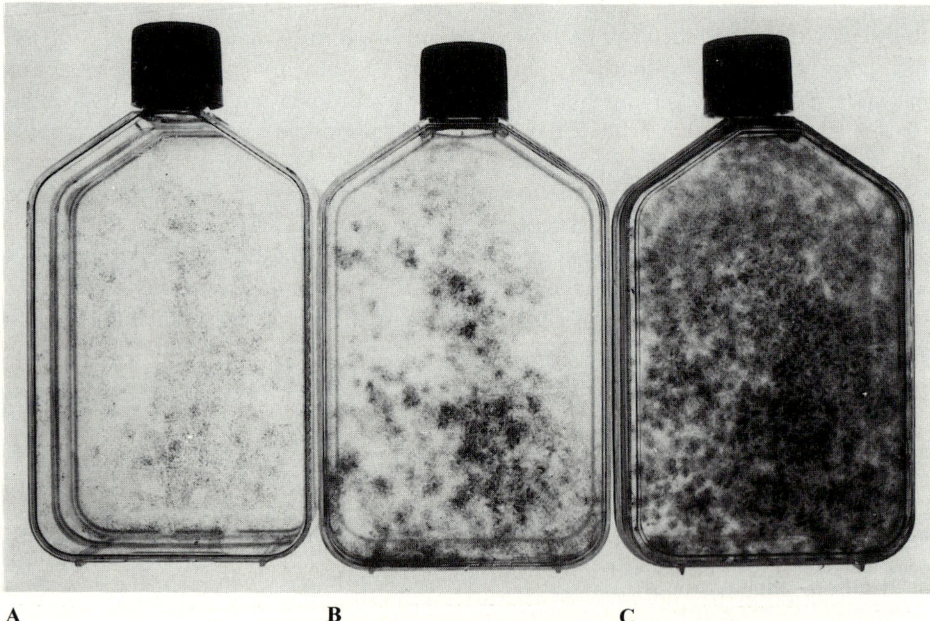

A                                    B                                    C

**Fig. 1A–C.** Immortalization of human embryonic lung fibroblasts (HEL). Flasks were inoculated with 250 cells per cm². The cells have been transfected with a plasmid containing either the MMTV promoter only (**A**) or the MMTV-LT fusion gene (**B, C**) at 25 µg per 20 ml of medium and were supplemented with fresh medium once a week. Medium in flask **C** contained $10^{-6}$ dexamethasone. Cells were stained after 6 weeks

**Table 3.** Properties of immortalized clones

| Clone | RNA molecules per cell | | Low density growth (% colony formation) | | 1% serum growth (% colony formation) | |
|---|---|---|---|---|---|---|
| | — | Dex + | — | Dex + | — | Dex + |
| HEL6-1 | 40 | 200 | 4.5 | 9 | 4.5 | 22 |
| HEL6-5 | 10 | 50 | <0.1 | 8 | <0.01 | 12 |
| HEL8-1 | 50 | 200 | 1 | 12 | n.d. | n.d. |
| HEL8-3 | 250 | 1800 | 10 | 15 | 24 | 25 |

Clonal derivatives of cells immortalized with pMTV—LT2 were analyzed for LT-mRNA by quantitative RNase protection, for clonal growth with 500 cells/60-mm dish, and for colony formation in medium with 1% fetal calf serum and cell input of $10^4$ cells/60-mm dish. Growth media were either without or with $10^{-6}$ M dexamethasone (STRAUSS et al. 1988).
n.d., no data.

cases transcription of the MMTV-LT hybrid gene is stimulated four- to seven-fold by $10^{-6}$ M dexamethasone. However, in the case of HEL8-3 the immortalization-specific growth properties are unrelated to the RNA level since the colony formation was not significantly increased by dexamethasone. Even more interestingly, in the case of the other three cell lines the immortalized phenotype could be reversed completely by removing dexamethasone from the medium (STRAUSS et al. 1988). This suggests that large T-antigen levels corresponding to 50–200 mRNA molecules per cell can induce immortalization dependent upon the continuous presence of large T-antigen expression. mRNA levels below 50 molecules per cell are not sufficient for this effect. In the case of the HEL8-3 clone, the irreversible state could be caused by the high basic level of expression. A similar clone was obtained after transfection pMT-LT2. Again, a high level expression (>1000 mRNA molecules/cell) correlated with an irreversible immortalization phenotype. However, we recently isolated a subclone of HEL8-3 which had lost large T-antigen expression and which was still immortalized. Therefore, it may be assumed that over-expression of the large T antigen (>200 mRNA molecules/cell) can lead to irreversible changes in the cell such that the immortalized state can be maintained even without expression of large T antigen.

More detailed analyses including screening of the viral content in metaphase chromosomes are necessary to define the molecular basis for reversible and irreversible immortalization.


# 4 Conclusions

The data presented here show that the $N$-terminal 40% of polyomavirus large T antigen, which contains both the DNA binding domain and the immortalizing domain (GUIZANI et al. 1987), is also responsible for the observed mutagenic activity. Our working hypothesis is that two mechanisms of immortalization by polyoma-virus large T antigen exist: (1) reversible changes, including specific cellular gene activation(s), linked to low or medium level expression of large T antigen. and (2) irreversible changes linked to high level expression. Mutagenesis would only occur in the latter situation or perhaps by additional activities induced including permanent stimulation of DNA synthesis. Thus, mutagenesis is most likely a consequence of gene activation involved in irreversible immortalization.

# References


Cerni C, Mougneau E, Zerlin M, Julius M, Marcu KB, Cuzin F (1986) c-*myc* and functionally related oncogenes induce both high rates of sister chromatid exchange and abnormal karyotypes in rat fibroblasts. Curr Top Microbiol Immunol 132: 193–201

Cowie A, De Villiers J, Kamen R (1986) Immortalization of rat embryo fibroblasts by mutant polyomavirus large T antigens deficient in DNA binding. Mol Cell Biol 6: 4344–4352

Dilworth S, Hansson HA, Darnfors C, Bjursell G, Streuli CH, Griffin BE (1986) Subcellular localization of the middle and large T antigens of polyoma virus. EMBO J 5: 491–499

Griffin BE, Maddock C (1979) New classes of viable deletion mutants in the early region of polyoma virus. J Virol 31: 645–656

Guizani I, Clertant P, Cuzin F (1987) Biochemical properties associated with the immortalizing domain of the large T protein of polyoma virus. Biochem Biophys Res Commun 144: 973–979

Griffin BE, Maddock C (1979) New classes of viable deletion mutants in the early region of polyoma virus. J Virol 31: 645–656

Kingston RE, Cowie A, Morimoto RI, Gwinn KA (1986) Binding of polyomavirus large T antigen to the human hsp70 promoter is not required for *trans* activation. Mol Cell Biol 6: 3180–3190

Lübbe L, Strauss M, Scherneck S, Geissler E (1982) The DNA tumor virus SV40 induces gene mutations in human cells. Reversion of HGPRT deficiency. Hum Genet 61: 236–241

Noda T, Satake M, Robins T, Ito Y (1986) Isolation and characterization of NIH3T3 cells expressing polyomavirus small T antigen. J Virol 60: 105–113

Rassoulzadegan M, Naghashfar Z, Cowie A, Grisoni M, Kamen R, Cuzin F (1983) Expression of the large T antigen of polyomavirus promotes the establishment in culture of "normal" rodent fibroblast cell lines. Proc Natl Acad Sci USA 80: 4345–4348

Strauss M, Lübbe L, Geissler E (1981) HGPRT structural gene mutation in Lesch-Nyhan-syndrome as indicated by antigenic activity and reversion of the enzyme deficiency. Hum Genet 57: 185–188

Strauss M, Kiessling U, Kähler R (1985) Rescue of transfected genes from mammalian cells by functional selection in *E. coli.* Mol Gen Genet 201: 277–281

Strauss M, Argani P, Mohr I, Gluzman Y (1987) Studies on the origin-specific DNA-binding domain of SV40 large T antigen. J Virol 61: 3326–3330

Strauss M, Lübbe L, Griffin BE (1988) Reversible immortalization of human embryonic fibroblasts by polyoma large T antigen. (submitted for publication)

Tevethia MJ (1984) Immortalization of primary mouse embryo fibroblasts with SV40 virions, viral DNA, and a subgenomic DNA fragment in a quantitative assay. Virology 137: 414–421

Theile M, Strauss M (1977) Mutagenesis by SV40. II. Mutat Res 45: 111–123

Theile M, Scherneck S, Geissler E (1976) Mutagenesis by SV40. I. Detection of mutations in Chinese hamster cell lines using different resistance markers. Mutat Res 37: 111–124

Theile M, Strauss M, Lübbe L, Scherneck S, Krause H, Geissler E (1980) SV40-induced somatic mutations: possible relevance to viral transformation. Cold Spring Harbor Symp Quant Biol 44: 377–382

Theile M, Krause H, Geissler E (1987) SV40 induced mutagenesis: action of the early viral region. J Gen Virol 68: 233–237

Tooze J (ed) (1981) DNA tumor viruses. Molecular biology of tumor viruses, p 2. Cold Spring Harbor Laboratory, Cold Spring Harbor

Wigler M, Silverstein S, Lee LS, Pellicer A, Chang T, Axel R (1977) Transfer of purified Herpes simplex virus thymidine kinase gene to cultured mouse cells. Cell 11: 223–232

# Part III: Transforming Functions of Papillomaviruses

# Introduction

The aurea of excitement about papillomavirus research is certainly due to the merger of interesting basic molecular biology and theoretical and, possibly, practical medicine. Indeed, papillomaviruses are not only most interesting model systems for cell and molecular biologists but are also considered to be etiological agents of animal and human diseases (Table 1). This will be made clear by the authors of the research papers collected in this part. They will discuss the evidence for the etiological role of papillomaviruses in the development of human cancers.

Papillomaviruses are found in a wide range of vertebrates, including amphibia, birds, and a wide variety of mammals. Most papillomavirus types have a single host and grow in the differentiating keratinocytes of the cutaneous or mucosal epithelium. For example, the specificity of host cell tropism is clearly apparent in human papillomaviruses as, in general, those viruses affecting external skin cells do not infect mucosal epithelial cells and vice versa. This extreme host cell specificity is the reason why papillomaviruses cannot be multiplied in tissue culture cells. But, as will be demonstrated below, molecularly cloned papillomavirus DNA can be transfected into some common cultured rodent fibroblasts (and other cultured cells), allowing an investigation of viral DNA replication, gene expression, and, most importantly in this context, cell transformation. However, no mature viral particles are produced in these cells as the late viral genes, coding for the structural proteins of the virions, are not expressed. The production of structural proteins and the assembly of virus particles are restricted to differentiating keratinocytes in intact cutaneous or mucosal tissues (reviews: PFISTER 1984; BROKER and BOTCHAN 1986).

**Table 1.** Association of human papillomaviruses with human diseases (summarized from articles in STEINBERG et al. 1987). The list includes only a fraction of the 50 or so known human papillomaviruses

| Human papillomavirus type | Disease |
| --- | --- |
| 2, 4 | Benign common warts |
| 5, 8, 14 | Epidermodysplasia verruciformis (malign in 30% of cases) |
| 6, 11 | Nasal and laryngeal papillomas; anogenital condylomata acuminata |
| 16, 18, 31, 33, 35 | Found in genital carcinomas |
| 30, 40 | Found in laryngeal carcinomas |
| 42 | Benign genital warts |

Current Topics in Microbiology and Immunology, Vol. 144
© Springer-Verlag Berlin · Heidelberg 1989

The genomes of papillomaviruses are circular, double-stranded DNA molecules, about 8000 base pairs (bp) in size. Like polyomavirus and SV40, the intranuclear viral DNA is organized as a minichromosome with stretches of DNA wrapped around nucleosomal cores. Minichromosomes possess a DNase I-hypersensitive, nucleosome-free region around regulatory sequences for transcription and replication (RÖSL et al. 1983).

But unlike SV40 or polyomavirus, all genes or open reading frames are encoded by only one of the two strands of papillomavirus DNA. The up to eight early genes overlap to a considerable extent using all three translational reading frames. The two late genes are tandemly aligned without overlap (Fig. 1).

A region between 400 and 900 bp long (in different papillomavirus types) begins with the stop codon of the last late gene and ends with the initiator codon of the first early gene. This region, variously referred to as the upstream regulatory region (URR), the long control region (LCR), or the noncoding region (NCR), includes enhancer and promoter elements as well as an origin of replication (reviews BROKER and BOTCHAN 1986; GIRI and DANOS 1986).

The general genetic organization is similar for all investigated papillomaviruses. However, the nucleotide sequences of different virus types can be highly divergent, with the exception of some genes like the open reading frame E1, coding for a protein involved in DNA replication, and the open reading frame L1, coding for the major capsid protein, which are quite similar for different papillomaviruses. In Fig. 1 are shown as examples the genetic maps of the papillomaviruses which will be discussed in the articles later on in this chapter. Here the functions of the early open reading frames will be briefly summarized.

*Open Reading Frame E1.* The function of this coding region was investigated using bovine papillomavirus type 1 (BPV) as a model system, but it is quite likely that similar results will be obtained for other papillomaviruses as the E1 coding region is highly conserved among them. The 3′-part of the E1 open reading frame has regions of homology with SV40 and polyomavirus large T antigens, particularly around the nucleotide binding site (SEIF 1984). Mutant analysis has shown that this 3′-part is absolutely necessary for the replication of BPV DNA, transfected into cultured mouse C127 fibroblasts. The 5′-part of the E1 region appears to code for a function, termed modulator, that maintains the number of extrachromosomal viral DNA at values of 100–200 copies/nucleus (in C127 cells). This part of the E1 open reading frame regulates the once-per-cell cycle mode of extrachromosomal papillomavirus DNA replication (LUSKY and BOTCHAN 1984; BERG et al. 1986). Available genetic evidence suggests that the activator protein, coded for by the 3′-part of E1, and the modulator function of the 5′-part are expressed as individual proteins.

*Open Reading Frame E2.* This reading frame codes for a protein that acts in *trans* to activate other early viral genes. The function of the E2-encoded protein is described below by HORWITZ et al., PHELPS et al., and by IFTNER et al.

*Open Reading Frame E5.* This small coding region is present in some, but not in all, papillomaviruses. It is absent, for example, in human papillomavirus type 8 (HPV-8) as discussed by IFTNER et al. (see also Fig. 1). However, in BPV-1, the

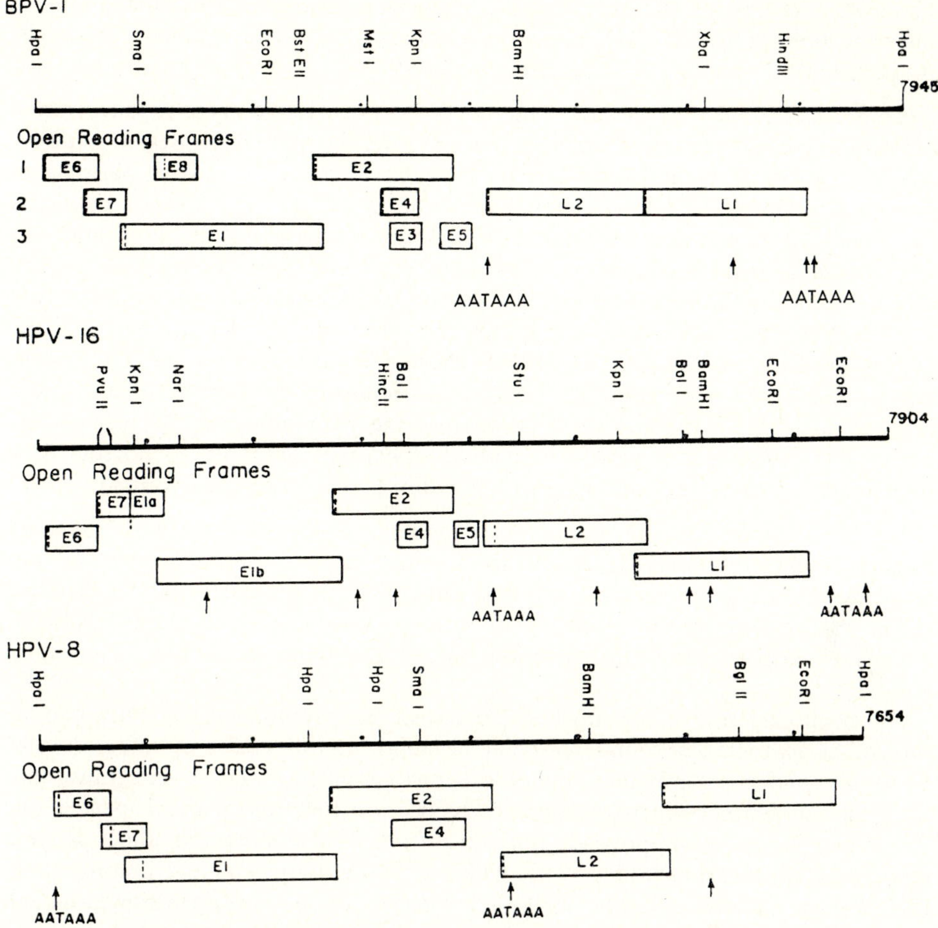



**Fig. 1.** Organization of some papillomavirus genomes. As examples we show the genetic maps of the virus types discussed in the articles of this chapter, namely bovine papilloma virus type 1 (*BPV-1*; see HORWITZ et al.), human papilloma virus type 16 (*HPV-16*; see PHELPS et al.) and type 8 (*HPV-8*; see IFTNER et al.). References concerning the nucleotide sequences of these genomes can be found in the respective articles of this chapter.

Papillomavirus genomes are double-stranded DNA circles, but it is usual to present them as linear maps. The BPV-1 genome, 7945 base pairs in size, was the first to be sequenced. Its *Hpa*I restriction site serves as a landmark for the start of the map. The maps of other papillomavirus genomes are colinearly arranged relative to the BPV-1 map.

The figure gives the sizes of the three virus genomes in base pairs, selected restriction sites, the location of polyadenylation signals (ATAAA, *vertical arrows*), and, most importantly, the location of potential coding regions in the three translational reading frames. The nucleosome-free upstream regulatory region is at the *right* beginning with the stop codon of the open reading frame L1. The known functions of the early open reading frames, *E1–E8*, are described in the text. The open reading frames *L2* and *L1* code for the minor and the major capsid protein, respectively. (Adapted from BROKER and BOTCHAN 1986)

E5 protein is one of the transforming proteins whose main function could be the induction of cellular DNA synthesis (GREEN and LOEWENSTEIN 1987). The properties of E5 are described by HORWITZ et al.

*Open Reading Frames E6 and E7.* The functions coded for by these reading frames are associated with the transforming activity of several papillomaviruses. A comparison of the deduced amino acid sequences shows regularly spaced patterns of cysteine residues (CXXC), and it has been suggested that E6 and E7 could be the results of a common evolutionary pathway starting with a duplication of a 33-amino acid coding sequence followed by genetic drift (DANOS and YANIV 1987). Except for this repeating pattern, E6 and E7 sequences are poorly conserved. This may be the reason why the E6 protein (and not the E7 protein) has the major transforming function of HPV-8 (IFTNER et al.) whereas the E7 protein has a transforming function of HPV-16 as reported by PHELPS et al. These authors made the intriguing observation that the papillomavirus E7 protein has remarkable functional as well as structural similarities with the adenovirus E1A protein known to be involved in gene activation (see Part IV, this volume).

*Other Early Open Reading Frames.* The protein coded for by the E4 reading frame is found in relatively large quantities in keratinocytes producing mature viral particles but not in transfected cultured cells. The E4 protein could be involved in the maturation or the assembly of virions (DOORBAR et al. 1986). Next to nothing is known about the functions coded for by open reading frames E3 and E8.

As mentioned above, up to 100–200 copies of papillomavirus DNA can be present as extrachromosomal episomal elements in the nuclei of infected cells. However, under experimental conditions, papillomavirus DNA integrates into the host cell genome when the positive replication function in the 3′-part of the E1 coding region is suppressed. But integration is also observed under in vivo conditions. Extrachromosomal papillomavirus DNA disappears and becomes integrated when infected precancerous cells are converted to fully transformed malign cancer cells. As discussed in more detail by PHELPS et al., integration does not occur at random sites on the virus DNA; the open reading frames E6/E7 usually remain intact and are found to be expressed. A major concern to tumor virologists was whether the integrated (and expressed) viral DNA elements are just blind passengers of the cancer cell, or whether they are important for the maintenance of the cancer phenotype. This problem has been addressed by KLEINHEINZ et al., who report in their contribution that the expression of the early viral genes in cancer cells may be important for the maintenance of the transformed state.

# References

Berg L, Lusky M, Stenlund A, Botchan MR (1986) Repression of bovine papilloma virus replication is mediated by a virally encoded *trans*-acting factor. Cell 46: 753–762
Broker TR, Botchan M (1986) Papilloma viruses: retrospective and prospective. Cancer Cells 4: 17–36

Danos O, Yaniv M (1987) E6 and E7 gene products evolved by amplification of a 33-aminoacid peptide with a potential nucleic acid binding structure. Cancer Cells 5: 145–150

Doorbar J, Campbell D, Grand RJA, Gattimore PH (1986) Identification of the human papilloma virus Ia E4 gene product. EMBO J 5: 355–372

Giri I, Danos O (1986) Papilloma-virus genomes: from sequence data to biological properties. Trends Genet 2: 227–232

Green M, Loewenstein PM (1987) Demonstration that a chemically synthesized BPV I oncoprotein and its C-terminal domain function do induce cellular DNA synthesis. Cell 51: 795–802

Lusky M, Botchan MR (1984) Characterization of the bovine papilloma-virus plasmid-maintenance sequence. Cell 36: 391–401

Pfister H (1984) Biology and biochemistry of papilloma viruses. Rev Physiol Biochem Pharmacol 99: 111–181

Rösl F, Waldeck W, Sauer G (1983) Isolation of episomal bovine papilloma-virus chromatin and identification of a DNase I hypersensitive region. J Virol 46: 567–574

Seif I (1984) Sequence homology between the larger tumor antigen of polyoma viruses and the putative E1 protein of papilloma viruses. Virology 138: 347–352

Steinberg BM, Brandsma JL, Taichman LB (eds) (1987) Papilloma viruses. Cancer Cells, vol 5. Cold Spring Harbor Laboratory, Cold Spring Harbor

# Structure, Activity, and Regulation
of the Bovine Papillomavirus E5 Gene
and Its Transforming Protein Product

B. H. Horwitz, J. Settleman, S. S. Prakash, and D. DiMaio

## 1 Introduction

Transfected bovine papillomavirus type 1 (BPV) DNA can induce tumorigenic
transformation of C127 cells, an established line of mouse cells (Dvoretsky
et al. 1980; Lowy et al. 1980). By assaying the activity of constructed viral
mutants, we and others have shown that an intact E5 gene is required for
efficient C127 cell focus formation (DiMaio et al. 1986; Groff and Lancaster
1986; Rabson et al. 1986). Moreover, expression of the E5 gene in the absence
of all other BPV genes is sufficient to induce focus formation (Schiller et al.
1986; Yang et al. 1985; Horwitz et al. 1988). Detailed mutational and biochemical analysis of this gene indicates that translation of the portion of
open reading frame (ORF) E5 downstream of the first methionine codon results in
the synthesis of a 7K polypeptide required for efficient transformation (DiMaio
et al. 1986; Schlegel et al. 1986; Burkhardt et al. 1987). The E5 protein
is predicted to be only 44 amino acids long and extremely hydrophobic (Fig. 1).

BOVINE PAPILLOMAVIRUS TYPE 1 ORF E5

```
ATG-CCA-AAT-CTA-TGG-TTT-CTA-TTG-TTC-TTG-GGA-CTA-GTT-GCT-GCA-
MET PRO ASN LEU TRP PHE LEU LEU PHE LEU GLY LEU VAL ALA ALA      Hydrophobic
                                                                 Portion
ATG-CAA-CTG-CTG-CTA-TTA-CTG-TTC-TTA-CTC-TTG-TTT-TTT-CTT-GTA-
MET GLN LEU LEU LEU LEU LEU PHE LEU LEU LEU PHE PHE LEU VAL

TAC-TGG-GAT-CAT-TTT-GAG-TGC-TCC-TGT-ACA-GGT-CTG-CCC-TTT-TAA      Hydrophilic
TYR TRP ASP HIS PHE GLU CYS SER CYS THR GLY LEU PRO PHE          Portion
```

Fig. 1. Coding sequence of the E5 gene and predicted amino acid sequence of the E5 protein.
(Reproduced from Horwitz et al. 1988)

Department of Human Genetics, Yale University School of Medicine, New Haven, CT 06510,
USA

Current Topics in Microbiology and Immunology, Vol. 144
© Springer-Verlag Berlin · Heidelberg 1989

Its *N*-terminal two-thirds consists almost exclusively of strongly hydrophobic amino acid residues, although a glutamine is present at position 17. In contrast, there are several charged and polar amino acid residues in the C-terminal third. There is no apparent similarity between the papillomavirus E5 proteins and other sequenced proteins.

An antiserum generated against the C-terminal third of the BPV E5 protein recognizes a 7K E5 protein in BPV-transformed cells (SCHLEGEL et al. 1986; BURKHARDT et al. 1987). As predicted from its hydrophobic composition, the E5 protein is recovered predominantly in the membrane fractions of cells. Moreover, it is isolated as a dimer that can be converted to monomer by treatment with dithiothreitol. To determine the importance of individual amino acid residues for the biological and biochemical properties of the E5 protein, we have subjected the E5 gene to extensive mutational analysis.

Although the transforming activity of the E5 gene was first detected by using stable transformation assays, acute assays of the effects of E5 expression must be used to identify the initial biochemical events which eventually culminate in stable transformation. We have developed an acute morphologic transformation assay in C127 cells that requires the expression of the E5 transforming gene.

Mutations outside of the E5 gene can also affect the efficiency of transformation. Mutations in the 5′ portion of ORF E2 can cause a substantial reduction in the ability of the viral DNA to induce foci in mouse C127 cells (LUSKY and BOTCHAN 1985; DIMAIO 1986; KLEINER et al. 1986; RABSON et al. 1986). However, ORF E2 has no detectable focus-forming activity when it is expressed in the absence of other BPV genes (SCHILLER et al. 1986; YANG et al. 1985). Because expression of the E2 gene can either stimulate or inhibit transcription from promoters linked to the BPV long control region (LCR), it has been suggested that the E2 gene influences the transformation efficiency of BPV by controlling the activity of viral transforming genes (SPALHOLZ et al. 1985; HAUGEN et al. 1987; LAMBERT et al. 1987). We demonstrate here that the E2 gene controls the expression of the E5 gene.

## 2 Materials and Methods

To generate mutations in the E5 gene, we have used a variety of directed mutagenesis procedures. Oligonucleotide-directed saturation mutagenesis (HORWITZ et al. 1988) was used to produce amino acid substitution mutations distributed throughout the C-terminal two-thirds of the E5 protein.

To assay the mutants for transforming activity, 100–200 ng of cloned viral DNA was digested with *Bam*HI to separate the viral genome from the bacterial vector and transfected into C127 cells as previously described (DIMAIO 1986). Foci were counted 2–3 weeks after transfection.

To generate virus stocks of the BPV/SV40 recombinant viral genome, circularized viral DNA from pPava-1 was transfected into CMT4 cells, a line of monkey cells which expresses SV40 large T antigen, thereby allowing high level DNA replication from the SV40 replication origin, efficient expression of the late capsid genes, and virion assembly (SETTLEMAN and DiMAIO 1988). High titer virus stocks were generated by serial passage on CMT4 cells.

Transient expression of the E5 protein was determined in CV-1 cells transfected with 20 µg of viral DNA (PRAKASH et al. 1988). These cells were metabolically labelled with [$^{35}$S]methionine and [$^{35}$S]cysteine 43–48 h after transfection. The E5 protein in cell extracts was detected by immunoprecipitation with rabbit E5 peptide-specific antiserum and SDS polyacrylamide gel electrophoresis as described (SCHLEGEL et al. 1986).

# 3 Results

## 3.1 Mutational analysis of the E5 Gene

Few specific amino acids in the $N$-terminal third of the E5 protein seem to be required for efficient focus formation or membrane association (DiMAIO et al. 1986). Numerous different in-frame mutations centered on amino acid position 3 do not significantly impair focus-forming activity, even though some of these mutations alter the charge of the protein or insert proline residues. Cells transformed with these mutants synthesize membrane-associated E5 proteins with altered mobility compared to the wild-type species (BURKHARDT et al. 1987). One transformation-competent mutant, E5XS-1, contains a substitution of amino acids 3–13. The substituted amino acids, like those they replace, are largely hydrophobic.

In the very hydrophobic middle third of the protein, all mutants that solely have hydrophobic amino acids substituted with different hydrophobic amino acids transform efficiently (HORWITZ et al. 1988). In contrast, all mutants containing lysine or arginine in this region are defective for focus formation. The most hydrophilic amino acid residue in this portion of the protein is glutamine 17, and the identical amino acid is present at this position in all sequenced fibropapillomaviruses. To test whether a hydrophobic amino acid is tolerated at this position, we changed the glutamine to a leucine. Preliminary analysis of this mutant indicates that it is transformation defective (HORWITZ and DiMAIO, unpublished data).

Missense mutations were also generated in the hydrophilic $c$-terminal segment of the E5 protein, and their effects on C127 cell focus formation were assessed (HORWITZ et al. 1988). At several positions, none of the mutations we isolated significantly inhibited transformation. At some of these positions, multiple mutants with different substitutions have been examined, and some of these mutations result in nonconservative amino acid substitutions. On the other hand, mutations that do interfere with transformation have been isolated at seven different positions (Fig. 2). At some of these positions, we have isolated other conservative mutations that do not inhibit transformation. Mutations that change either of the cysteines to serine or arginine result in transformation defects, as does a double mutation in which both cysteines were replaced by serine.

In collaboration with Anne L. BURKHARDT and Richard SCHLEGEL, we examined the E5 protein synthesized in cell lines established with each of the defective

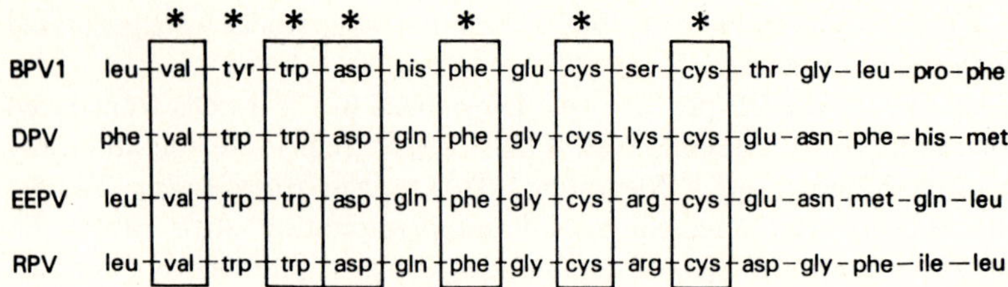



**Fig. 2.** Comparison of the *C*-terminal amino acid sequences of bovine papillomavirus (BPV), deer papillomavirus (DPV), European elk papillomavirus (EEPV), and reindeer papillomavirus (RPV). Amino acids that are identical in all four sequences are *boxed*. Amino acids in BPV that have been changed in defective mutants are *starred*. (Reproduced from HORWITZ et al. 1988)

mutants to insure that the defects were not due to lack of expression of the mutant protein or to its inability to associate with membranes (HORWITZ 1988). In all cases examined, substantial amounts of membrane-associated E5 protein were detected. Although these cell lines were selected for transformation, a procedure which may have resulted in secondary changes in the stability of the transforming protein, these results suggest that the mutant proteins are not markedly unstable. Almost all of the wild-type E5 protein can be isolated in the dimeric form, but the defective mutant which has both cysteines changed to serines does not detectably form dimers.

Genetic mapping experiments demonstrate that the mutations identified by sequencing of the E5 gene caused the transformation defects. We also showed that these defects were not due to altered expression of viral genes other than E5 by demonstrating that the mutations severely inhibited focus formation when inserted into an LTR-driven E5 construct. These results strongly suggest that these mutations in the E5 gene cause transformation defects by directly affecting the transforming ability of the E5 protein.

## 3.2 Acute Transformation by the E5 Protein

To introduce the E5 gene efficiently into cultured cells, we have generated a BPV/ SV40 recombinant virus which contains BPV ORFs E2, E3, E4, and E5 in place of the SV40 T-antigen coding sequences (SETTLEMAN and DiMAIO 1988). The resulting plasmid, pPava-1, contains the intact SV40 late region and origin of replication, the BPV sequences linked in the correct orientation to the SV40 early promoter, and the complete pBR322 genome. Transfected pPava-1 induces the formation of stably transformed foci in mouse C127 cells. Virus particles were generated from pPava-1 as described in Methods. Low molecular weight DNA isolated from Pava-1-infected monkey CMT4 cells contained a substantial amount of re-plicated, unrearranged viral DNA.

To test whether the BPV-1 E5 gene is expressed in Pava-infected cells, CMT4 cells were infected and metabolically labelled, and cell extracts were subjected to immunoprecipitation with an E5-specific antiserum (SCHLEGEL et al. 1986). Pava-1-infected cells expressed large amounts of the 7K E5 protein, whereas mock-infected

CMT4 cells and cells infected with a mutant virus containing an E5 frameshift mutation expressed no E5 protein (Settleman and DiMaio 1988). Infection of mouse C127 cells with Pava-1 resulted in a dramatic morphology change characteristic of tumorigenic transformation within 1–2 days, whereas mock-infected cells remained unchanged (Fig. 3). In addition to appearing morphologically transformed, the infected cells reached a saturation density several-fold higher than mock-infected cells and rapidly acidified the culture medium. Incubation of Pava-1 with SV40-neutralizing antibody abolished this effect. We tested the activity of a viral mutant containing a frameshift mutation in the E5 gene. Cells infected with this mutant remain flat and are morphologically indistinguishable from mock-infected cells. The amounts of wild-type and mutant virus used in these infections were equivalent as assayed by their expression of the BPV E2 gene, which is also

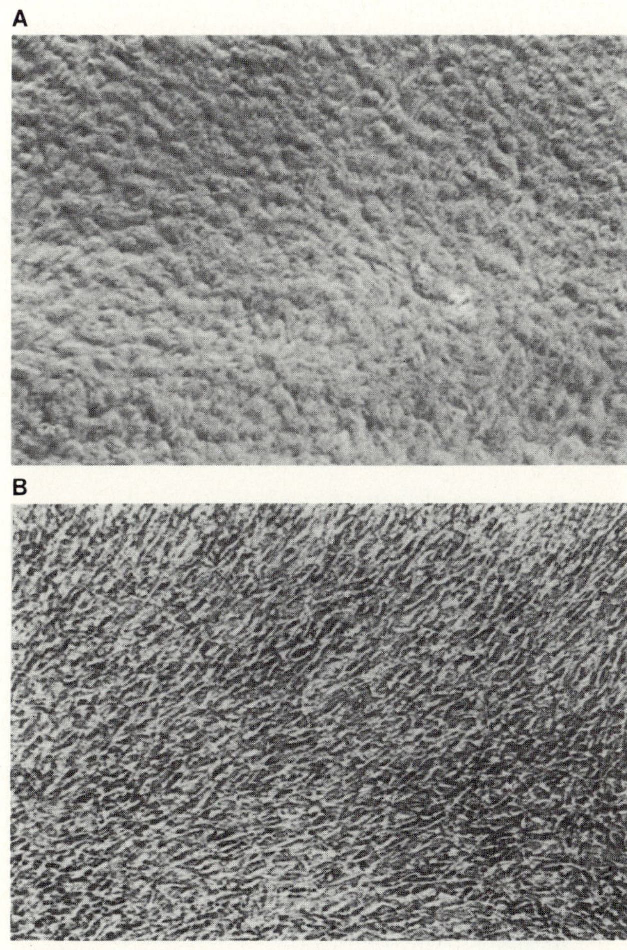

**Fig. 3A, B.** Acute morphological transformation of C127 cells. The cells were mock-infected (**A**) or infected with a high titer stock of Pava-1 virus (**B**). Cells were photographed 48 h after infection.

expressed by these viruses. Thus, the rapid morphology change caused by Pava-1 infection requires expression of the E5 gene product.

## 3.3 Effect of ORF E2 Mutations on ORF E5 Expression

To examine the regulation of the E5 gene, the amount of 7K E5 polypeptide in CV-1 cells was determined after they were transfected with either wild-type BPV DNA or with a mutant containing a deletion in the 5'-end of ORF E2 (PRAKASH et al. 1988). The E5-specific antiserum was used to immunoprecipitate metabolically labelled E5 protein from extracts prepared from cells 2 days after transfection. As shown in Fig. 4, the 7K E5 transforming protein is readily detectable in cells transfected with wild-type viral DNA, but it is not detectable in cells transfected with carrier DNA alone or with an ORF E5 frameshift mutant. The amount of the protein is also markedly reduced in cells transfected with the E2 mutant. However, as shown in Fig. 4, lane 4, cotransfection with the mutant plasmid and with one expressing the wild-type E2 gene results in the production of the E5 protein. The E2 expression plasmid by itself does not produce the E5 protein. We also examined the expression of a chloramphenicol acetyl transferase gene fused in-frame immediately downstream of the E5 initiation codon in the context of the full-length BPV genome (PRAKASH et al. 1988). In this system, E5 expression was reduced by mutations that inactivate the E2 gene, removal of the LCR that contains binding sites for the E2 protein (ANDROPHY et al. 1987; MOSKALUK and BASTIA 1987), or expression of the E2 repressor activity (LAMBERT et al. 1987). These results demonstrate that production of the E5 protein is regulated by the expression of the E2 gene.

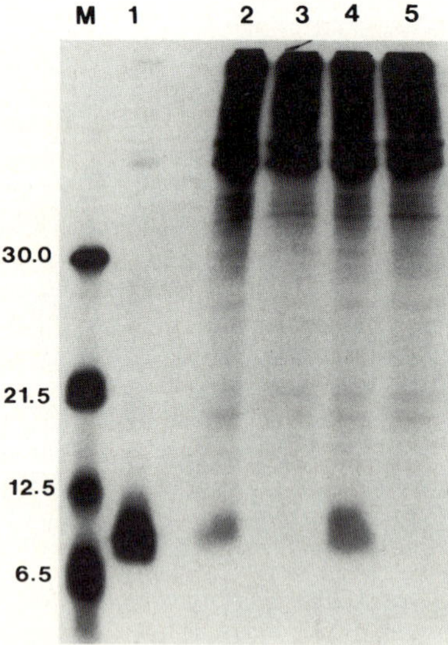

Fig. 4. Bovine papillomavirus (BPV) E5 protein synthesized in transfected cells. CV-1 cells were transfected with 20 µg of cloned viral DNA by using the calcium phosphate method. After metabolic labelling with $^{35}$S-labelled amino acids 43–48 h after infection, extracts were prepared and analyzed by immunoprecipitation and gel electrophoresis. *1*, E5 protein marker; *2*, CV-1 cells transfected with wild-type BPV DNA; *3*, CV-1 cells transfected with an E2 mutant; *4*, CV-1 cells transfected with an E2 mutant plus an E2 expression plasmid; *5*, CV-1 cells transfected with an E5 mutant plus an E2 expression plasmid. (Reproduced from PRAKASH et al. 1988)

# 4 Discussion

Our mutational analysis indicates that the overall hydrophobic nature of the first two-thirds of the E5 protein is essential for efficient focus formation, but a precise amino acid sequence is not required. Although many variant hydrophobic sequences are compatible with membrane association and efficient focus formation, the introduction of a strongly basic amino acid into the middle of the protein inhibits transformation. These results suggest that the very hydrophobic middle third of the E5 protein is involved in membrane association and that this interaction is essential for transformation.

Although by crude cell fractionation techniques the defective E5 proteins with new basic residues are membrane associated, more detailed analysis may reveal differences from the wild-type localization or orientation. On the other hand, the insertion of an arginine or a lysine may interfere with an activity of the protein other than membrane association. The preliminary analysis of the mutant with a leucine substituted for glutamine 17 suggests that there is a requirement for a polar amino acid at this position. The occupation of this position by histidine in a competent mutant and by glycine in an incompetent one is consistent with this idea.

The occurrence of numerous hydrophilic and charged amino acids in the C-terminal third of the BPV E5 protein suggests that it plays a different role in the function of the protein than does the hydrophobic portion. In fact, at several positions the introduction of a nonconservative amino acid (e.g., substitution of a lysine for a glutamic acid) does not result in transformation defects. Although it can be inferred that the wild-type amino acids are not absolutely required at these positions, we have not determined whether all amino acids would be tolerated. However, diverse mutations at seven positions in the C-terminal segment do cause transformation defects, apparently without grossly destabilizing the protein. Conservative changes are tolerated at some of these positions, although all mutations examined so far at the cysteine residues result in transformation defects and impaired dimerization. The amino acids we have identified as being important for the transforming activity of E5 protein are highly conserved in the predicted E5 proteins encoded by the other fibropapillomaviruses that have similar biological properties to BPV (Fig. 2). In contrast, there is considerable variability at the six positions at which mutations have not caused defects.

These results indicate that certain amino acids in the C-terminal segment of the E5 protein are essential for its transforming activity. This is consistent with the results of recent microinjection experiments indicating that C-terminal E5 peptides express a mitogenic activity (GREEN and LOWENSTEIN 1987). The essential amino acids probably play an important role in the function of the E5 protein, perhaps by contacting another protein, by maintaining an active configuration of the E5 protein, or by contributing to an enzymatic function. However, it is unlikely that a catalytic site is composed exclusively of E5 amino acids because of the small size of the protein and the large proportion of positions that can accommodate amino acid changes. The E5 protein may interact with larger proteins and influence their activity. Potential targets for productive interaction with the viral protein are cellular proteins regulating cell growth and differentiation. Elucidation of the mechanisms by which the E5 protein causes altered cell growth and metabolism will require biochemical and genetic identification and characterization of these proposed intracellular targets.

To facilitate this biochemical analysis, we have constructed a BPV/SV40 recombinant virus that allows the efficient introduction and expression of the BPV E2 and E5 genes in cultured cells. Efficient expression of the BPV genes in this virus is sufficient to cause rapid morphological changes and altered growth characteristics typical of tumorigenic cells. Although the generation of stably transformed foci by transfected BPV DNA is relatively inefficient, virtually all C127 cells expressing these genes can be acutely morphologically transformed. The ORF E5 missense mutants described above can be used to determine whether acute and stable papillomavirus transformations are genetically separable. Moreover, this is a novel system for analyzing the acute biochemical responses elicited by expression of the BPV transforming and regulatory genes and may thereby provide insight into the mechanism of their action.

A regulatory circuit that controls E5 expression has also been described. Our results show that the E2 gene acts in *trans* to control expression of the E5 transforming gene. This response requires a *cis*-acting element in the viral LCR, which is located about 4000 base pairs away from the E5 initiation codon. These data are consistent with the hypothesis that the E2 protein affects the level of ORF E5 mRNA, perhaps by binding to the E2-responsive enhancer in the LCR, and thereby influences the activity of viral promoters. Elimination of E2 transactivation activity, removal of the LCR, or increased E2 repressor synthesis cause dramatic decreases in transient expression of the E5 gene and severely inhibit transformation. These results suggest that E2-mediated transactivation and repression dictate the level of expression of the E5 transforming gene and hence the efficiency of focus formation.

*Acknowledgements*. We thank R. Schlegel and A. L. Burkhardt for a productive and enjoyable collaboration. We also thank A. West, D. Reise, K. Neary, T. Zibello, D. Guralski, J. T. Schiller, B. Spalholz, P. Howley, Y. Gluzman, and L. Norkin for help and for essential reagents. This work was supported in part by a grant from the National Cancer Institute (CA37157). D.D. is the recipient of a Mallinckrodt Scholar Award.

# References

Androphy EJ, Lowy DR, Schiller JS (1987) Bovine papillomavirus E2 *trans*-activating gene product binds to specific site in papillomavirus DNA. Nature 325: 70–73

Burkhardt A, DiMaio D, Schlegel R (1987) Genetic and biochemical definition of the bovine papillomavirus E5 transforming protein. EMBO J 8: 2381–2385

DiMaio D (1986) Nonsense mutation in open reading frame E2 of bovine papillomavirus DNA. J Virol 57: 475–480

DiMaio D, Guralski D, Schiller JT (1986) Translation of open reading frame E5 of bovine papillomavirus is required for its transforming activity. Proc Natl Acad Sci USA 83: 1797–1801

Dvoretzky I, Shoher R, Chattopadhyay SK, Lowy DH (1980) A quantitative focus forming assay for bovine papilloma virus Virology 103: 369–375

Green M, Lowenstein PM (1987) Demonstration that a chemically synthesized BPV1 oncoprotein and its C-terminal domain function to induce cellular DNA synthesis. Cell 51: 795–802

Groff DE, Lancaster WD (1986) Genetic analysis of the 3' early region transformation and replication functions of bovine papillomavirus type 1. Virology 150: 221–230

Haugen TH, Cripe TP, Ginder GD, Karin M, Turek LP (1987) *Trans*-activation of an upstream early gene promoter by bovine papillomavirus 1 by a product of the viral E2 gene. EMBO J 6: 145–152

Horwitz BH, Burkhardt AL, Schlegel R, DiMaio D (1988) 44-Amino acid E5 transforming protein of bovine papillomavirus requires a hydrophobic core and specific carboxyl-terminal amino acids. Mol Cell Biol 8: 4071–4078

Kleiner E, Dietrich W, Pfister H (1986) Differential regulation of papillomavirus early gene products in transformed fibroblasts and carcinoma cell lines. EMBO J 5: 1945–1950

Lambert PF, Spalholz BA, Howley PM (1987) A transcriptional repressor encoded by BPV-1 shares a common carboxy terminal domain with the E2 transactivation Cell 50: 69–78

Lowy DR, Dvoretzky I, Shober R, Law MF, Engel L, Howley PM (1980) In vitro tumorigenic transformation by a defined subgenomic fragment of bovine papilloma virus DNA. Nature 287: 72–74

Lusky M, Botchan MR (1985) Genetic analysis of bovine papillomavirus type 1 *trans*-acting replication factors. J Virol 53: 955–965

Moskaluk C, Bastia D (1987) The E2 "gene" of bovine papillomavirus encodes an enhancer binding protein. Proc Natl Acad Sci USA 84: 1215–1218

Prakash SS, Horwitz BH, Zibello T, Settleman J, DiMaio D (1988) Bovine papillomavirus E2 gene regulates the expression of the viral E5 transforming gene. J. Virology 62: 3608–3613

Rabson MS, Yee C, Yang Y-C, Howley PM (1986) Bovine papillomavirus type 1 3' early region transformation and plasmid maintenance functions. J Virol 60: 626–634

Schiller JT, Vass WC, Vousdan KH, Lowy DR (1986) The E5 open reading frame of bovine papillomavirus type 1 encodes a transforming gene. J Virol 57: 1–6

Schlegel R, Wade-Glass M, Rabson MS, Yang Y-C (1986) The E5 transforming gene of bovine papillomavirus encodes a small hydrophobic polypeptide. Science 233: 464–467

Settleman J, DiMaio D (1988) Efficient *trans*activation and morphologic transformation by bovine papillomavirus genes expressed from a BPV/SV40 recombinant virus. Proc Natl Acad Sci USA 85: 9007–9011

Spalholz BA, Yang YC, Howley PM (1985) *Trans*activation of a bovine papillomavirus transcriptional regulatory element by the E2 gene product. Cell 42: 183–191

Yang Y-C, Rabson M, Howley PM (1985) Dissociation of bovine papillomavirus transforming and *trans*activating functions. Nature 318: 575–577

# Functional and Sequence Similarities Between HPV16 E7 and Adenovirus E1A

W. C. PHELPS, C. L. YEE, K. MÜNGER, and P. M. HOWLEY

## 1 Introduction

Recent molecular and epidemiological data have established a strong association between certain human papillomaviruses (HPVs) and some types of human anogenital cancers. More than a dozen different HPV types have now been isolated from epithelial tumors of the genital region. In general, benign genital condyloma acuminata have been associated with the presence of HPV types 6 and 11, whereas precancerous lesions such as moderate to severe cervical dysplasia and carcinoma in situ, as well as invasive cervical carcinoma, have been associated with HPV types 16, 18, 31, 33, and 35 (ZUR HAUSEN and SCHNEIDER 1987). Of the HPVs which have been associated with anogenital malignancies, HPV-16 has been detected most frequently ($>60\%$) in biopsies from cervical carcinoma.

Molecular studies of benign lesions infected with a variety of papillomaviruses have revealed that the viral genome is normally maintained within the host cell as a multicopy, extrachromosomal plasmid. Analysis of HPV-16 and HPV-18 DNA in genital neoplasia has indicated that benign and precancerous lesions (e.g., condyloma, bowenoid papulosis, Bowen's disease, and cervical intraepithelial neoplasia) generally contain extrachromosomal HPV DNA, although frequently co-existent with a minor population of integrated viral genomes (DÜRST et al. 1983; DiLUCA et al. 1986; SHIRASAWA et al. 1986, 1988). In contrast, examination of HPV-16 and HPV-18 DNA in genital malignancies has revealed that the viral DNA is preferentially integrated within the host cell genome (DÜRST et al. 1983; BOSHART et al. 1984) suggesting that integration of the viral genome may play a direct role in the process of malignant progression. In addition, integrated and transcriptionally active copies of HPV-16 or HPV-18 DNA are present in cell lines established from cervical carcinomas including SiHa, Caski, HeLa, and C4-1 (BOSHART et al. 1984, YEE et al. 1985), as well as the more recently derived lines SKG-I, II, and III (TAKEBE et al. 1987), and HX151c, HX156c, and HX160c (SPENCE et al. 1988). Examination of both cervical carcinoma tissues and derived cell lines indicates that integration frequently occurs in the viral E1/E2 region, leading to substantial deletion or disruption of the viral genome in that region (SCHWARZ et al. 1985; MATSUKURA et al. 1986; BAKER et al. 1987). Disruption of the E2 open reading frame (ORF) by integration presumably results in the deregulation of the viral

---

Laboratory of Tumor Virus Biology, National Cancer Institute, Bethesda, Maryland 20892, USA

Current Topics in Microbiology and Immunology, Vol. 144
© Springer-Verlag Berlin · Heidelberg 1989

promoters normally under E2 regulation. The HPV-16 E2 ORF encodes a transcriptional *trans*-regulatory factor (PHELPS and HOWLEY 1987) capable of modulating the expression of a viral promoter upstream of the E6 and E7 ORFs through regulatory elements in the viral long control region (LCR). In addition, the HPV-16 LCR contains E2-independent enhancer elements which are keratinocyte specific and glucocorticoid responsive (CRIPE et al. 1987; GLOSS et al. 1987). Therefore, HPV gene regulation appears to involve a complex interaction between viral gene products and cellular transcriptional factors.

RNA analyses of cervical carcinoma tissues and derived cell lines have revealed that the E6 and E7 ORFs are generally expressed, suggesting that their gene products may be necessary for the maintenance of the malignant phenotype (SCHNEIDER-GADICKE and SCHWARZ 1986; SMOTKIN and WETTSTEIN 1986; BAKER et al. 1987; TAKEBE et al. 1987). Morphological transformation of established rodent cells has been described for HPV-16 (YASUMOTO et al. 1986; TSUNOKAWA et al. 1986; KANDA et al. 1987) and HPV-18 (BEDELL et al. 1987), and this transforming activity has been localized to the E6/E7 region (BEDELL et al. 1987; KANDA et al. 1988; K. MÜNGER, unpublished data). In addition, the HPV-16 E6/E7 region has been shown to encode a function capable of cooperating with an activated *ras* oncogene in the transformation of rat embryo fibroblasts (MATLASHEWSKI et al. 1987).

In this study, we have mapped the HPV-16 immortalization function to the E7 ORF and have extended the analysis of the E1A-like activities of HPV-16 E7. We demonstrate that it has transcriptional *trans* activation properties analogous to Ad E1A in that E7 can affect heterologous promoters including the Ad E2 promoter. The Ad E2 promoter sequences required for E1A stimulation and HPV-16 E7 activation are coincident. Furthermore, we have demonstrated that stimulation of transfected primary baby rat kidney cells with hydrocortisone obviates the requirement for expression of HPV-16 E7 from a strong surrogate promoter and thus permits an examination of the control of the homologous HPV-16 promoters in transformed cells. Finally, comparison of the amino acid sequences of HPV-16 E7 and the adenovirus E1A proteins reveals regions of significant amino acid similarity which are well conserved within the E7 proteins of other genital-associated papillomaviruses.


## 2 Methods and Experimental Procedures


### 2.1 Cells

African green monkey kidney CV-1 cells were maintained in Dulbecco modified Eagle medium (Gibco laboratories) with 10% fetal bovine serum and supplemented with penicillin (100 units/ml) and streptomycin (100 µg/ml). Primary baby rat kidney (BRK) cells were prepared from 5-day-old Fischer rats as previously described (RULEY 1983). The kidneys were finely minced and digested with collagenase and dispase (Boehringer Mannheim) for several hours. Approximately $3–5 \times 10^5$ cells per 60-mm dish were plated in Dulbecco modified Eagle media as above. Hydrocortisone (Sigma, St. Louis, MO) was added to the media 24 h after the addition of the DNA at a concentration of 1 µg/ml.

## 2.2 Plasmids

Adenovirus type 2 pE1A (Imperiale et al. 1983), Ad5 pE2CAT, and 5'-deletion mutants pE2-97, pE2-79, pE2-70, and pE2-59 were kindly provided by Joe Nevins and have been previously described (IMPERIALE and NEVINS 1984). The plasmid pEJRAS expressing the activated *ras* oncogene was a gift of Robert Weinberg and has been described (PARADA et al. 1982).

The HPV-16 expression plasmids were constructed utilizing the original HPV-16 clone (p769) described by DÜRST et al. (1983). Fragments of the HPV-16 early region were placed downstream of the SV40 early promoter at nucleotides 79 (p1059), 505 (p858), and 2711 (p859). Translation termination linkers were inserted at various locations in the early region as described (PHELPS et al. 1988). Bacterial transformation of *Escherichia coli* K-12, HB101, or DH5 (Bethesda Research Laboratories), recombinant screening, propagation, and nucleic acid manipulations were performed by conventional methods (MANIATIS et al. 1982).

## 2.3 Mammalian Cell Transfections

DNA transfection for transient CAT assays were performed by calcium phosphate coprecipitation as previously described (PHELPS and HOWLEY 1987) using a total of 10 µg per 60 mm dish. Then 4 h after addition of the DNA, the cells were treated with 15 % glycerol for 1 min. The cell monolayer was washed twice with fresh medium and incubated in complete medium containing 5 m$M$ sodium butyrate (pH 7.0) for 48 h. Primary BRK cells were transfected 2 days after plating with 5 µg of each DNA for a total of 10 µg per 60-mm dish. Precipitates remained on the cells overnight, and the plates were fed the next day in complete medium with or without hydrocortisone.

## 2.4 CAT Assays

CAT assays were performed as previously described (SPALHOLZ et al. 1985). Briefly, cell extracts were normalized for total protein concentration and incubated with [$^{14}$C]chloramphenicol (50 mCi/mole; Amersham Corp., Arlington Heights, IL) and 4 m$M$ acetyl coenzyme A (Pharmacia, Piscataway, N. J.) in 250 m$M$ Tris-HCl (pH 7.8) at 37 °C, and the products of the acetylation reaction were separated by ascending thin-layer chromatography in chloroform-methanol (95:5). Fractioned products were localized by autoradiography and exised for quantitation by liquid scintillation.

# 3 Results

## 3.1 HPV-16 *Trans*-Activation of the Ad E2 Promoter

Previously, it has been shown that the papillomavirus E2 ORF expresses transcriptional *trans*-regulatory functions with activation properties (SPALHOLZ et al. 1985; PHELPS and HOWLEY 1987; HIROCHIKA et al. 1987; HAUGEN et al. 1987; THIERRY et al. 1987; CRIPE et al. 1987) and repressor functions (LAMBERT et al. 1987). In order to identify and further characterize transcriptional modulators encoded by HPV-16, a series of plasmids expressing various portions of the HPV-16 early region were analyzed for their effect on heterologous promoters.

The well-characterized adenovirus early E2 promoter was chosen as one of the group of viral and cellular genes whose transcription is responsive to the adenovirus E1A proteins. Different segments of the HPV-16 early region, expressed from the SV40 early promoter, were assayed by cotransfection into CV-1 monkey cells for their effect on the expression of the Ad E2 promoter linked to the bacterial CAT gene (pE2CAT). As is shown in Fig. 1, the expression of Ad E2CAT was stimulated by cotransfection with pE1A, or either of two different HPV-16 expression plasmids, p1059 and p858, indicating that the HPV-16 early region encodes a factor capable of augmenting the expression of the Ad E2 promoter. The HPV-16 expression plasmid, p859, encoding only the 3′ ORFs, E2, E4, and

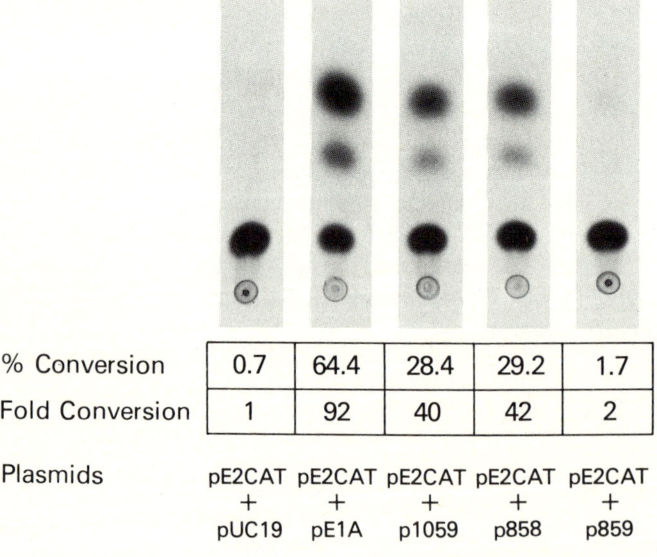



Fig. 1. HPV-16 *trans*-activation of Ad E2CAT. Autoradiograph of a typical CAT analysis. CV-1 monkey cells were transfected with 10 μg plasmid DNA per 60-mm dish. Routinely, 5 μg of of pE2CAT was cotransfected with 5 μg of either pUC19 or a test plasmid: pE1A, p1059, p858, or p859. Cells were harvested after 48 h, and 15 μg total protein was analyzed for CAT activity in a 30-min assay. Acetylated and unacetylated spots were localized by autoradiography, excised from the TLC plate, and quantitated by liquid scintillation

E5, was unable to stimulate the expression of CAT from the Ad E2 promoter, suggesting that the upstream ORFs E6, E7, or E1 were required. This *trans*-activation function has been previously described, and it has been shown by primer extension analysis of transient CAT RNA that HPV-16-mediated stimulation of the Ad E2 promoter was manifest at the level of steady state RNA (PHELPS et al. 1988).

## 3.2 Localization of the HPV-16 E7 *Trans*-Activation Function

Identification of the HPV-16 early gene responsible for Ad E2 *trans*-activation was accomplished by analysis of a series of plasmids in which different portions of the early region were positioned downstream of the SV40 early promoter (PHELPS et al. 1988). As shown in Fig. 1, expression of the E7 or E1 upstream ORFs (p1059 and p858) was required for activity. Therefore, utilizing p858 as a parental plasmid, a series of specific mutations were created by insertion of a translation termination linker (TTL) into a variety of restriction enzyme sites to interrupt translation of particular ORFs. Interruption of translation of the E2, E4, or E1 ORF had no quantitative or qualitative effect on *trans*-activation of Ad E2CAT as demonstrated by cotransfection with p1359, p1301, p1028, p1016, p1360, p1194, p1194, p1196, or p1198 (Fig. 2). However, interruption of the HPV-16 E7 ORF in p1197 or p1199 completely eliminated *trans*-activation; thus, the E7 gene product is required for HPV-16-mediated *trans*-activation of the Ad E2 promoter; and further, this *trans*-acting function was clearly distinct from the previously described HPV-16 E2 transcriptional *trans*-activation function (PHELPS and HOWLEY 1987).

## 3.3 Localization of the Target Sequences for E7 *Trans*-Activation

A series of *Bal*31-generated upstream deletion mutants of Ad E2CAT which have been previously described (IMPERIALE and NEVINS 1984) were cotransfected into CV-1 cells with either pE1A or p858. Previous analysis of this series of plasmids indicated that deletion of Ad E2 promoter sequences to −70 upstream of the major RNA start site significantly impaired E1A responsiveness (IMPERIALE and NEVINS 1984). As shown in Fig. 3, the same deletion mutants which were activated by pE1A responded to cotransfection with the HPV-16 plasmid, p858. Deletion of promoter sequences between −79 and −70 completely abrogated both E1A- and HPV-16-mediated *trans*-activation of the Ad E2 promoter. These data indicate that the targets for Ad E1A and HPV-16 E7 were coincident within the Ad E2 promoter and further suggest that activation of this promoter may occur through a similar mechanism, possibly by modification of host transcriptional factors.

## 3.4 Cooperativity with *ras* in the Transformation of BRK Cells

In experiments designed to evaluate the HPV-16 early region for other oncogenic properties which have been attributed to the multifunctional Ad E1A and E1B regions, primary BRK cells from 5-day-old Fischer rats were transfected with genomic

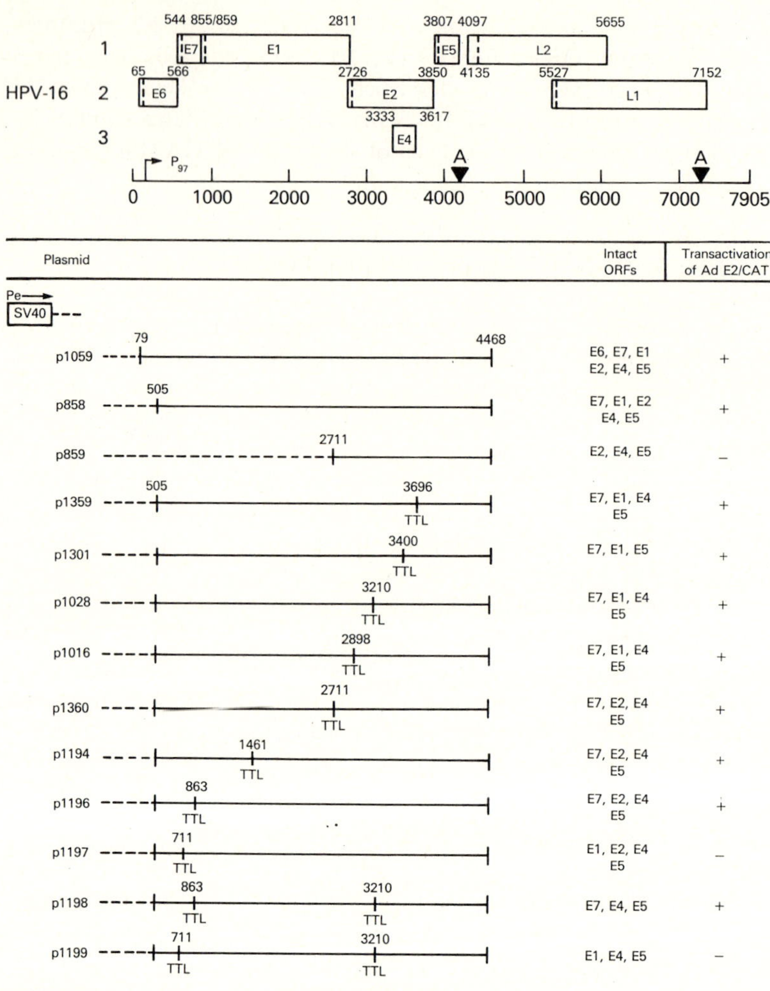

Fig. 2. Localization of the HPV-16 *trans*-acting function. The HPV-16 genome is schematically represented at the *top* of the figure with the major early and late ORFs indicated. The first and last nucleotides of each ORF are indicated according to the sequence data of SEEDORF et al. (1985). The diagram depicts the E1 ORF as being intact since it is thought that this structure more accurately reflects the wild-type arrangement. However, the HPV-16 plasmid used, p769, is the original clone described by DÜRST et al. (1983), and contains a frame shift at nucleotide 1138 (BAKER et al. 1987). The early and late polyadenylation sites (*A*) and the $P_{97}$ promoter are noted. Translation termination linkers (*TTL*) were inserted into various restriction enzyme sites in the HPV-16 early region to inactivate specifically individual ORFs (PHELPS and HOWLEY 1987). HPV-16 sequences were placed downstream of the SV40 early promoter to facilitate efficient expression in CV-1 cells. The intact ORFs are indicated, and the ability to activate pE2CAT is indicated on the *right*. Positive transactivation (+) normally resulted in a stimulation of CAT activity 15–35-fold with a background (−) of 0.5%–3.0% acetylation

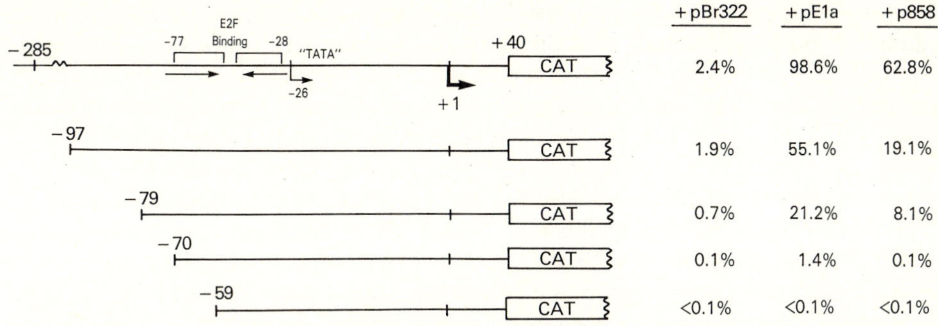


**Fig. 3.** Localization of the E7 target sequences in Ad E2CAT. *Bal*31 deletions with end points at −97, −79, −70, and −59 derived from the −285 to +40 Ad E2CAT plasmid were kindly provided by Joe NEVINS (KOVESDI et al. 1986). Five micrograms of each Ad E2CAT plasmid together with 5 µg of either pBR322, pE1A, or p858 were cotransfected into CV-1 monkey cells and assayed for CAT activity. The locations of the previously described E2F binding sites (KOVESDI et al. 1986) and the major and minor RNA initiation sites are shown

HPV-16 DNA (p769) and many of the plasmids described in Fig. 2. These plasmids were transfected either alone or in combination with plasmids encoding Ad E1A or the activated *ras* oncogene. Cell monolayers were observed for up to 6 weeks for evidence of morphological transformation. Transfection of HPV-16 plasmids alone or in combination with Ad E1A failed to induce morphologically transformed foci. However, transfection of the HPV-16 expression plasmids, p1059 or p858, together with pEJRAS readily induced transformation of BRK cells (Table 1). Cotransfection of BRK cells with p859 encoding the HPV-16 3′ ORFs E2, E4, and E5 failed to induce any apparent morphological alterations. Further, transformation of BRK cells by cotransfection with pEJRAS and p1226 or p1225 indicated that expression of HPV-16 E7 was sufficient for *ras* cooperativity. Cotransfection of p858 with a plasmid expressing the polyomavirus middle T antigen

**Table 1.** Cooperativity with *ras* in the transformation of baby rat kidney cells

| Plasmid | Intact ORFS | Transcriptional promoters | BRK cell transformation | |
|---|---|---|---|---|
| | | | with E1A | with *ras* |
| p769 | all | Homologous | — | — |
| p769 and hydrocortisone[a] | all | Homologous | — | + |
| p1059 | E6, E7, E1, E2 E4, E5 | SV40 | — | + |
| p858 | E7, E1, E2 E4, E5 | SV40 | — | + |
| p859 | E2, E4, E5 | SV40 | — | — |
| p1226 | E6, E7 | SV40 | — | + |
| p1224 | E6 | SV40 | — | — |
| p1225 | E7 | SV40 | — | + |

[a] Hydrocortisone (1 µg/ml) was added to the medium 24 h after the addition of the calcium phosphate-DNA

was also observed to transform BRK cells efficiently in the absence of pEJRAS (data not shown), indicating that the HPV-16 immortalizing function is not specific for the *ras* oncoprotein. Transfection of the whole genome (p769) was incapable of cooperating with *ras* under standard conditions, possibly due to inefficient expression of the viral genes from the homologous viral promoter in rodent cells. However, as noted above, the HPV-16 LCR contains a glucocorticoid-responsive element; we therefore assayed the genomic clone for *ras* cooperativity with the inclusion of hydrocortisone in the growth medium. Under these conditions, cooperativity was noted with the genomic HPV-16 clone in the absence of a surrogate promoter.

Individual BRK transformed foci were expanded and assessed for anchorage-independent growth in soft agar and for tumorigenicity in nude mice. HPV-16/*ras*-transformed BRK cells containing the HPV-16 plasmids p858, p1016, p1028, p1196, or p1198 readily formed colonies in soft agar and were tumorigenic in nude mice within 10 days.

## 3.5 Structural Similarity Between Ad E1A and HPV-16 E7

Comparison of the amino acid sequences of the E1A proteins from various adenovirus serotypes has revealed that these polypeptides are comprised of alternating regions of high and relatively low sequence conservation (KIMELMAN et al. 1985). Three highly conserved regions have been identified, and subsequent mutational analyses have mapped transformation and *trans*-activation functions differentially to these domains. Conserved domains 1 and 2 in E1A are shared by the 243- and 289-amino acid polypeptides translated from the 12*S* and 13*S* mRNAs, respectively. These two domains are important for transcriptional repression and for cellular immortalization functions, including cooperation with *ras* in the transformation of primary rat cells (LILLIE et al. 1986; MORAN et al. 1986; ZERLER et al. 1987; SCHNEIDER et al. 1987). Domain 3 is the 46-amino acid segment which is unique to the 289-amino acid product of the 13*S* mRNA transcript and has been shown to contain determinants both necessary and sufficient for E1A-mediated transcriptional *trans*-activation (LILLIE et al. 1986, 1987). Examination of the amino acid sequence of HPV-16 E7 revealed that a single domain in E7 between residues 2 and 37 contains striking sequence similarity to domains 1 (residues 37–49) and 2 (residues 116–137) of Ad E1A (Fig. 4). Although HPV-16 E7 does not appear to contain any other primary amino acid similarity to Ad E1A, it is interesting to note that the *C*-terminal half of HPV-16 E7 contains two Cys-x-x-Cys motifs analogous to those found in the conserved domain 3 of Ad E1A. Such motifs have been implicated in metal binding by several eukaryotic transcription factors (BERG 1986; MILLER et al. 1985; KADONAGA et al. 1987).

A number of point mutations in domains 1 and 2 of Ad E1A have been shown to abrogate specifically the capacity of E1A to cooperate with an activated *ras* oncogene in the transformation of primary rodent cells. Mutations in domain 1 described by ZERLER et al. (1987) which inactivate *ras* cooperativity occur within a region of Ad E1A that does not appear to have a similar counterpart in HPV-16 E7. However, mutations in domain 2 at Cys 124, Glu 126, Ser 132, and Glu 135 (LILLIE et al. 1986; ZERLER et al. 1987),

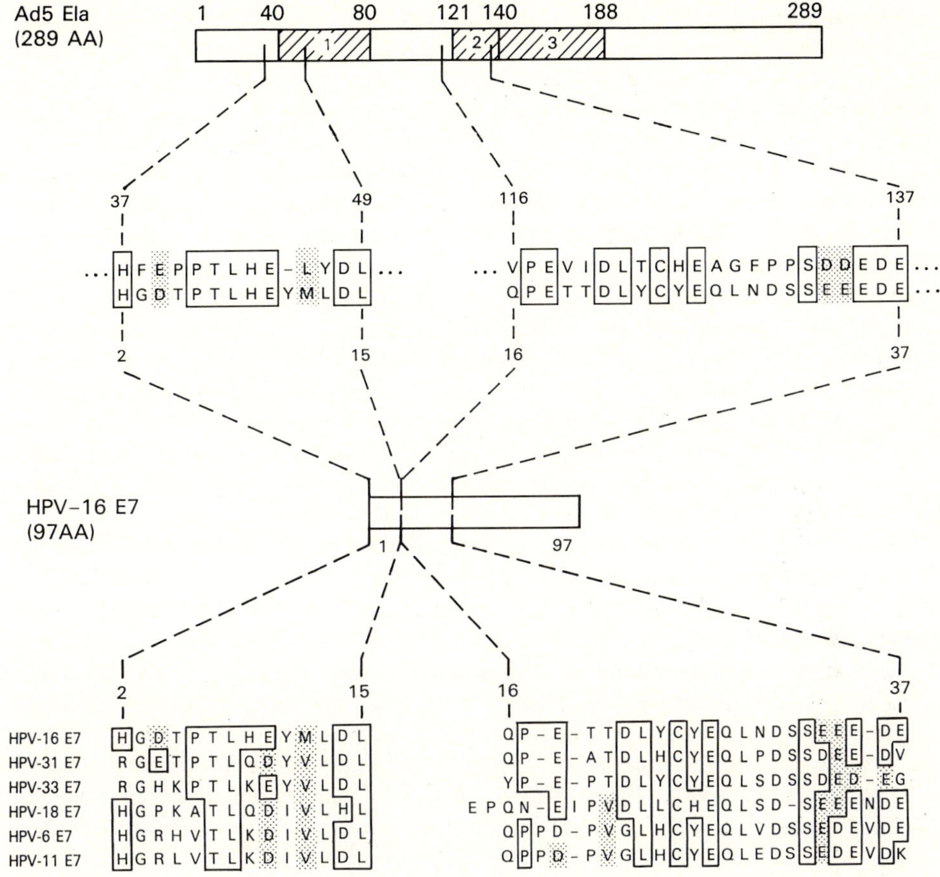

Fig. 4. Structural similarity between Ad E1A and HPV-16 E7. A schematic diagram of the Ad5 E1A region is shown at the *top* of the figure with the conserved amino acid domains *1–3* indicated (KIMELMAN et al. 1985). Homologous (*boxed*) and isofunctional (*stippled*) amino acids of domains 1 and 2 from E1A which are found in HPV-16 E7 are located within the *N*-terminal 37 residues. The predicted amino acid sequences for this region of the E7 proteins of the other genital-associated HPVs are shown *below*

which inactivate cooperativity with *ras*, are conserved in the region of similarity in HPV-16 E7 (Fig. 4). The predicted amino acid sequences of the E7 proteins among the other genital-associated HPVs (HPVs 6, 11, 18, 31, and 33) are well-conserved, suggesting that similar activities may be encoded by the E7 genes of other HPVs. Indeed, the critical amino acids in domain 2 of Ad E1A at Cys 124, Glu 126, Glu 135, and Ser 132 are well-conserved throughout the genital-associated HPVs (Fig. 4). In contrast, these conserved amino acids are not present in the predicted sequence of the BPV-1 E7 protein, which also does not appear to possess similar transcriptional modulation or transforming properties (B. SPALHOLZ, C. YEE, unpublished data).

## 4 Discussion

Studies concerning the transcriptional program of the human papillomaviruses have frequently been modeled after the detailed genetic analyses of BPV-1. The E2 ORFs of BPV-1, CRPV, and of the several HPVs studied have been demonstrated to encode gene products involved in transcriptional regulation mediated through a specific DNA binding site for E2, ACCN$_6$GGT (SPALHOLZ et al. 1985, 1987; DARTMANN et al. 1986; ANDROPHY et al. 1987; HIROCHIKA et al. 1987; HAUGEN et al. 1987; MCBRIDE et al. 1988; HAWLEY-NELSON et al. 1988). As with BPV-1, the HPV-16 full-length E2 gene product encodes a transcriptional *trans*-acting factor capable of activating a conditional enhancer element in the viral LCR (PHELPS and HOWLEY 1987; CRIPE et al. 1987). In this study, we have examined the properties of an additional transcriptional modulator encoded by the HPV-16 E7 ORF. The E7 protein appears to be multifunctional, possessing both transcriptional modulatory and cellular transforming properties analogous to those described for the adenovirus E1A proteins (PHELPS et al. 1988). This activity is thus far unique to HPV-16 among the papillomaviruses in that similar transcriptional and immortalizing properties have not been detected in BPV-1. Based on the extensive homology among the E7 proteins of the genital-associated HPVs, it seems likely that these other HPVs may encode similar functions.

Analysis of *Bal*31-generated Ad E2 promoter deletions suggested that the target sites for Ad E1A and HPV-16 E7 may be similar and that they may, therefore, operate through a common mechanism. *Trans*-activation by Ad E1A does not appear to involve the direct binding of the viral protein to the promoter sequences (FELDMAN et al. 1982; FERGUSON et al. 1985) but rather may be mediated through interaction with, or modification of, cellular transcription factors. Analysis of DNA/protein interactions within the Ad E2 promoter has revealed the presence of two binding sites for a cellular factor, E2F, whose activity is augmented in adenovirus-infected cells (KOVESDI et al. 1986). E2F is a 54K protein which binds at two sites in the Ad E2 promoter between −33 and −74 to the sequence element TTTCGCGC (JONES et al. 1988). In addition, binding of another factor between −70 and −79 has implicated the ATS or CRE sequence motif found in promoters induced by cAMP, suggesting that E1A may interact with the cAMP binding protein (CREB) in the activation of the adenovirus early genes (LEE et al. 1987).

The mechanism by which HPV-16 E7 *trans*-activates the Ad E2 promoter is not known. The similarity in the target sites in the Ad E2 promoter suggests that the E7 protein might interact with E2F or with CREB. Alternatively, HPV-16 E7 *trans*-activation could be mediated through a pathway independent of both factors, possibly by direct binding to the Ad E2 promoter, or through interaction with other transcriptional factors (JALINOT et al. 1987). Presumably, E7 transcriptional modulation of HPV-16 expression must be mediated through homologous HPV promoters; however, the normal target for E7 regulation has not yet been identified although the occurrence of E2F and CREB binding motifs in HPV-16 may provide useful landmarks in the analysis of E7 *trans*-regulation.

Analysis of cervical carcinoma tissues and cell lines harboring HPV-16 or HPV-18 has indicated that there is active viral gene expression principally from the E6 and E7 ORFs (Schwarz et al. 1985; Schneider-Gadicke and Schwarz 1986; Baker et al. 1987). In this study, we present evidence that the HPV-16 E7 gene encodes both transcriptional *trans*-regulatory and cellular transformation properties. Cellular transformation and modulation of heterologous promoters are viral functions that are likely to play a direct role in the complex, multistep process of carcinogenic progression. The continued expression of these functions, therefore, would be consistent with HPV playing an active role in the maintenance of the malignant phenotype of these cells.

Human foreskin keratinocytes can be immortalized by transfection with HPV-16 DNA (Pirisi et al. 1987; Dürst et al. 1987). Utilizing techniques for electroporation of DNA, we have begun a genetic analysis of the HPV-16 requirements for immortalization of human epithelial cells to determine whether functions described for E7 are involved in the arrest of differentiation and stimulation of proliferation associated with the immortalization of keratinocytes.

*Acknowledgements.* We would like to thank Nan Freas for preparation of this manuscript. W.C.P. was supported by NRSA postdoctoral fellowship CA07713 from the National Cancer Institute.

# References

Androphy EJ, Lowy DR, Schiller JT (1987) Bovine papillomavirus E2 *trans*-acting gene product binds to specific sites in papillomavirus DNA. Nature 325: 70–73

Baker CC, Phelps WC, Lindgren V, Braun MJ, Gonda MA, Howley PM (1987) Structural and transcriptional analysis of human papillomavirus type 16 sequences in cervical carcinoma cell lines. J Virol 61: 962–971

Bedell MA, Jones KH, Laimins LA (1987) The E6–E7 region of human papillomavirus type 18 is sufficient for transformation of NIH 3T3 and Rat-1 cells. J Virol 61: 3635–3640

Berg JM (1986) Potential metal-binding domain in nucleic acid binding proteins. Science 232: 484–487

Boshart M, Gissmann L, Ikenberg H, Kleinheinz A, Scheurlen W, zur Hausen H (1984) A new type of papillomavirus DNA and its presence in genital cancer biopsies and in cell lines derived from cervical cancer. EMBO J 3: 1151–1157

Cripe TP, Haugen TH, Turk OP, Tabatabai F, Schmid PG, Dürst M, Gissmann L, Roman A, Turek LP (1987) Transcriptional regulation of the human papillomavirus-16 E6–E7 promoter by a keratino-cyte-dependent enhancer, and by viral E2 *trans*-activator and repressor gene products: implications for cervical carcinogenesis. EMBO J 6: 3745–3753

Dartmann K, Schwarz E, Gissmann L, zur Hausen H (1986) The nucleotide sequence and genome organization of human papillomavirus type 11. Virology 151: 124–130

Di Luca D, Pilotti S, Stefanon B, Rotola A, Monini P, Tognon M, De Palo G, Rilke F, Cassai E (1986) Human papillomavirus type 16 DNA in genital tumors: a pathological and molecular analysis. J Gen Virol 67: 583–589

Dürst M, Gissmann L, Ikenberg H, zur Hausen H (1983) A papillomavirus DNA from a cervical carcinoma and its prevalence in cancer biopsy samples from different geographic regions. Proc Natl Acad Sci USA 80: 3812–3815

Dürst M, Dzarlieva-Petrusevska RT, Boukamp P, Fusenig NE, Gissmann L (1987) Molecular and cytogenetic analysis of immortalized human primary keratinocytes obtained after transfection with human papillomavirus type 16 DNA. Oncogene 1: 251–256

Feldman LT, Imperiale MJ, Nevins JR (1982) Activation of early adenovirus transcription by the herpes virus immediate early gene, evidence for a common cellular control factor. Proc Natl Acad Sci USA 79: 4952–4956

Ferguson B, Kripple B, Andrisani O, Jones N, Westphal H, Rosenberg M (1985) Ela 13*S* and 12*S* mRNA products made in *Escherichia coli* both function as nucleus localized transcription activators but do not directly bind to DNA. Mol Cell Biol 5: 2653–2661

Gloss B, Bernard HU, Seedorf K, Klock G (1987) The upstream regulatory region of the human papilloma virus-16 contains an E2 protein-independent enhancer which is specific for cervical carcinoma cells and regulated by glucocorticoid hormones. EMBO J 6: 3735–3743

Guis D, Grossman S, Bedell MA, Laimins LA (1988) Inducible and constitutive enhancer domains in the noncoding region of human papillomavirus type 18. J Virol 62: 665–672

Haugen TH, Cripe TP, Ginder GD, Karin M, Turek LP (1987) Transactivation of an upstream early gene promoter of bovine papillomavirus-1 by a product of the viral E2 gene. EMBO J 6: 145–152

Hawley-Nelson P, Androphy EJ, Lowy DR, Schiller JT (1988) The specific DNA recognition sequences of the bovine papillomavirus E2 protein is an E2-dependent enhancer. EMBO J 7: 525–531

Hirochika H, Broker TR, Chow LT (1987) Enhancers and *trans*-acting E2 transcriptional factors of papillomaviruses. J Virol 61: 2599–2606

Imperiale MJ, Nevins JR (1984) Adenovirus 5 E2 transcription unit: an Ela-inducible promoter with an essextial element that functions independently of position or orientation. Mol Cell Biol 4: 875–882

Imperiale MJ, Feldman LT, Nevins JR (1983) Activation of gene expression by adenovirus and herpesvirus regulatory genes acting in *trans* and by a *cis*-acting adenovirus enhancer element. Cell 35: 127–136

Jalinot P, Devaux B, Kedinger C (1987) The abundance and in vitro DNA binding of three cellular proteins interacting with the adenovirus EIIa early promoter are not modificd by the E 1a gene products. Mol Cell Biol 7: 3806–3817

Jones NC, Rigby PWJ, Ziff EB (1988) *Trans*-acting protein factors and the regulation of eukaryotic transcription: lessons from studies on DNA tumor viruses. Genes Dev 2: 267–281

Kanda T, Watanabe S, Yoshiike K (1987) Human papillomavirus type 16 transformation of rat 3Y1 cells. Jpn J Cancer Res (GANN) 78: 103–108

Kanda T, Watanabe S, Yoshiike K (1987) Human papillomavirus type 16 open reading frame E7 encodes a transforming gene for rat 3Y1 cells. J Virol 62: 610–613

Kadonga IT, Carner KR, Masiarz FR, Tjian R (1987) Isolation of cDNA encoding transcription factor SpI and functional analysis of the DNA binding domain. Cell 51: 1079–1090

Kimelman D, Miller JS, Porter D, Roberts BE (1985) E 1a regions of the human adenoviruses and of the highly oncogenic simian adenovirus 7 are closely related. J Virol 53: 399–409

Kovesdi I, Reichel R, Nevins JR (1986) Identification of a cellular transcription factor involved in Ela *trans*-activation. Cell 45: 219–228

Lambert PF, Spalholz BA, Howley PM (1987) A transcriptional repressor encoded by BPV-1 shares a common carboxy terminal domain with the E2 transactivator. Cell 50: 69–78

Lee KAW, Hai TY, SivaRaman L, Thimmappaya B, Hurst HC, Jones NC, Green MR (1987) A cellular transcription factor ATF activates transcription of multiple E1A-inducible adenovirus early promoters. Proc Natl Acad Sci USA 84: 8355–8359

Lillie JW, Green M, Green MR (1986) An adenovirus Ela protein region required for transformation and transcription repression. Cell 46: 1043–1051

Lillie JW, Loewenstein PM, Green MR, Green M (1987) Functional domains of adenovirus type 5 Ela proteins. Cell 50: 1091–1100

Maniatis T, Fritsch EF, Sambrook J (1982) Molecular cloning: a laboratory manual. Cold Spring Harbor Laboratory, Cold Spring Harbor, NY

Matlashewski G, Schneider J, Banks L, Jones N, Murray A, Crawford L (1987) Human papillomavirus type 16 DNA cooperates with activated *ras* in transforming primary cells. EMBO J 6: 1741–1746

Matsukura T, Kanda T, Furuno A, Yoshikawa H, Kawana T, Yoshiike K (1986) Cloning of monomeric human papillomavirus type 16 DNA integrated within cell DNA from a cervical carcinoma. J Virol 58: 979–982

McBride AA, Schlegel R, Howley PM (1988) The carboxy-terminal domain shared by the bovine papillomavirus E2 transactivator and repressor proteins contains a specific DNA binding activity. EMBO J 7: 533–539

Miller J, McLachlan AD, Klug A (1985) Repetitive zinc-binding domains in the protein transcription factor IIIA from *Xenopus* oocytes. EMBO J 4: 1609–1614

Moran E, Zerler B, Harrison TM, Mathews MB (1986) Identification of separate domains in the adenovirus E 1a gene for immortalization activity and the activation of virus early genes. Mol Cell Biol 6: 3470–3480

Parada LF, Tabin CJ, Shih C, Weinberg RA (1982) Human EJ bladder carcinoma is a homologue of Harvey sarcoma virus *ras* gene. Nature 297: 474–478

Phelps WC, Howley PM (1987) Transcriptional transactivation by the human papillomavirus type 16 E2 gene product. J Virol 61: 1630—1638

Phelps WC, Yee CL, Münger K, Howley PM (1988) The human papillomavirus type 16 E7 gene encodes transactivation and transformation functions similar to those of adenovirus E1A. Cell 53: 539–547

Pirisi L, Yasumoto S, Feller M, Doniger J, DiPaolo JA (1987) Transformation of human fibroblasts and keratinocytes with human papillomavirus type 16 DNA. J Virol 61: 1061–1066

Ruley HE (1983) Adenovirus early region 1A enables viral and cellular transforming genes to transform primary cells in culture. Nature 304: 602–606

Schneider JF, Fisher F, Goding CR, Jones NC (1987) Mutational analysis of the adenovirus E1a gene: the role of transcriptional regulation in transformation. EMBO J 6: 2053–2060

Schneider-Gadicke A, Schwarz E (1986) Different human cervical carcinoma cell lines shown similar transcription patterns of human papillomavirus type 18 early genes. EMBO J 5: 2285–2292

Schwarz E, Freese UK, Glasmann L, Mayer W, Roggenbuck B, Stremlau A, zur Hausen H (1985) Structure and transcription of human papillomavirus sequences in cervical carcinoma cells. Nature 314: 111–114

Seedorf K, Krammer G, Dürst M, Suhai S, Rowekamp W (1985) Human papillomavirus type 16 DNA sequence. Virology 145: 181–185

Shirasawa H, Tomita Y, Kubota K, Kasai T, Sekiya S, Takamizawa H, Simizu B (1986) Detection of human papillomavirus type 16 DNA and evidence for integration into the cell DNA in cervical dysplasia. J Gen Virol 67: 2011–2015

Shirasawa H, Tomita Y, Kubota K, Kasai T, Sekiya S, Takamizawa H, Simizu B (1988) Transcriptional differences of the human papillomavirus type 16 genome between precancerous lesions and invasive carcinomas. J Virol 62: 1022–1027

Smotkin D, Wettstein FO (1986) Transcription of human papillomavirus type 16 early genes in a cervical cancer and a cancer-derived cell line and identification of the E7 protein. Proc Natl Acad Sci USA 83: 4680–4684

Spence RP, Murray A, Banks L, Kelland LR, Crawford L (1988) Analysis of human papillomavirus sequences in cell lines recently derived from cervical cancers. Cancer Res 48: 324–328

Spalholz BA, Lambert PF, Yee CL, Howley PM (1987) Bovine papillomavirus transcriptional regulation: localization of the E2-responsive elements of the long control region. J Virol 61: 2128–2137

Spalholz BA, Yang Y-C, Howley PM (1986) Transactivation of a bovine papillomavirus transcriptional regulatory element by the E2 gene product. Cell 42: 183–191

Takebe N, Tsunokawa Y, Nozawa S, Terade M, Sugimura T (1987) Conservation of E6 and E7 regions of human papillomavirus types 16 and 18 present in cervical cancers. Biochem Biophys Res Commun 143: 837–844

Thierry F, Heard JM, Dartmann K, Yaniv M (1987) Characterization of a transcriptional promoter of human papillomavirus 18 and modulation of its expression by simian virus 40 and adenovirus early antigens. J Virol 61: 134–142

Tsunokawa Y, Takebe N, Kasamatsu T, Terada M, Sugimura T (1986) Transforming activity of human papillomavirus type 16 DNA sequences in a cervical cancer. Proc Natl Acad Sci USA 83: 2200–2203

Yasumoto S, Burkhardt AL, Doniger J, DiPaolo JA (1986) Human papillomavirus type 16 DNA-induced malignant transformation of NIH 3T3 cells. J Virol 57: 572–577

Yee CL, Krishnan-Hewlett I, Baker CC, Schlegel R, Howley PM (1985) Presences and expression of human papillomavirus sequences in human cervical carcinoma cell lines. Am J Pathol 119: 3261–3266

Zerler B, Roberts RJ, Mathews MB, Moran E (1987) Different functional domains of the adenovirus Ela gene are involved in regulation of host cell cycle products. Mol Cell Biol 7:821–829

zur Hausen H, Schneider A (1987) The role of papillomaviruses in human anogenital cancer. In: Howley PM, Salzman N (eds) The papovaviridae: the papillomaviruses. Plenum, New York pp 24–263

# Two Independently Transforming Functions
# of Human Papillomavirus 8

T. Iftner, P. G. Fuchs, and H. Pfister

## 1 Introduction

The human papillomaviruses (HPVs) cause a variety of benign tumors of the skin and mucosa and are of particular medical interest due to their association with human cancers (Pfister 1984). A large group of HPVs is associated with epidermodysplasia verruciformis (ev), a lifelong disease, which is characterized by disseminated flat warts and macular skin lesions that develop into squamous cell carcinomas in 30%–50% of the sufferers (Orth 1987). More than 90% of these cancers are persistently infected with HPV-5 or HPV-8, which were also discovered in skin carcinomas of immunosuppressed transplant recipients (Bunney et al. 1987). These HPV types are therefore assumed to represent particular risk factors for the development of skin cancer.

An in vitro test system for oncogenic papillomaviruses was first established with bovine papillomavirus (BPV) 1, which induces fibropapillomas in cattle and readily transforms rodent fibroblasts in culture (Pfister 1984). Two independently transforming genes, E5 and E6, were identified, which act synergistically to induce the fully transformed phenotype: altered morphology, anchorage independence, reduced serum requirement, and tumorigenicity in nude mice (Schiller et al. 1984, 1986; Yang et al. 1985).

More recently HPV-1, 5, 16, and 18 have been shown to transform mouse or rat fibroblast cell lines (Watts et al. 1984; Green et al. 1986; Yasumoto et al. 1986; Bedell et al. 1987). They are usually less efficient than BPV-1 and induce fewer foci that take longer to develop. In the case of HPV-16 and 18, which are associated with cervical cancer, transforming activity could be assigned to open reading frame (ORF) E7 (Kanda et al. 1988; Watanabe and Yoshiike 1988).

Our study was designed to identify the transforming functions of HPV-8, which was isolated from an ev patient. Its genome has been sequenced (Fuchs et al. 1986) and is organized similar to other papillomaviruses. No ORF could be identified, however, with homology to E5, which accounts for the major transforming activity of BPV-1.

Institut für Klinische und Molekulare Virologie, Universität Erlangen—Nürnberg, Erlangen, FRG

Current Topics in Microbiology and Immunology, Vol. 144
© Springer-Verlag Berlin · Heidelberg 1989

## 2 Results

We subcloned all early ORFs of HPV-8 into the *Bam*HI site of the retroviral expression vector pZipNeo-SV(X)1, hereafter termed SVX (Cepko et al. 1984). The vector contains two large terminal repeats of the Moloney murine leukemia virus, which are necessary for the initiation and polyadenylation of HPV-8 transcripts, a neomycin phosphotransferase gene for selection in eukaryotic cells, and splicing signals that allow the processing of primary transcripts (Fig. 1).

Mouse C127 cells (Lowy et al. 1978) were transfected with the HPV-8-SVX constructs using the calcium phosphate coprecipitation method (Graham and van der Eb 1973). The cells were split at a ratio of 1:6 1 or 2 days after transfection and then fed twice weekly without further subdividing. No foci appeared during an observation period of 8 weeks.

In a second approach cell lines were established with all constructs after selection for G418 (400 µg/ml) resistance. These lines were labeled with C8 for C127 and HPV8, followed by the number of the cloned HPV-8 ORF. Southern blot analysis of DNA from all cell lines identified integrated HPV-8-SVX constructs at a low copy number, and restriction enzyme cleavage data confirmed the integrity of the HPV-8 inserts and the relevant transcription units (Iftner et al. 1988). Consequently, HPV-8-specific transcripts could be demonstrated by Northern blot analysis. Mouse C127 cells, which were transfected with the expression vector only and selected for G418 resistance, served as a negative control for further experiments. Among all lines established, only C86 and C867 showed an altered morphology when compared with C127 or C127-SVX. The cells appeared elongated and spindle-

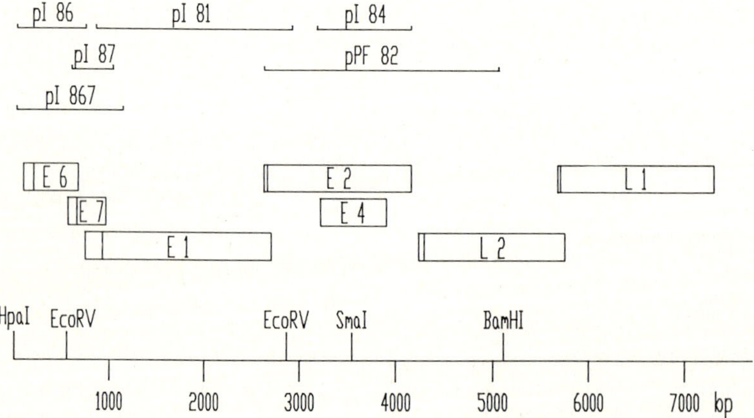



**Fig. 1.** Map position of HPV-8 DNA fragments, which were cloned into the expression vector pZipNeo-SVX. The HPV-8 ORFs are represented by *open bars*, with *vertical lines* indicating the first translational start codons of each. Blunt-ended HPV-8 DNA fragments were generated by digestions with two or more endonucleases, ligated to *Bam*HI linkers, cleaved with *Bam*HI, and cloned into the single *Bam*HI site of pZipNeo-SVX. The correct orientation of the inserts was determined by restriction analysis of analytical plasmid preparations from transformed *E. coli* HB101 clones

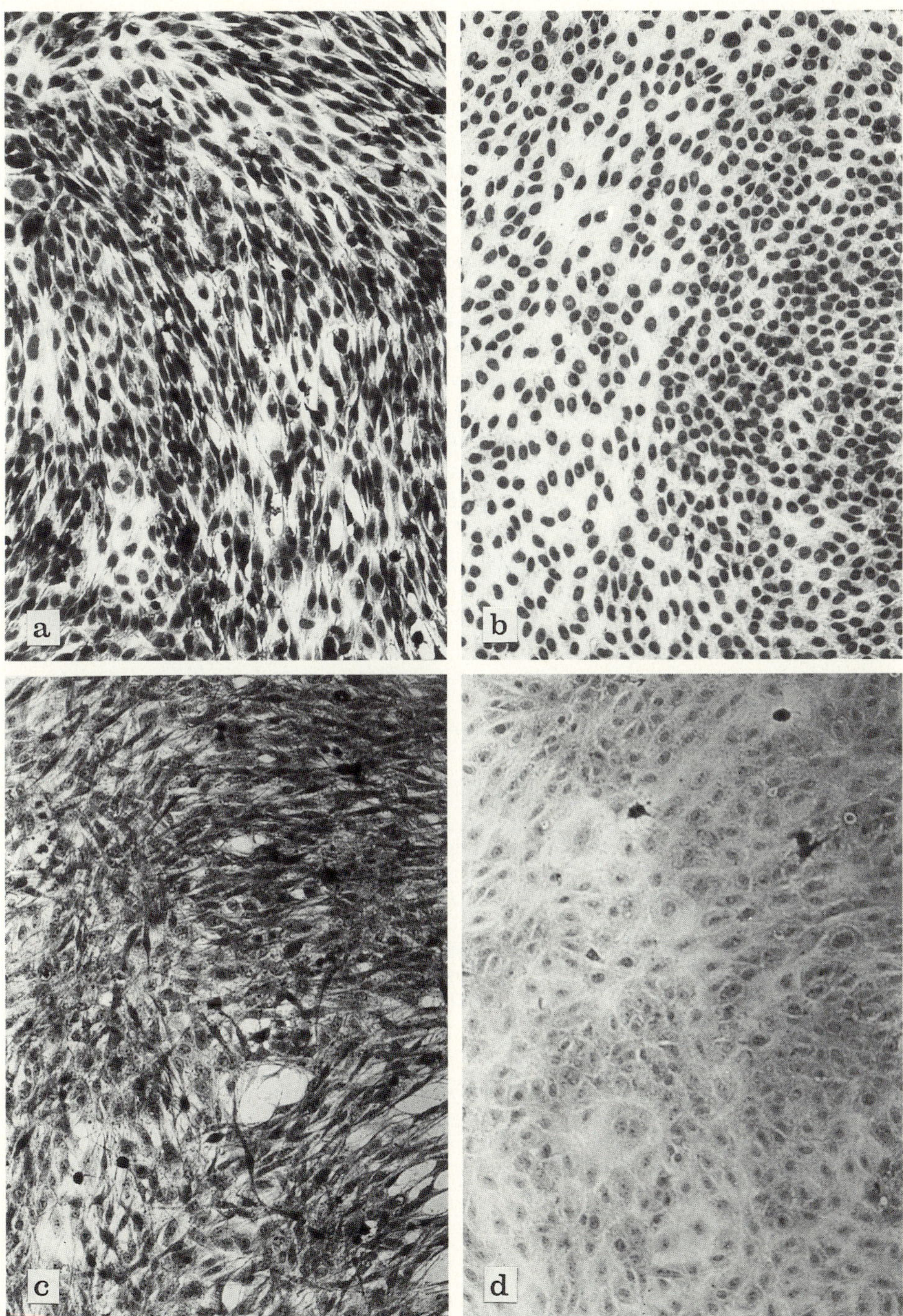

**Fig. 2a–d.** Morphology of C127 (**a, b**) and Rat1 (**c, d**) cells established after transfection with pZipNeo-SVX (**d**), pI86 (**a, c**) or pI82 (**b**). Cells were stained with Giemsa

shaped and grew on top of each other (Fig. 2). This morphological transformation could be reproduced in 11 independent assays.

Both NIH 3T3 (JAINCHILL et al. 1969) and Ratl cells (BEDELL et al. 1987) were transfected with SVX or pI86 and tested for the appearance of foci. No foci were detectable with both lines. The G418-resistant Rat1 line, however, transfected with pI86 appeared clearly transformed (Fig. 2), whereas NIH 3T3-derived N86 cells did not morphologically differ from SVX-transfected controls (data not shown).

Additional parameters of transformation were studied with C8 cell lines. The growth rate was determined in minimum essential medium containing either 5% or 0.5% fetal calf serum (Table 1). C86 and C867 cells grew at almost the same rate in both media whereas doubling of C81, C87, C127-SVX, and the parental C127 cells took more than twice as long in the low-serum medium. C82 cells surprisingly grew at virtually the same rate in 0.5% serum and in 5% serum, like the lines with the morphologically transformed phenotype.

This finding was substantiated by the anchorage-independent growth of C82 in soft agar. About 50000 cells were seeded with 0.3% agarose type VII (Sigma, St. Louis, MO) into a 60-mm plate. The cultures were fed twice weekly and colonies scored 2–3 weeks later (Table 2). As a positive control we used BPV-1 wild-type transformed C127/B81 cells (KLEINER et al. 1986) that formed large

**Table 1.** Growth of HPV8/pZipNeo-SVX transfected C127 cells in medium with high and low serum content

| Cell line | Generation time (h) | |
|---|---|---|
| | 5% serum | 0.5% serum |
| C81 | 20 | >100 |
| C82 | 20 | 22 |
| C86 | 20 | 24 |
| C867 | 20 | 21 |
| C87 | 20 | 46 |
| C127/pZipNeoSVX | 20 | 46 |
| C127 | 24 | 62 |

**Table 2.** Anchorage-independent growth of HPV8/pZipNeo-SVX-transfected cells in semisolid medium

| Cell line | Cloning efficiency (%) | Colony size (number of cells) |
|---|---|---|
| C127/B81 | 5.3 | 1500 |
| C81 | 0.0 | — |
| C82 | 4.5 | 200–300 |
| C86 | 6.4 | 1000 |
| C867 | 2.6 | 1000 |
| C87 | <0.1 | 100 |
| C127 pZipNeo-SVX | 0.0 | — |

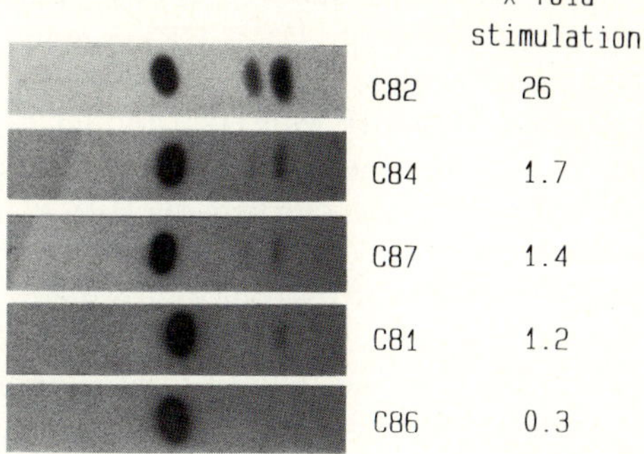



**Fig. 3.** Autoradiogram of a thin layer chromatography plate showing CAT enzyme activity in extracts from various C8 cells transfected with an HPV-8-SV40 promoter-CAT construct (p33 CAT+, Seeberger et al. 1987). *Spots to the left* are due to unacetylated chloramphenicol and the *two spots to the right* of each lane represent monoacetylated forms. Stimulation was calculated from the enzyme activity in the cell line shown versus enzyme activity in C127 cells transfected with the same CAT construct

macroscopically visible colonies within 3 weeks. C86, C867, and C82 were also able to grow in soft agar with cloning efficiencies comparable to cells transformed by BPV-1 wild-type. The colonies took longer, however, to reach the same size as colonies of C127/B81. C81 and C127 cells containing only the vector backbone did not form colonies in soft agar.

C82 cells contain besides ORF E4 and part of ORF L2 the ORF E2, which was shown to encode an efficient *trans*-activator of viral transcription with all papillomaviruses tested so far (Spalholz et al. 1985; Hirochika et al. 1987; Phelps and Howley 1987) except for HPV-18 (Thierry and Yaniv 1987). We have previously described that an enhancer element of HPV-8 is activated in *trans* by the BPV-1 E2 protein (Seeberger et al. 1987). So we used an enhancer-dependent expression vector for chloramphenicol acetyltransferase (CAT) with the HPV-8 enhancer cloned in sense orientation upstream of CAT to test for *trans*-activation in C8 cells. The CAT activity was clearly enhanced in C82 when compared with the enzyme activity in C127 or any other C8 line (Fig. 3). This indicates that HPV-8 E2 also acts as a *trans*-activator of virus-specific transcription.

# 3 Discussion

Oncogenic DNA viruses frequently encode more than one gene involved in cell transformation (Green 1985). Different genes usually act independently and trigger different pathways but cooperate to induce the fully transformed phenotype characteristic for the respective virus. Data presented here introduce HPV-8 as another example with two transforming genome regions. In contrast to BPV-1 only one gene could be identified that induces morphological transformation of rodent fibroblasts. This is in line with the lack of an HPV-8 ORF homologous to the BPV-1 oncogene E5 (Fuchs et al. 1986). The 3'-moiety of the early genome part of HPV-8 was able, however, to stimulate C127 cells to grow in soft agar and in the presence of low serum. Growth in semisolid medium without any apparent morphological changes or tendency to pile up was also observed with NIH 3T3 and 3YI cells transfected with cloned HPV-16 DNA (Noda et al. 1988).

Expression of the HPV-8 ORF E6 under the control of a retroviral large terminal repeat was sufficient to induce morphological transformation, anchorage independence, and reduced serum requirement of C127 fibroblasts. Similar to BPV-1 E6 (SCHILLER et al. 1984) HPV-8 E6 cannot transform NIH 3T3 cells. It was interesting to note that ORF E7 of HPV-8 showed no transforming activity in contrast to findings with HPV-16 and 18, where ORF E7 encodes the only transforming protein identified so far (KANDA et al. 1988; WATANABE and YOSHIIKE 1988). The discrepancies between HPV-8 and BPV-1 on one side and HPV-16 and 18 on the other may be understood with regard to the similarities of ORFs E6 and E7. These ORFs code for proteins with four or two cysteine-X-X-cysteine motifs, respectively, that represent potential zinc ion binding sites (DANOS and YANIV 1987). The E6 and E7 genes were interpreted as having evolved by repeated duplication of a 33-amino acid module followed by genetic drift. This could explain why functions switch from E6 to E7. On the other hand one cannot exclude the fact that BPV-1 and HPV-8 transform cells by a mechanism different from that of HPV-16 and 18. The BPV-1 E6 protein was detected in the nucleus and associated with nonnuclear membranes (ANDROPHY et al. 1985), whereas HPV-16 E7 represents a cytoplasmic phosphoprotein (SMOTKIN and WETTSTEIN 1987). HPV-16 E7 furthermore transforms NIH 3T3 cells (K. H. VOUSDEN and D. R. LOWY, personal communication) in contrast to BPV-1 and HPV-8 E6, which lack this activity. Both the findings on intracellular localization and on function favor the idea of differently acting proteins.

The second transforming genome segment of HPV-8 covers ORFs E2, E4, and part of L2. Assigning the transforming activity to a specific ORF has to await genetic experiments, which are presently being carried out in our laboratory. It is tempting to speculate, however, that *trans*-activation of transcription by an ORF E2 product might play a role in transformation.

The role of in vitro transforming functions of HPV-8 in naturally occurring infections remains to be established. The 5'-part of the early region corresponding to ORFs E6 and E7 was found to be transcribed in a skin carcinoma persistently infected with the closely related HPV-5 (ORTH 1987). An HPV-17 DNA-positive skin cancer on the other hand showed viral transcripts predominantly from the 3'-moiety of the early region (YUTSUDO and HAKURA 1987). This may indicate that persistent expression of one transforming function or the other is important for the growth properties of the HPV-associated skin carcinoma cells.

*Acknowledgements.* We are indebted to Dr. R. C. Mulligan, who provided the expression vector pZipNeo-SVX. This work was supported by the Deutsche Forschungsgemeinschaft (Fu 175/1–2).

# References

Androphy E, Schiller JT, Lowy DR (1985) Identification of the protein encoded by the E6 transforming gene of bovine papillomavirus. Science 230: 442–445
Bedell MA, Katherine HJ, Laimins AL (1987) The E6—E7 region of human papillomavirus type 18 is sufficient for transformation of NIH 3T3 and Rat 1 cells. J Virol 61: 3635–3640
Bunney MH, Barr BB, MacLaren K, Smith TW, Benton EC, Anderton JL, Hunter JAA (1987) Human papillomavirus type 5 and skin cancer in renal allograft recipients. Lancet ii: 151–152

Cepko CL, Roberts BE, Mulligan RC (1984) Construction and applications of a highly transmissible murine retrovirus shuttle vector. Cell 37: 1055–1062

Danos O, Yaniv M (1987) E6 and E7 gene products evolved by amplification of a 33-amino-acid peptide with a potential nucleic-acid-binding structure. Cancer Cells 5: 145–150

Fuchs PG, Iftner T, Weninger J, Pfister H (1986) Epidermodysplasia verruciformis-associated human papillomavirus 8: genomic sequence and comparative analysis. J Virol 58: 626–634

Graham FL, van der Eb AJ (1973) A new technique for the assay of infectivity of human adenovirus 5 DNA. Virology 52: 456–467

Green M (1985) Transformation and oncogenesis: DNA viruses. In: Fields BN (ed) Virology. Raven, New York, pp 183–234

Green M, Brackmann KH, Loewenstein PM (1986) Rat embryo fibroblast cells expressing human papillomavirus 1a genes exhibit altered growth properties and tumorigenicity. J Virol 60: 868–873

Hirochika H, Broker TR, Chow LT (1987) Enhancers and *trans*-acting E2 transcriptional factors of papillomaviruses. J Virol 61: 2599–2606

Iftner T, Bierfelder S, Csapo Z, Pfister H (1988) Involvement of human papillomavirus 8 genes E6 and E7 in transformation and replication. J Virol (submitted)

Jainchill JL, Aaronson SA, Todaro GJ (1969) Murine sarcoma and leukemia viruses: assay using clonal lines of contact inhibited mouse cells. J Virol 4: 549–553

Kanda T, Furuno A, Yoshiike K (1988) Human papillomavirus type 16 open reading frame E7 encodes a transforming gene for rat 3Y1 cells. J Virol 62: 610–613

Kleiner E, Dietrich W, Pfister H (1986) Differential regulation of papillomavirus early gene expression in transformed fibroblasts and carcinoma cell lines. EMBO J 5: 1945–1950

Lowy DR, Rands E, Scolnick EM (1978) Helper-independent transformation by unintegrated Harvey sarcoma virus DNA. J Virol 26: 291–298

Noda T, Yajima H, Ito Y (1988) Progression of the phenotype of transformed cells after growth stimulation of cells by a human papillomavirus type 16 gene function. J Virol 62: 313–324

Orth G (1987) Epidermodysplasia verruciformis. In: Salzmann NP, Howley PM (eds) The papovaviridae. Plenum, New York, pp 129–244

Pfister H (1984) Biology and biochemistry of papillomaviruses. Rev Physiol Biochem Pharmacol 99: 111–181

Phelps WC, Howley PM (1987) Transcriptional *trans*-activation by the human papillomavirus type 16 E2 gene product. J Virol 61: 1630–1638

Schiller JT, Vass WC, Lowy DR (1984) Identification of a second transforming region in bovine papillomavirus DNA. Proc Natl Acad Sci USA 81: 7880–7884

Schiller JT, Vass WC, Vousden KH, Lowy DR (1986) The E5 open reading frame of bovine papillomavirus type 1 encodes a transforming gene. J Virol 57: 1–6

Seeberger R, Haugen T, Turek L, Pfister H (1987) An enhancer of human papillomavirus type 8 is *trans*-activated by the bovine papillomavirus type 1 E2 function. Cancer Cells 5: 33–38

Smotkin D, Wettstein FO (1987) The major human papillomavirus protein in cervical cancers is a cytoplasmic phosphoprotein. J Virol 61: 1686–1689

Spalholz BA, Yang Y-C, Howley PM (1985) Transactivation of a bovine papilloma virus transcriptional regulatory element by the E2 gene product. Cell 42: 183–191

Thierry F, Yaniv M (1987) The BPV1-E2 *trans*-acting protein can be either an activator or a repressor of the HPV18 regulatory region. EMBO J 63: 3391–3397

Watanabe S, Yoshiike K (1988) Transformation of rat 3Y1 cells by human papillomavirus type 18 DNA. Int J Cancer (in press)

Watts SL, Phelps WC, Ostrow RS, Zachow KR, Faras AJ (1984) Cellular transformation by human papillomavirus DNA in vitro. Science 225: 634–636

Yang YC, Spalholz BA, Rabson MS, Howley PM (1985) Dissociation of transforming and trans-activation functions for bovine papillomavirus type 1. Nature 318: 575–577

Yasumoto S, Burkhardt AL, Doniger J, DiPaolo JA (1986) Human papillomavirus type 16 DNA-induced malignant transformation of NIH 3T3 cells. J Virol 57: 572–577

Yutsudo M, Hakura A (1987) Human papillomavirus type 17 transcripts expressed in skin carcinoma tissue of a patient with epidermodysplasia verruciformis. Int J Cancer 39: 586–589

# Human Papillomavirus Early Gene Products and Maintenance of the Transformed State of Cervical Cancer Cells In Vitro

A. KLEINHEINZ[1], M. VON KNEBEL DOEBERITZ[1], T. P. CRIPE[2], L. P. TUREK[2], and L. GISSMANN[1]

There are several lines of evidence from epidemiological as well as from laboratory experimental work linking human papillomaviruses (HPV) to the development of squamous cell cancer of the uterine cervix:

1. The association of a sexually transmissible infectious agent is strongly suggested by epidemiological observations (for review, see DOLL 1986)
2. Cervical intraepithelial neoplasias (CIN), considered as putative precursor lesions of cervical cancer, are regularly associated with HPVs. Recently, the development of this kind of lesion was induced by infection of normal human epithelium in vitro and subsequent grafting into nude mice (KREIDER et al. 1985). From such lesions papillomavirus particles can be purified which are efficient in a second round of infection. Thus the Koch's postulate for a causal relationship of the infectious agent and the disease have been fulfilled.
3. HPVs exhibit a transforming potential in vitro. Primary or established rodent cells can be transformed by transfection with the HPV early gene E7 (BEDELL et al. 1987; KANDA et al. 1988; L. LAIMINS, personal communication), and a synergistic effect was observed with the activated human *ras* oncogene (MATLASHEWSKI et al. 1987). In addition, human keratinocytes, which represent the natural target cells for papillomavirus infection, are immortalized by HPV-16 or HPV-18 DNA, respectively (DÜRST et al. 1987; PIRISI et al. 1987; P. KAUR and J. MCDOUGALL, personal communication).

Although in the majority of cases cervical cancer biopsies as well as established cell lines (e.g., HeLa, C4-1, SiHa) contain integrated HPV DNA which is transcribed and translated (for review, see GISSMANN et al. 1987), it could be argued that the virus is only required initially in order to induce the malignant phenotype of the cell. In that case its detectable traces reflect its passenger state rather than its necessary activity for the maintenance of transformation. Two sets of experiments were therefore undertaken to investigate the proliferation of cervical carcinoma cells in vitro after modulation of early HPV-16 or HPV-18 gene expression.

As a first approach we attempted to influence the expression of E6 and E7 proteins by the use of anti-sense RNA-producing plasmids (IZANT and WEINTRAUB 1985). The respective region of the HPV-18 genome was cloned in sense or anti-sense orientation downstream of the human cytomegalovirus (HCMV) immediate

---

[1] Deutsches Krebsforschungszentrum, Heidelberg, FRG
[2] University of Iowa, Iowa City, USA

Current Topics in Microbiology and Immunology, Vol. 144
© Springer-Verlag Berlin · Heidelberg 1989

early promoter-enhancer element (VON KNEBEL DOEBERITZ et al. 1988). The plasmids, which also contain the pSV2*neo* gene, were transfected into the HPV-18-positive cervical cancer cell line C4-1 or into HPV-negative mouse cells (C127), and G418-resistant colonies were selected. In C4-1 cells the number of G418-resistant colonies obtained after transfection of the sense construct was twice as high as with the anti-sense plasmid. No difference was found in the HPV-negative C127 cells. This result suggests that the expression of anti-sense RNA (and presumably the absence or reduced abundance of the endogenous HPV-18 mRNA) interferes with survival of the HPV-positive cells.

In order to investigate this effect more closely the HCMV promoter-enhancer element of the anti-sense construct was replaced by the long terminal repeat (LTR) of the mouse mammary tumor virus (MMTV), which is normally silent but can be activated by the addition of dexamethasone to the culture medium. C4-1 cells were transfected as described above, and G418-resistant colonies were obtained. Three of these colonies (K1-3) were expanded and analyzed in more detail. In all three subclones the induction of anti-sense RNA was observed after treatment of the cells with dexamethasone. The growth rate (measured by incorporation of tritiated thymidine) of normal C4-1 cells or cells transfected only with pSV2*neo* DNA was increased by a factor of 5–6 upon dexamethasone treatment. Although we do not know which cellular or viral genes are responsible for this increase in DNA synthesis, it is possible that in C4-1 cells an overexpression of the HPV-18 E6-E7 genes contributes to this effect, since transcription from the HPV E6-E7 promoter can be stimulated by steroid hormones (GLOSS et al. 1987). Correspondingly, the level of HPV-18 mRNA and of the E7 protein was elevated in C4-1 or C4-1-neo cells. In contrast, the subclones K1, K2, and K3, transfected with the inducible anti-sense construct, did not show an increased growth rate, and E6-E7 mRNA and E7 protein levels were reduced or unchanged after dexamethasone treatment. It is thus possible that the induced expression of the E6-E7 anti-sense RNA counteracts the increase in E6-E7 mRNA. The correlation between the level of HPV-18 early gene expression and growth activity observed in C4-1 cells and their different derivatives suggests that in established cervical carcinoma cells HPV genes play a role in cell growth regulation (VON KNEBEL DOEBERITZ et al. 1988).

In a similar approach the effect of the E2 ORF was analyzed. Due to specific integration of the circular HPV-16 or HPV-18 DNA into the host genome as observed in cervical cancer cell lines (SCHWARZ et al. 1985; BAKER et al. 1987), this particular region is inactivated. It has been postulated that this specificity of integration is a prerequisite to ensure the constitutive expression of E6-E7 required for cell transformation (SCHWARZ et al. 1985). In fact recent experiments by CRIPE et al. (1987) demonstrate that one of the at at least two E2 proteins (short E2: sE2) can repress transcription of downstream sequences when attached to short palindromes (E2P) present in the upstream regulatory region of different papillomaviruses (DARTMANN et al. 1986; ANDROPHY et al. 1987; HAUGEN et al. 1987).

Cotransfection of the HPV-16-positive cervical cancer cell line SiHa with pSV2*neo* plus increasing amounts of HPV-16 sE2 driven by the Rous sarcoma virus LTR [pRSV(16)-sE2] leads to a decreasing number of G418-resistant colonies. To study this effect in more detail the pRSV(16)-sE2 or pRSV(BPV-1)-sE2 DNA was microinjected into SiHa cells. As early as 90 min after injection the first cells underwent a morphological change (rounding up and shrinking of the nuclei, see Fig. 1). At 2 and 3 h after injection, between 90% and 100% of the injected cells became moribund and were detached from the glass slide after further incubation. This is not a simple toxic effect of some poisonous contaminant within the particular DNA preparation since primary human fibroblasts which were injected with the same construct survived this treatment. The specificity of the effect conferred by the sE2 gene product was further underlined by injecting frame shift mutants of pRSV(16)-sE2 or pRSV(BPV-1)-sE2 constructed by linker insertion which failed to kill the SiHa cells. The injection of the complete E2 ORF of HPV-16 under RSV LTR control led to the same result as sE2. Because a construct expressing the complete early region or the E6-7 ORF of HPV-16 had no effect on SiHa cells, it is unlikely that the *trans*-activation of the potential oncogenic E6-E7 genes by the long E2 (SPALHOLZ et al. 1985) was responsible for the lethal effect. THIERRY and YANIV (1987) argue that the close spatial relationship of the E2P and the TATA box prevents the binding of cellular transcription factors if the E2P sites are occupied, thus switching off E6-E7 transcription. This would explain why the complete and the truncated form of E2 give rise to the same effect.

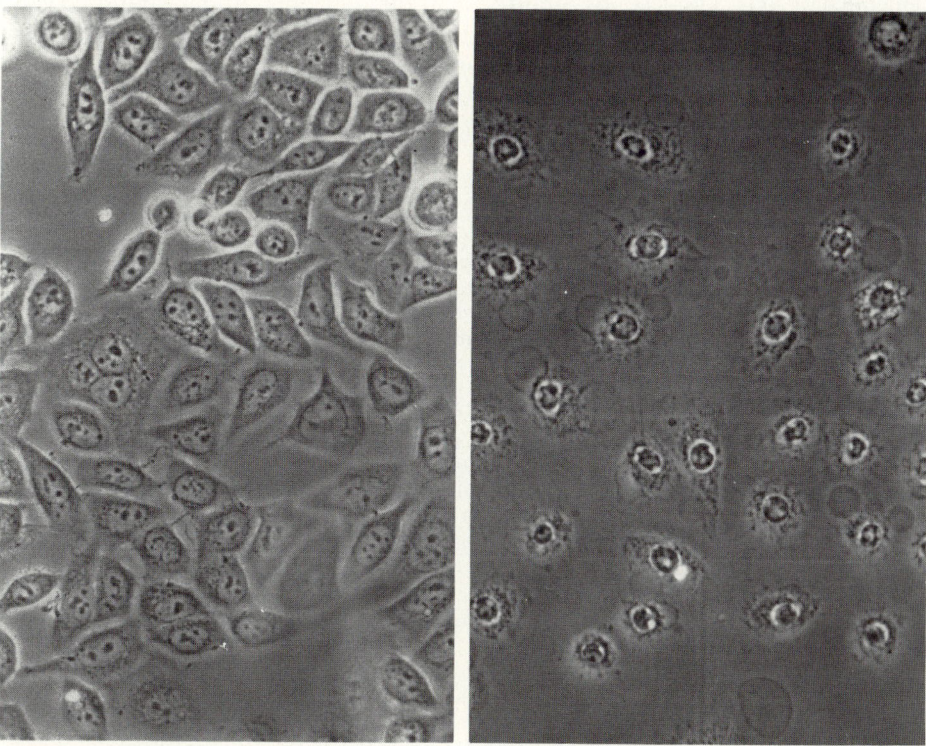

**Fig. 1a, b.** SiHa cells before (**a**) and 5 h after (**b**) microinjection of about 100 copies of the pRSV(16)-sE2 plasmid

The injection of an RSV-lacZ construct (kindly provided by G. SCHÜTZ) demonstrated the immediate expression of the respective protein: when the cells were fixed and treated with X-Gal 30% of the injected cells stained blue 30 min after injection, 80%–100% after 1 h. Since a certain amount of protein molecules per cell is required for a positive staining one can anticipate that the first protein molecules are produced shortly after microinjection of the plasmid. It is thus possible that the E2 proteins are also expressed and downregulate HPV-16 E6-E7 gene expression very rapidly, particularly if the short half-life of HPV-16 E7 protein is taken into account (SMOTKIN and WETTSTEIN 1987). It is not clear, however, whether the observed killing effect on SiHa cells is due to interference of the E2 gene products with the endogenous HPV-16 E6-E7 gene expression or to some other mechanism. Recent experiments have demonstrated that the E2 proteins are strong transcriptional regulatory factors exerting an effect on a wide range of promoters (T. H. HAUGEN et al., unpublished data). In situ hybridization experiments are presently underway to determine whether a decrease of E6-E7 gene expression precedes the lethal effect on the SiHa cells.

*Acknowledgements.* We are grateful to Dr. H. zur Hausen for stimulating discussions and constant support and to Dr. M. Pawlita for critical suggestions on this manuscript. The skillful technical assistance of A. Henn and M. Fiedler is also gratefully acknowledged. This work was supported by the Deutsche Forschungs-gemeinschaft (Gi 128/2-1), by grants from the Veterans Administration (VA), National Institutes of Health, and the University of Iowa Diabetes and Endo-crinology Research Center. L.P.T. receives a VA Career Award.

# References

Androphy EJ, Lowy DR, Schiller JT (1987) A peptide encoded by the bovine papillomavirus E2 *trans*-activating gene binds to specific sites in papillomavirus DNA. Nature 325: 70–73

Baker CC, Phelbs WC, Lindgren V, Braun MJ, Gonda MA, Howley PM (1987) Structural and transcriptional analysis of human papillomavirus type 16 sequences in cervical carcinoma cell lines. J Virol 61: 962–971

Bedell MA, Jones KH, Laimins LA (1987) The E6-E7 region of human papillomavirus type 18 is sufficient for transformation of NIH 3T3 and rat cells. J Virol 61: 3635–3640

Cripe TP, Haugen TH, Turk JP, Tabatabai F, Schmid PS, Dürst M, Gissmann L, Roman A, Turek LP (1987) Transcriptional regulation of the human papillomavirus-16 E6-E7 promoter by a keratinocyte dependent enhancer, and by viral E2 *trans*-activator and repressor gene products: implications for cervical carcinogenesis. EMBO J 12: 3745–3753

Dartmann K, Schwarz E, Gissmann L, zur Hausen H (1986) The nucleotide sequence and genome organization of human papillomavirus type 11. Virology 151: 124–130

Doll R (1986) Implications of epidemiological evidence for future progress. In: Peto R, zur Hausen H (eds) Viral etiology of cervical cancer, Banburry report 21. Cold Spring Harbor Laboratory, Cold Spring Harbor, pp 321–326

Dürst M, Dzarlieva-Petrusevska RT, Boukamp P, Fusenik N, Gissmann L (1987) Molecular and cytogenic analysis of immortalized human primary keratinocytes obtained after transfection with human papillomavirus type 16 DNA. Oncogene 1: 251–256

Gissmann L, Dürst M, Oltersdorf T, von Knebel Doeberitz M (1987) Human papillomaviruses and cervical cancer. In: Steinberg BM, Bradsma JL, Taichman LB (eds) Cancer cells, vol 5. Cold Spring Harbor Laboratory, Cold Spring Harbor, pp 275–280

Gloss B, Bernard HU, Seedorf K, Klock G (1987) The upstream regulatory region of the human papillomavirus 16 contains an E2 protein-independent enhancer which is specific for cervical carcinoma cells and regulated by glucocorticoid hormones. EMBO J 6: 3735–3743

Haugen TH, Cripe TP, Ginder GD, Karin M, Turek LP (1987) *Trans*-activation of an upstream early gene promoter of bovine papillomavirus 1 by a product of the viral E2 gene. EMBO J 6: 145–152

Izant JG, Weintraub H (1985) Constitutive and conditional suppression of exogenous and endogeneous genes by anti-sense RNA. Science 229: 345–352

Kanda T, Furuno A, Yoshiike K (1988) Human papillomavirus type 16 open reading frame E7 encodes a transforming gene for rat 3Y1 cells. J Virol 62: 610–613

Kreider JW, Howett MK, Wolfe SA, Bartlett GL, Zaino RJ, Sedlacek M, Mortel R (1985) Morphological transformation of human uterine cervix in vivo with papillomavirus from condylomata acuminata. Nature 317: 639–641

Matlashewski G, Schneider J, Banks L, Jones N, Murray A, Crawford L (1987) Human papillomavirus type 16 DNA cooperates with activated *ras* in transforming primary cells. EMBO J 6: 1741–1746

Pirisi L, Yasumoto S, Feller M, Doninger J, diPaolo JA (1987) Transformation of human fibroblasts and keratinocytes with human papillomavirus type 16 DNA. J Virol 61: 1061–1066

Schwarz E, Freese UK, Gissmann L, Mayer W, Roggenbuck B, Stremlau A, zur Hausen H (1985) Structure and transcription of human papillomavirus sequences in cervical carcinoma cells. Nature 314: 111–114

Smotkin D, Wettstein FO (1987) The major human papillomavirus protein in cervical cancers is a cytoplasmic phosphoprotein. J Virol 61: 1686–1689

Spalholz BA, Yang YC, Howley PM (1985) Transactivation of a bovine papilloma virus transcriptional regulatory element by the E2-gene product. Cell 42: 183–191

Thierry F, Yaniv M (1987) The BPV1-E2 *trans*-acting protein can be either an activator or a repressor of the HPV 18 regulatory protein. EMBO J 6: 3391–3397

von Knebel Doeberitz M, Oltersdorf T, Schwarz E, Gissmann L (1988) Modulation of human papillomavirus gene expression in C4-1 cervical carcinoma cells is correlated with altered growth properties. Cancer Res 38: (in press)

# Part IV: Transformation by Adenoviruses

# Introduction

The human adenoviruses contain double-stranded linear DNA chromosomes about 36000 base pairs in length. They undergo markedly different infectious processes in human as compared with rodent cells. In human cells, the entire viral chromosome is expressed in an ordered and highly regulated fashion. A set of early viral genes are activated rapidly after infection.

The viral E1A gene products play a key role in this activation. They stimulate gene expression at the level of transcription, apparently by influencing the activities of multiple cellular transcription factors. Concomitant with the onset of viral DNA replication, a second, late set of viral genes are expressed.

Human cells are eventually killed by adenovirus infection, and progeny virus is generated. Rodent cells are much less permissive than human cells for growth of human adenoviruses: Generally, only early viral genes are expressed. Although the majority of individual rodent cells infected with adenoviruses eventually die, some survive and become oncogenically transformed. The E1A and E1B genes, comprising only 11% of the viral chromosome, contain all of the information required for transformation of cultured cells.

Human adenoviruses have been extensively employed as models for analysis of gene expression and its regulation in human cells and for analysis of oncogenic transformation in rodent cells. The following four contributions describe mechanisms underlying and regulating gene expression in virus-infected and transformed cells.

SCHAACK and SHENK describe a new function for the adenovirus-coded terminal protein which has long been known to act as a primer for viral DNA replication. This protein, which is covalently attached to the 5′ ends of viral DNA molecules, begins very early after infection to facilitate timely activation of viral transcription. Terminal protein presumably carries out this role by mediating attachment of the viral chromosome to the nuclear scaffold.

LILLIE et al. investigate the mechanism by which virus-coded E1A proteins and host factors interact to activate transcription of viral genes. First, they show that the E1A transcriptional activating domain fused to the DNA binding domain of the yeast *gal*4 protein can increase transcriptional activity of genes containing a *gal*4 binding site more efficiently than genes which lack the binding site in their promoter region. This raises the possibility that E1A proteins act to influence transcription directly at the promoter even though they don't exhibit sequence-specific DNA binding. Next, LILLIE et al. report the purification of ATF and AP-1 which bind very similar DNA sequences. These two cellular factors very likely play a role in E1A-mediated activation of adenovirus expression. So far, they have

Current Topics in Microbiology and Immunology, Vol. 144
© Springer-Verlag Berlin · Heidelberg 1989

found no evidence for E1A-induced physical modifications of the factors. Given the E1A-*gal*4 result, it is tempting to suggest that E1A modulates the function of ATF and AP-1 by entering into a complex with these factors at the transcriptional initiation site.

VAN DER EB et al. show that expression of a number of cellular genes, including c-*myc*, is inhibited in rat cells transformed by human adenoviruses. This result extends earlier studies which have shown expression of class I major histocompatability antigen to be inhibited in some transformed cells. If one assumes that the c-*myc* gene product plays a key role in the control of cellular proliferation, its reduced level of expression may mean that this normal cellular control mechanism is bypassed in adenovirus-transformed cells. VAN DER EB et al. raise the possibility that c-*myc* function is assumed by the viral E1A gene in transformed cells.

The adenovirus chromosome or segments of the chromosome are covalently attached to host cell DNA within transformed cells. DOERFLER et al. describe a series of experiments designed to recapitulate integration-specific events in cell-free extracts. A cloned segment of hamster DNA was used as a target for integration. This segment is known to have previously served as an integration site in vivo during transformation of hamster cells. The hamster DNA was mixed with fragments of adenovirus DNA in nuclear extracts of hamster cells. Hamster sequences were recovered via their attached plasmid DNAs by transformation of bacterial cells and screened for the presence of viral sequences. A variety of recombinant molecules were identified. As yet, it is not clear whether the specific hamster DNA sequence employed was more recombinogenic than a random DNA sequence. Nevertheless, this in vitro system holds great promise for dissection of the recombination process undergone by the adenovirus chromosome during oncogenic transformation of hamster cells.

# Adenovirus Terminal Protein Mediates Efficient and Timely Activation of Viral Transcription

J. Schaack and T. Shenk

## 1 Introduction

The adenovirus terminal protein (TP) is found covalently attached to the 5′-ends of viral DNA molecules (Robinson et al. 1973; Robinson and Bellett 1974; Rekosh et al. 1977). The linkage is through a phosphodiester bond formed between the β-hydroxyl group of a serine residue within the *C*-terminal domain of the TP and the 5′-hydroxyl of the terminal deoxycytosine residue on the viral DNA (Desiderio and Kelly 1981; Smart and Stillman 1982). The protein initially attached to the viral chromosome is termed the preterminal protein (pTP), and it is cleaved to form the TP during virion maturation (Challberg and Kelly 1981; Stillmann et al. 1981). The functional significance of the cleavage is unclear.

This pTP functions during initiation of adenovirus DNA replication. Initiation begins with the formation of a covalent linkage between dCMP and pTP (reviewed in Stillman 1983; Kelly 1984). Formation of this complex requires the participation of both viral and cellular replication proteins and is dependent on the presence of adenovirus DNA. Once the dCMP-pTP complex is formed, elongation of the new daughter strand proceeds by the addition of nucleotides at the 3′-OH group of the dCMP residue. Thus, pTP primes adenovirus DNA replication.

It is coded for by the adenovirus E2B gene (Smart and Stillman 1982). A series of mutant viruses that contain small linker insertions within the region of the E2B gene encoding pTP has been isolated (Freimuth and Ginsberg 1986). One of these mutants, H5*sub*100, contains a net 3-bp insert within the region (sequence position 10,564) encoding the N-terminus of pTP. This mutant is temperature-sensitive for DNA replication and viral growth (Freimuth and Ginsberg 1986).

We now report on two additional components to the *sub*100 phenotype. First, the mutant was found to be markedly defective in activation of E1A transcription. This defect could not be complemented in *trans*. Second, whereas the majority of wild-type viral DNA became tightly associated with the nuclear scaffold very soon after infection, *sub*100 DNA was readily extracted from the scaffold compartment. These observations suggest that the TP mediates an association between the infecting adenovirus chromosome and the nuclear scaffold and that this association is required for efficient and timely activation of adenovirus gene expression.

---

Department of Biology, Princeton University, Princeton, NJ 08544, USA

Current Topics in Microbiology and Immunology, Vol. 144
© Springer-Verlag Berlin · Heidelberg 1989

# 2 Results and Discussion

## 2.1 TP Facilitates Activation of the E1A Transcription Unit

A role of the TP in the activation of adenovirus transcription was demonstrated by monitoring the accumulation of E1A mRNAs subsequent to *sub*100 infection. HeLa cells were infected at a multiplicity of 25 pfu/cell with either *sub*100 (Freimuth and Ginsberg 1986; Fig. 1) or a phenotypically wild-type virus, *dl*309 (Jones and Shenk 1979). Total cytoplasmic RNA was prepared 4 h after infection and assayed by the RNase protection procedure (Melton et al. 1984) using a $^{32}$P-labeled E1A-specific probe RNA (Fig. 2). Cells infected with *sub*100 contained

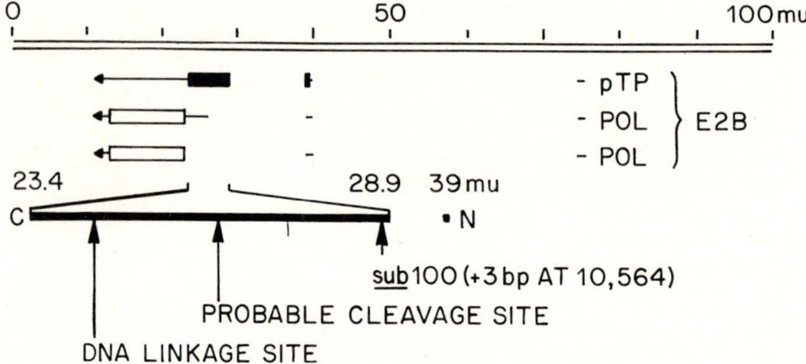



**Fig. 1.** Diagram of the adenovirus E2B transcription unit. Map locations are shown in map units (*mu*). Three E2B-coded mRNAs are transcribed in the direction indicated by *arrowheads*. Exons are designated by *lines* or *rectangles*, and introns are indicated by *spaces*. Coding regions for preterminal protein (*pTP*) and DNA polymerase (*POL*) are represented by *rectangles*. The coding region for pTP is shown expanded at the *bottom* of the figure. The positions of the *DNA linkage site*, *probable cleavage site*, and *sub*100 mutation are indicated on the pTP

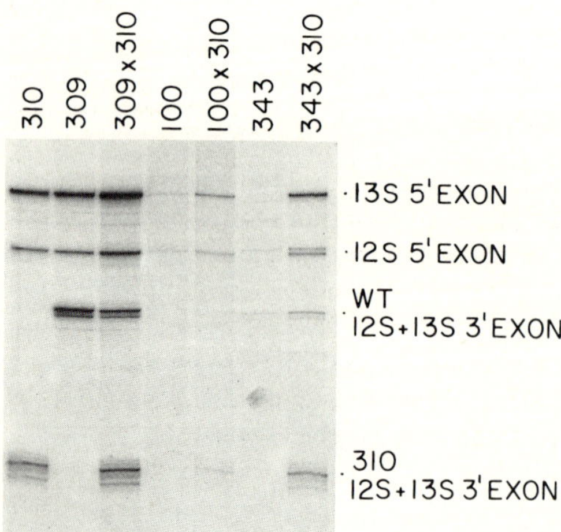



**Fig. 2.** RNase protection analysis of E1A-specific mRNAs produced in cells infected with wild-type (*WT*) and mutant viruses. RNA was prepared 4 h after infection of HeLa cells and analyzed using a $^{32}$P-labeled RNA probe corresponding to the E1A coding region. The bands corresponding to wild-type and mutant exons of E1A mRNAs are labeled

substantially less E1A-specific mRNA (Fig. 2, lane 100) than *dl*309-infected cells (Fig. 2, lane 309).

If the failure of *sub*100 to produce E1A-coded mRNAs was due to its covalently attached mutant TP, then the defect should not be complemented in *trans* by a virus with wild-type TP. This prediction was tested using *dl*310 (Jones and Shenk 1979) as the complementing virus. Its 12S and 13S E1A mRNAs can be distinguished from those of other adenoviruses, including *sub*100, in an RNase protection assay. E1A mRNAs encoded by *dl*310 lack a segment of their 3' exon. Wild-type probe RNAs are cleaved by RNase when hybridized to *dl*310-coded RNA molecules, generating a shortened band corresponding to the common 3' exon of 12S and 13S mRNAs. As a result, bands corresponding to the E1A-specific 3' exon encoded by *dl*310 can be readily distinguished from those representing wild-type exons, and this is demonstrated in an experiment in which cells were coinfected with *dl*309 (wild-type 3' exon) and *dl*310 (deleted 3' exon) (Fig. 2, lane 309 × 310). Coinfection with *dl*310 did not lead to an increase in the level of the band corresponding to the *sub*100 E1A-specific 3' exon (Fig. 2, lane 100 × 310), indicating thet the *sub*100 transcriptional defect could not be complemented.

As a control, the ability of *dl*310 to complement E1A mRNA synthesis by a *trans*-complementable mutant, *dl*343, was tested. Mutant *dl*343 lacks 2 base pairs within the E1A coding region (Hearing and Shenk 1985). It is unable to produce E1A gene products that are required, in turn, for efficient expression of E1A mRNAs. As expected, cells infected with *dl*343 contained very low levels of E1A-specific mRNAs (Fig. 2, lane 343), and coinfection with *dl*310 led to an increase in intensity of the band corresponding to the *dl*343-coded 3' exon (Fig. 2, lane 343 × 310).

The transcriptional analysis displayed in Fig. 2 was performed at 37 °C. Since *sub*100 has been reported to be temperature-sensitive for growth (Freimuth and Ginsberg 1986), the analysis was repeated at both the permissive (32 °C) and nonpermissive (39,5 °C) temperatures (data not shown). The transcriptional defect was observed at all temperatures tested. Presumably the pTP encoded by *sub*100 displays a temperature-dependent defect in its DNA replication function and a temperature-independent defect in its early transcriptional function.

In other experiments, *sub*100 was shown to be defective for transcriptional activation of both the E2A gene and the RNA polymerase III-transcribed VA RNA genes, even in the presence of normal levels of E1A gene products (data not shown).

Thus, *sub*100 was unable to activate transcription of its early genes rapidly and efficiently. This defect could not be complemented in *trans*, consistent with a lesion in the covalently attached TP which functions in *cis* on infecting viral chromosomes.


## 2.2 TP Mediates Nuclear Matrix Association of Adenovirus DNA

Several early events in the infectious process were monitored to investigate the basis for the transcriptional defect displayed by *sub*100. More *sub*100 than *dl*309 reached the nucleaus when equal multiplicities of infection were used, indicating a high particle/pfu ratio for *sub*100. The DNAs became sensitive to DNaseI, with similar kinetics (data not whown). Therefore, it is very likely that *sub*100 DNA reaches the nucleus and is unpacked normally.

The association of mutant and wild-type viral DNAs with the nuclear matrix was next tested using the lithium diiodosalicylate (LIS) extraction procedure of MIRKOVITCH et al. (1984). Normally, in this procedure, nuclei are prepared and extracted with LIS to remove the bulk of histone and many other nonhistone proteins, and then nuclear DNA is digested with a restriction endonuclease. DNA sequences that remain with the nucleus after digestion are defined as nuclear scaffold associated. Accordingly, HeLa cell nuclei were prepared and extracted with LIS 4 h after infection with either *sub*100 or *dl*309. Extracted DNAs were bound to nitrocellulose using a slot blot apparatus and assayed for the presence of Ad sequences by hybridization with $^{32}$P-labeled adenovirus DNA (Fig. 3). Although DNAs in extracted nuclei were digested with a restriction enzyme, this step was not necessary to observe a dramatic difference in scaffold association between *sub*100 and *dl*309 DNAs. The majority of *sub*100 DNA was released during the LIS extraction step while *dl*309 DNA was not. This difference was observed at both 32 °C and 39.5 °C, indicating that the scaffold-association phenotype of *sub*100 was not temperature dependent.

The intensities of bands in Fig. 3 were quantified by densitometric scanning of appropriate autoradiographic exposures. The ratios of the amount of viral DNA released by LIS extraction to that remaining with the nuclear scaffold were determined. The averages for the 32 °C and 39.5 °C experiments are shown in Table 1. The ratio of LIS supernatant DNA to LIS pellet DNA was 5 for *sub*100 and 0.17 for *dl*309.

These results establish a new function for TP. It starts very early after infection to active viral transcription. Apparently, it performs its transcriptional role by mediating an association between the nuclear scaffold and the viral chromosome. A great deal of circumstantial evidence supporting a role for nuclear matrix attachment of DNA in active transcription has been accumulated (e.g., PARDOLL and VOGELSTEIN 1980; NELKIN et al. 1980; JACKSON et al. 1981; COOK et al.

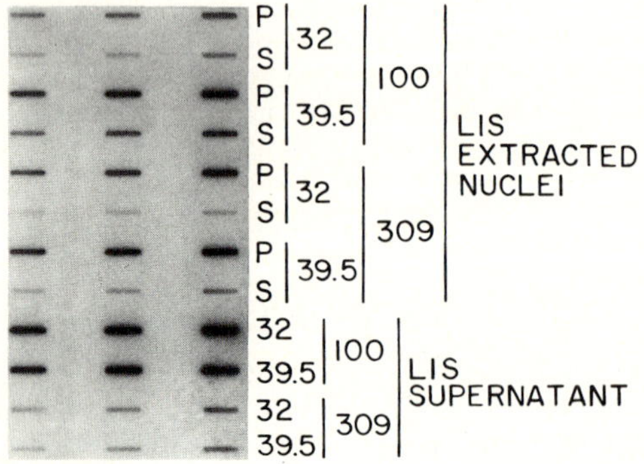


**Fig. 3.** Slot blot analysis of wild-type and mutant viral DNAs fractionated by the LIS extraction procedure. DNAs from the LIS supernatant or from residual LIS-extracted nuclei (*P*, pellet; *S*, supernatant after *Hind*III cleavage of nuclear DNA) were applied to nitrocellulose. In each case three samples of each DNA were applied which contained two-fold increments in total DNA concentration (least DNA in left-hand slots). The bound DNAs were then analyzed by hybridization with $^{32}$P-labeled adenovirus DNA. The numbers *32* and *39.5* designate the temperatures (in °C) at which infected cells were maintained prior to analysis

**Table 1.** Qauntification of viral DNAs released by LIS extraction

| Virus | LIS supernatant | LIS pellet | Supernatant/pellet |
|---|---|---|---|
| *sub*100 | 1 | 0.2 | 5 |
| *dl*309 | 0.05 | 0.3 | 0.17 |

The density of the *sub*100 band obtained for the LIS supernatant was arbitrarily set to 1, and the other intensities are presented relative to this value. The densities are each the average of two independent experimental determinations, one at 32 °C and the other at 39.5 °C.

1982; ROBINSON et al. 1983; NELSON et al. 1986). Our results with the adenovirus TP provide the first experimental evidence that altering the strength of the DNA-nuclear matrix association affects the efficiency of transcription.

It is interesting to note that the mutation in *sub*100 is located in the region of the E2B gene encoding the N-terminal domain of pTP (Fig. 1). This places the altered amino acids within the portion of pTP which is no longer covalently attached to viral DNA after the cleavage that occurs during virion maturation. Since the *sub*100 mutation affects a function which occurs after the cleavage event during a subsequent round of infection, it is very likely that the N-terminal portion of pTP remains associated with TP after the cleavage occurs.

*Acknowledgements.* We thank P. Freimuth and H. Ginsberg for the gift of *sub*100. This work was supported by a grant from the National Institutes of Health (CA 41086). J. Schaack was a postdoctoral fellow of the Jane Coffin Childs Memorial Fund for Medical Research, and T. Shenk is an American Cancer Society Research Professor.

# References

Challberg MD, Kella TJ (1981) Processing of the adenovirus terminal protein. J Virol 38: 272–277

Cook PR, Lang J, Hayday A, Lania L, Fried M, Chiswell DJ, Wyke JA (1982) Active viral genes in transformed cells lie close to the nuclear cage. EMBO J 4: 447–452

Desiderio SV, Kelly TJ (1981) The structure of the linkage between adenovirus DNA and the 55,000 dalton terminal protein. J Mol Biol 145: 319–337

Freimuth PI, Ginsberg HS (1986) Codon insertion mutants of the adenovirus terminal protein. Proc Natl Acad Sci USA 83: 7816–7820

Hearing P, Shenk T (1985) Sequence-independent autoregulation of the adenovirus type 5 E1A transcription unit. Mol Cell Biol 5: 3214–3221

Jackson DA, McCready SJ, Cook PR (1981) RNA is synthesized at the nuclear cage. Nature 292: 552–555

Jones N, Shenk T (1979) Isolation of Ad5 host range deletion mutants defective for transformation of rat embryo cells. Cell 17: 683–689

Kelly TJ (1984) Adenovirus DNA replication. In: Ginsberg H (ed) The Adenoviruses. Plenum, New York, pp 271–307

Melton DA, Krieg PA, Rebagliati MR, Maniatis T, Zinn K, Green MR (1984) Efficient in vitro synthesis of biologically active RNA and RNA hybridization probes from plasmids containing a bacteriophage SP6 promoter. Nucleic Acids Res 12: 7035–7056

Mirkovitch J, Mirault ME, Laemmli UK (1984) Organization of the higher-order chromatin loop: specific DNA attachment sites on nuclear scaffold. Cell 39: 223–232

Nelkin BD, Pardoll DM, Vogelstein B (1980) Mapping sequences in loops of nuclear DNA by their progressive detachment from the nuclear cage. Nucleic Acids Res 8: 2895–2908

Nelson WG, Plenta KJ, Barrack ER, Coffey DS (1986) The role of the nuclear matrix in the organization and function of DNA. Annu Rev Biophys Biophys Chem 15: 457–475

Pardoll DM, Vogelstein B (1980) Sequence analysis of nuclear matrix associated DNA from rat liver. Exp Cell Res 128: 466–470

Rekosh DMK, Russell WC, Bellet AJD, Robinson AJ (1977) Identification of a protein linked to the ends of adenovirus DNA. Cell 11: 283–295

Robinson AJ, Bellett AJD (1974) A circular DNA-protein complex from adenoviruses and its possible role in DNA replication. Cold Spring Harbor Symp Quant Biol 39: 523–531

Robinson AJ, Younghusband HB, Bellett AJD (1973) A circular DNA-protein complex from adenoviruses. Virology 56: 54–69

Robinson SI, Small D, Idzerda R, McKnight GS, Vogelstein B (1983) The association of transcriptonally active genes with the nuclear matrix of the chicken oviduct. Nucleic Acids Res 11: 5113–5130

Smart JE, Stilman BW (1982) Adenovirus terminal protein precursor: partial amino acid sequence and the site of covalent linkage to virus DNA. J Biol Chem 257: 13499–13506

Stillman B (1983) The replication of adenovirus DNA with purified proteins. Cell 35: 7–9

Stillman BW, Lewis JB, Chow LT, Mathews MB, Smart JE (1981) Identification of the Gene and mRNA for the adenovirus terminal protein precursor. Cell 23: 497–508

# Transcriptional Activation of Adenoviral Early Genes

J. W. Lillie, T. Hai, W. J. Coukos, K. A. W. Lee, K. J. Martin, and M. R. Green

## 1 Introduction

Adenovirus E1A proteins play an essential role in both productive infection and cellular transformation. As an oncogene, E1A induces cellular DNA synthesis in growth-arrested cells, and in cooperation with other oncogenes transforms non-permissive primary cells (for a review, see Graham 1984). During productive infection, E1A is required to activate transcription of early viral genes (reviewed in Berk 1986). E1A proteins also regulate the transcription of some cellular genes. The expression of some cellular genes is activated, while the expression of others is repressed. Transcriptional regulation of cellular genes is thought to account for E1A's biological effects (see Kingston et al. 1985).

The mechanism by which E1A activates transcription is unknown. E1A activates transcription of genes with divergent promoter sequences transcribed by both polymerase II and III. There is evidence that E1A might activate transcription by modifying a sequence-specific DNA binding protein (Kovesdi et al. 1986), but many promoters responsive to E1A induction do not contain binding sites for this factor. Some genes transcriptionally activated by E1A do, however, have common promoter sequences: for example, early viral genes E2, E3, and E4 have ATF sites (Lee et al. 1987). In addition, certain TATA-box sequences may be sufficient for transcriptional activation by E1A (Green et al. 1983; Wu et al. 1987; Simon et al. 1988). But in contrast to most cellular transcriptional activators (for a review, see Ptashne 1986), E1A appears to function without directly interacting with specific promoter sequences (reviewed in Berk 1986).

## 2 Two Major Early E1A Proteins

One potential source of confusion is the possibility that E1A activates transcription by more than one mechanism. There are two major early E1A proteins, 243- and 289-amino acid (AA), nuclear-localized phosphoproteins, which have identical N- and C-termini and differ only by 46 amino acids unique to the 289-AA protein (Perricaudet et al. 1979). Both proteins contain two short distinct stretches of amino acids conserved among adenovirus subgroups and species (Ormondt et al. 1980;

Department of Biochemistry and Molecular Biology, Harvard University, Cambridge, MA, USA

Current Topics in Microbiology and Immunology, Vol. 144
© Springer-Verlag Berlin · Heidelberg 1989

KIMELMAN et al. 1985) referred to as regions 1 and 2 (see MORAN and MATHEWS 1987). In addition, a third conserved region, region 3, is specific to the 289-AA protein. Region 3 is necessary and sufficient for transcriptional activation of early viral genes by E1A (LILLIE et al. 1986, 1987; MORAN et al. 1986). Although regions 1 and 2 do not appear to be required for E1A-mediated activation of early viral gene transcription, the transcription of some other genes appears to be activated by the 243-AA protein (see, for example, ZERLER et al. 1987). Since regions 1 and 2 share no sequences with region 3, it seems likely that transcriptional activation by these regions proceeds by a different mechanism.

## 3 E1A-GAL4 Fusions

Although it seems unlikely that the E1A region 3 contains a sequence-specific DNA binding activity, this region could interact with other transcription factors at the promoter either by binding directly to one or more transcription factors or by binding DNA non-specifically (CHATTERJEE et al. 1988). Once localized at the promoter, E1A might become a stable component of the transcription complex or might transiently induce the modification of a transcription factor. To test whether E1A could function directly at a specific promoter we asked whether increasing its concentration would improve its ability to activate transcription from that promoter. To this end we have constructed a plasmid that expresses E1A amino acids 121–223 (including region 3) fused to the sequence-specific DNA-binding domain of the yeast *GAL*4 protein. We reasoned that fusing E1A to a sequence-specific DNA binding protein would increase its concentration at a promoter bearing the appropriate DNA binding site. In fact, in mammalian cells the hybrid E1A-*GAL*4 protein induces much higher levels of transcription from target promoters containing *GAL*4 sites than from the promoters lacking *GAL*4 sites. Mutations of E1A that impair its normal ability to activate transcription also decrease activation by E1A-*GAL*4, suggesting that the wild-type and fusion proteins function by the same mechanism. These preliminary results suggest that E1A may function directly at the promoter. Experiments are underway to determine whether the fusion protein becomes an integral component of the transcription complex.

## 4 Transcription Factor ATF

Although no specific DNA sequence or transcription factor that mediates E1A inducibility has yet been identified, we have continued our efforts to define factors required for transcription of early viral genes. We have recently identified a cellular factor, ATF, that is required for transcription of multiple E1A-inducible adenovirus promoters (E4, E2A, and E3; LEE et al. 1987). ATF binds to sequences containing the conserved CGTCA motif, and ATF binding sites are required for transcription of the E4, E2A, and E3 promoters. For the adenovirus E4 promoter, binding sites for ATF are important for E1A inducibility. In addition to E1A-inducible promoters, ATF binding sites are also found in cAMP-inducible promoters. Furthermore, ATF binding sites have been implicated in the cAMP inducibility of those genes (see LIN and GREEN 1988 and references therein). Therefore, a common *cis*-acting element, the ATF binding site, is involved in both E1A- and cAMP-inducible transcription.

As a step toward understanding the regulation of ATF at the biochemical level, we purified it (HAI et al. 1988). In previous studies, ATF had been identified as a single 43–45K polypeptide (MONTMINY and BILEZSIKJIAN 1987; HURST and JONES 1987; LIN and GREEN 1988). In contrast, our purification yields not a single protein but rather a series of polypeptides. The highly purified ATF fraction contains multiple polypeptides which migrate in two groups on a gradient SDS-polyacrylamide gel: one group (consisting of two bands) migrates with apparent molecular weights around 47K and the other group (also consisting of two bands) migrates with apparent molecular weights around 43K. Several lines of evidence lead to the conclusion that those polypeptides represent multiple forms of ATF and another transcription factor, AP-1, whose recognition sequence, GTGAGT $\frac{C}{A}$ A, differs from the ATF consensus, GTGACGT $\frac{C}{A}$ A, by the absence of a cytosine residue (see HAI et al. 1988 and references therein).

Our results indicate that the upper bands in each set are closely related and appear to represent two forms of ATF, whereas the lower bands in both sets are closely related and appear to be different forms of AP-1. These conclusions are based on the observations that the upper bands and the lower bands have preferential affinities for their cognate binding sites, have differential reactivities toward the AP-1-specific pep2-specific antiserum (from P. Vogt), and have distinctive abilities to be substrates for the cAMP-dependent protein kinase (protein kinase A) in vitro. Only the upper bands can be efficiently phosphorylated by protein kinase A. We referred to the upper bands as ATF-43 and ATF-47, and the lower bands as AP-1–43 and AP-1–47.

The copurification of AP-1 with ATF is presumably due to the cross-binding of AP-1 to the ATF site, which is present at high concentrations on the affinity column. The similar ATF and AP-1 DNA recognition sites suggest a possible similarity between their DNA binding domains. We have investigated their relationship through an immunological assay and have found that ATF reacts with the antiserum against the DNA binding domain of AP-1 (c-jun-specific antiserum, from M. Karin). The immunological cross-reactivity of ATF and AP-1 suggests that they contain similar amino acid sequences in the DNA binding domains. Therefore, ATF and AP-1 can be grouped together as a family of transcription factors although their binding sites are found in promoters that respond to different regulatory signals. Proof of this point, however, must await cloning and sequencing of the gene that encodes ATF.

We have further investigated the structural relationship between the two forms of ATF. When ATF-43 and ATF-47 are $^{32}$P-labeled in vitro with protein kinase A followed by complete digestion with trypsin, both yield a single, labeled peptide. These two peptides fractionate identically on a high performance liquid chromatography column and have the same mobility on an SDS-polyacrylamide gel. These results indicate that ATF-43 and ATF-47 probably contain at least one common tryptic peptide and may have the same phosphorylation site(s) for protein kinase A.

We have also noted that the electrophoretic mobility difference between the two forms of ATF (the upper bands) is similar to that between the two forms of AP-1 (the lower bands). Thus, the different ATF and AP-1 proteins may be generated by a similar mechanism, such as alternative splicing or posttranslational modification. Alternatively, each ATF and AP-1 polypeptide may be encoded by a separate gene. It is of major interest to understand the functional significance of the multiple forms of ATF and AP-1. For example, ATF binding sites are found in both cAMP- and E1A-inducible promoters (LIN and GREEN 1988 and references therein). Perhaps the different forms of ATF mediate these distinct transcriptional regulatory responses.

It has been suggested that the induction of transcription by cAMP involves phosphorylation of ATF by protein kinase A (MONTMINY and BILEZSIKJIAN 1987). To test whether the induction of transcription by cAMP and/or E1A involves phosphorylation of ATF by protein kinase A, we analyzed the mobility of ATF by two-dimensional gel electrophoresis, involving isoelectric focusing in the first dimension and SDS-polyacrylamide gel electrophoresis in the second dimension. In contrast to a previous report no differences in phosphorylation of either ATF-43 or ATF-47 were detected from cells induced by cAMP or E1A. Therefore, how E1A or cAMP directly or indirectly affects ATF remains to be determined.

## 5 Other Factors

Currently, we are also attempting to identify other factors with which E1A interacts. In vitro assays have been developed for E1A transcriptional activation. The preliminary results indicate that a synthetic peptide consisting of the 49 amino acids of region 3 activates transcription of class II and III genes. This peptide can be radio-labeled in vitro, enabling its direct interactions with transcription factors to be examined. In addition, wild-type and mutant E1A proteins expressed and isolated from *Escherichia coli* are also being tested in the in vitro transcription assay.

## References

Berk AJ (1986) Adenovirus promoters and E1a transactivation. Annu Rev Genet 20: 45–79
Chatterjee PK, Bruner M, Flint SJ, Harter ML (1988) DNA-binding properties of an adenovirus 289R E1a protein. EMBO J 7: 835–842
Graham FL (1984) Transformation by and oncogenicity of human adenoviruses. In: Ginsberg H (ed) The Adenoviruses. Plenum, New York, pp 339–398
Green MR, Treisman R, Maniatis T (1983) Transcriptional activation of cloned human β-globin genes by viral immediate-early gene products. Cell 35: 137–148
Hai T, Liu F, Allegretto EA, Karin M, Green MR (1988) A family of immunologically related transcription factors that includes ATF and AP-1. Genes Dev 2: 1216–17
Hurst HC, Jones N (1987) Identification of factors that interact with the E1a-inducible adenovirus E3 promoter. Genes Dev 1: 1132–1146.
Kimelman D, Miller JS, Porter D, Roberts BE (1985) E1a regions of the human adenoviruses and of the highly oncogenic simian adenovirus 7 are closely related. J Virol 53: 399–409

Kingston RE, Baldwin AS, Sharp PA (1985) Transcription control by oncogenes. Cell 41: 3–5
Kovesdi I, Reichel R, Nevins JR (1986) Identification of a cellular transcription factor involved in E1A *trans*-activation. Cell 45: 219–228
Lee, KAW, Hai TY, SivaRaman L, Thimmappaya B, Hurst HC, Jones NC, Green MR (1987) A cellular protein, activating transcription factor, activates transcription of multiple E1a-inducible adenovirus early promoters. Proc Natl Acad Sci USA 84: 8355–8359
Lillie J, Green M, Green MR (1986) An adenovirus E1a protein region required for transformation and transcriptional repression. Cell 46: 1043–1051
Lillie JW, Loewenstein PM, Green MR, Green M (1987) Functional Domains of adenovirus type 5 E1a proteins. Cell 50: 1091–1100
Lin YS, Green MR (1988) Interaction of a common transcription factor, ATF, with regulatory elements in both E1a- and cyclic AMP-inducible promoters. Proc Natl Acad Sci USA 85: 3396–3400
Montminy MR, Bilezsikjian LM (1987) Binding of a nuclear protein to the cyclic AMP response element of the somatostatin gene. Nature 328: 175–178
Moran E, Zerler B, Harrison TM, Mathews MB (1986) Identification of separate domains in the adenovirus E1A gene for immortalization activity and the activation of virus early genes. Mol Cell Biol 6: 3470–3480
Moran E, Mathews MB (1987) Multiple functional domains in the adenovirus E1a gene. Cell 48: 177–178
Ormondt van H, Maat J, Dijkema R (1980) Comparison of nucleotide sequences of the early E1a regions for subgroups A, B, and C of human adenoviruses. Gene 12: 63–76
Perricaudet M, Akusjarvi G, Virtanen A, Rettersson U (1979) Structure of two spliced mRNAs from the transforming region of human subgroup C adenoviruses. Nature 281: 694–696
Ptashne M (1986) Gene regulation by proteins acting nearby and at a distance. Nature 322: 697–701
Simon MC, Fisch TM, Benecke BJ, Nevins JR, Heintz N (1988) Definition of multiple, functionally distinct TATA elements, one of which is a target in the hsp70 promoter for E1a regulation in preparation
Wu L, Rosser DSE, Schmidt MC, Berk A (1987) A TATA box implicated in E1a transcriptional activation of a simple adenovirus 2 promoter. Nature 326: 512–515
Zerler B, Roberts RJ, Mathews MB, Moran M (1987) Different functional domains of the adenovirus E1a gene are involved in regulation of host cell cycle products. Mol Cell Biol 7: 821–829

# Suppression of Cellular Gene Activity in Adenovirus-Transformed Cells

A. J. VAN DER EB, H. T. M. TIMMERS, R. OFFRINGA, A. ZANTEMA, S. J. L. VAN DEN HEUVEL, J. A. F. VAN DAM, and J. L. BOS

## 1 Introduction

The transforming and oncogenic potential of adenoviruses is localized in the early region 1 (E1), one of the regions of the viral genome expressed in the early phase of the lytic infection. E1 is approximately 4000 base pairs long and consists of two transcriptional units, E1A and E1B. In transformed cells E1A codes for two coterminal mRNAs specifying two related proteins, whereas E1B codes for one mRNA which is translated into two unrelated proteins (for reviews, see: BERNARDS and VAN DER EB 1984; BRANTON et al. 1985).

The molecular mechanism of transformation by adenoviruses is still unknown. One of the properties characteristic of transformed cells in general is that their control mechanisms of cell proliferation are disturbed to a greater or lesser extent. As this definition also applies to Ad-transformed cells, one could predict that regulation of cell growth is abnormal in these cells. TIMMERS et al. (1988a) have shown that Ad-transformed cells indeed are abnormal in their growth factor requirement: Ad5-transformed cells are capable of proliferating in the complete absence of serum and hence independently of growth factors (GF), whereas Ad12-transformed cells require insulin or insulin-like growth factor-I (IGF-I) but not other factors. The (partial) lack of GF requirement can be explained in two ways: either the transformed cells produce their own GFs (autocrine growth stimulation) or the transforming genes somehow can mimic the effects that are brought about by GFs. Since we have been unable to find evidence for the production of GFs by Ad-transformed cells (unpublished observations), we tested whether the GF-induced signal transduction pathways are still functional. In an initial approach we examined the expression of the serum-inducible genes JE, c-*myc*, and c-*fos*. Upon addition of serum or certain GFs to resting cells, expression of these genes is induced within 1–3 h, reaches a maximum, and then returns to a stable lower expression level except for c-*fos* which returns to zero. Since the induction of expression of these genes represents a late effect of signal transduction, alterations in this mechanism would be reflected in changes in the expression of these genes. As an extension of these studies we have also examined the expression of other genes that are known or suspected to have effects on the oncogenic phenotype of transformed cells. Finally, we report on the identification of a gene whose expression is completely

Department of Medical Biochemistry, Sylvius Laboratories, P.O. Box 9503, 2300 RA Leiden, The Netherlands

Current Topics in Microbiology and Immunology, Vol. 144
© Springer-Verlag Berlin · Heidelberg 1989

suppressed in cells that are oncogenic in immunocompetent animals but is normally expressed in other transformed cells, and on the effect on oncogenicity of the formation of a complex between the Ad5 55K E1B protein and the cellular p53 antigen.


## 2  Materials and Methods

### 2.1  Cells and Oncogenicity Assay

The following cell cultures have been used: primary cultures of baby rat kidney (BRK) cells, baby mouse kidney (BMK) cells, and human embryonic retinoblast (HER) cells, and the established rat cell lines NRK and 3Y1 and the mouse NIH 3T3 cell line, as well as the corresponding adenovirus E1- or E1A-transformed derivatives. The cells were grown in minimum essential medium (MEM), with or without extra amino acids and vitamins, supplemented with 8% newborn calf serum (NCS; BRK cells only) or 8% fetal calf serum (FCS; all other cell types). Oncogenicity was tested by injecting $3 \times 10^6$–$10^7$ cells subcutaneously into either nude mice (nu/nu) or syngeneic immunocompetent rats.


### 2.2  Northern Blotting and Run-on Analysis

Northern blotting analysis was carried out with total cytoplasmic RNA or poly(A)-selected RNA as described by SCHRIER et al. (1983). Nuclear run-on analysis was performed according to VAESSEN et al. (1987).


### 2.3  Immunoprecipitations and Quantitative Immunofluorescence

Immunoprecipitation of the 22K and 27K proteins was carried out according to ZANTEMA et al. (1985b). Quantitative immunocytochemical staining of the p53 T antigen was carried out by means of cytofluorometry using an MPVII microgluoro-meter (Leitz, Wetzlar). The measuring diaphragm was adjusted in such a way that the fluorescein isothiocyanate (FITC) immunofluorescence of the nucleus and the fluorescence of other regions in the cell were measured independently.


### 2.4  Clones and Antibodies

DNA fragments used as hybridization probes were derived from the following clones:

pR11.6 containing the genomic rat c-*myc* sequence was obtained from G. Klein, Stockholm.

pRGAPDH-13 containing the rat glyceraldehyde 3-phosphate dehydrogenase cDNA was obtained from J.-H. Blanchard, Montpellier.

pBC-JE containing the mouse JE cDNA was provided by C. Stiles, Boston.

pTR-1 containing a 1.6-kb fragment of the rat transin (stromelysin) cDNA was obtained from R. Breathnach, Strasbourg.

cDNAs spanning the entire coding region of the human fibronectin gene were obtained from F. E. Baralle, Oxford.

pHS208 containing the human hsp27 cDNA was obtained from E. Hicky and L. A. Weber, Tampas.

UMH213 containing the mouse *hsp*70 cDNA was provided by H. Pelham, Cambridge, UK.

The hybridoma cell line producing the rat monoclonal antibody 2C8 was isolated by A. Zantema (ZANTEMA et al. 1985b) as was the hybridoma line producing antibody against the Ad5-specific 55K-E1B protein (unpublished data).

The hybridoma cell line producing the p53-specific monoclonal antibody 122 was originally isolated by E. G. Gurney and was obtained from the American type culture collection.


# 3 Results


## 3.1 Expression of the Serum-Inducible Genes JE, c-*myc* and c-*fos*


As mentioned in the introduction, adenovirus (Ad)-transformed cells are partially or almost completely independent of serum GFs for their proliferation (TIMMERS et al. 1988a). Since no evidence has been found that this behaviour is caused by an autocrine mechanism, an alternative explanation could be that the viral transforming genes somehow activate an intermediate step in the GF-induced signalling pathway, which would also render the cells independent of GF. To test whether the signal transduction path way is constitutively activated in Ad-transformed cells, we examined the expression of the GF-inducible genes JE, c-*myc*, and c-*fos*. The proto-oncogenes c-*myc* and c-*fos* are (transiently) activated when cells are exposed to serum or PDGF (KELLY et al. 1983; GREENBERG et al. 1983). JE is a gene isolated by STILES and coworkers (COCHRAN et al. 1983) as a PDGF-inducible gene. It is also activated by poly(IC) as well as by interferon.

To investigate whether JE and c-*myc* mRNA are expressed in growing Ad-transformed cells Northern blotting analysis was carried out with primary cultures of BRK, BMK, and HER cells and with the established NRK and 3T3 cell lines, as well as with the corresponding Ad5- and Ad12-transformed derivatives. It was found that the JE mRNA was greatly reduced or undetectable in the Ad-transformed cells, whereas it was present in the untransformed parental cells (TIMMERS et al. 1988b). Similar results were obtained with the c-*myc* gene, although the residual levels of c-*myc* RNA were somewhat higher and the reduction was variable in the Ad12-transformed cells (TIMMERS et al. 1988b). The results are summarized in Table 1. Since the suppression also occurred in E1A-transformed cells, the effects must be due to the activity of the E1A region. These results show that two genes active in growing cells are not expressed in proliferating Ad-transformed

**Table 1.** Expression of JE and c-*myc* mRNA in primary cultures of BRK, BMK, and HER cells, and the established cell lines NRK, NIH 3T3, and 3Y1, as well as in the Ad5 and Ad12 E1-transformed derivatives

| Cell type | Untransformed | Transformed by | |
|---|---|---|---|
| | | Ad5 E1 | Ad12 E1 |
| *JE expression* | | | |
| BRK | + + | — | — |
| BMK | + + | — | — |
| HER | —[a] | — | — |
| NRK | + + + | + /— | + /— |
| NIH 3T3 | + + | + /— | n.d. |
| 3Y1 | — | — | n.d. |
| *c-myc expression* | | | |
| BRK | + + | + /— | + /— or + + |
| BMK | + + | + /— | + /— or + + |
| HER | + + | + /— | + /— or + + |
| NRK | + + | + /— | + + |
| NIH 3T3 | + + | + /— | n.d. |
| 3Y1 | + + | + /— | n.d. |

—, not detectable; + /—, +, + +, + + +, indicate relative expression levels; n.d., not done; BRK, baby rat kidney; BMK, baby mouse kidney; HER, human embryonic retinoblast.

[a] The lack of a signal in the human cells might be due to a low level of sequence similarity between the human gene and the mouse JE probe.

cells and hence apparently are dispensable. If one assumes that the c-*myc* protein (and possibly the JE protein) has an essential function in the control of cell proliferation, this result implies that in Ad-transformed cells this function is taken over by the viral E1A gene. The inhibition of JE and c-*myc* mRNA expression can be explained in at least two ways. Firstly, E1A blocks the signal transduction pathway(s) somewhere between the GF-receptor and JE or c-*myc*. As a result, no growth-stimulating signals will reach the genes, and therefore they remain inactive. Secondly, the signal transduction pathway(s) are unaffected, but E1A somehow inhibits expression of JE, c-*myc*, and possibly other genes with a similar function. In both cases, E1A must be able to replace functionally the genes that are inhibited in expression.

Further studies using nuclear run-on assays show that inhibition of JE expression probably occurs at the level of transcription initiation. To investigate whether expression of JE mRNA could still be induced with serum or the tumor-promoting agent (TPA) after a period of serum starvation, Ad-transformed NRK cells were incubated in medium containing 0.5% serum for 16 h, after which FCS or TPA was added to final concentrations of 8% and 100 ng/ml, respectively. It was found that expression of JE mRNA could still be induced, although at a very low level (Fig. 1). Basically similar results were obtained with the c-*myc* gene. Further work showed that after serum starvation the c-*fos* gene could be activated to the

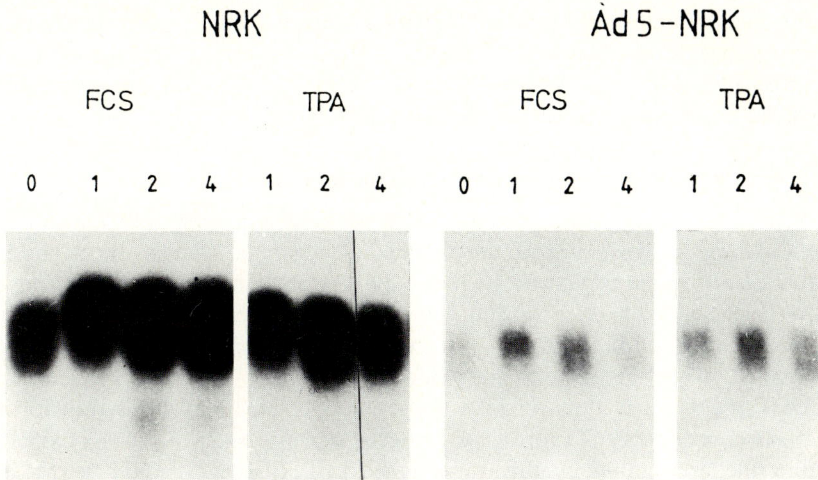


**Fig. 1.** Induction of JE mRNA expression by serum or TPA in *NRK* cells and *Ad5* E1-transformed *NRK* cells. Freshly confluent monolayers were cultured in medium containing 0.5% fetal calf serum for 24 h. Then, 8% fetal calf serum or TPA 100 ng/ml was added. RNA was isolated after *1*, *2*, and *4* h, and 20 µg of each RNA sample was analyzed for the presence of JE RNA by Northern blotting, using a radiolabeled JE cDNA clone as a probe

same extent in Ad-transformed cells as in untransformed cells (results not shown). Apparently, c-*fos* induction is not affected in Ad-transformed cells, indicating that the signal transduction pathways responsible for activation of c-*fos* are still functional.

## 3.2 Expression of the Fibronectin and the Stromelysin Genes

Previous studies have shown that oncogenicity of Ad12-transformed cells in immunocompetent animals can be explained, at least in part, by the suppression of class I MHC antigen expression (SCHRIER et al. 1983; BERNARDS et al. 1983; VAN DER EB and OOSTRA 1985). Here, we discuss other genes whose expression is modulated in Ad-tranformed cells and which may also contribute to the neoplastic behavior or the transformed phenotype.

### 3.2.1 Expression of the Fibronectin Gene

Fibronectin (FN) is an extracellular matrix protein which is produced in large quantities in normal fibroblasts and other untransformed cells. FN expression is reduced in most tumor cells, which may be one of the reasons why such cells often attach poorly to tissue-culture dishes. In Ad-transformed rat cells, FN mRNA expression is reduced approximately five-fold, but in cells transformed by Ad5 E1A + activated EJ*ras* we found that expression of FN mRNA is undetectable in nuclear run-on and Northern blotting analyses. These latter cells show a rounded morphology, attach poorly to plastic culture dishes, and are extremely oncogenic, both in nude mice and in immunocompetent syngeneic rats (JOCHEMSEN et al. 1988). It appeared of interest, therefore, to study the effect of re-expression of FN on the transformed and oncogenic phenotype of these cells. Preliminary results have shown that FN-expressing Ad5 E1A + EJ*ras* cells can efficiently attach to plastic dishes, but that the cells have not lost their oncogenicity in nude mice.

### 3.2.2 Expression of the Stromelysin or Transin Gene

Stromelysin (sml) is an excreted protease which presumably has a function in modulating the structure of the extracellular matrix. Similar functions have been attributed to other proteases, such as collagenases. Untransformed normal cells have low levels of expression of the *sml* gene, whearease *ras*- or *src*-transformed cells produce large amounts of the *sml* gene product. *Sml* gene expression is transiently activated in serum-starved, untransformed cells by serum or certain GFs (EGF, PDGF, FGF) as well as by TPA (MATRISIAN et al. 1986). Cells transformed by an activated *ras* oncogene express the *sml* gene constitutively, a process mediated by the autocrine production by the transformed cells of the GF TGF-α (MCKAY et al. 1986).

In contrast to *ras*-transformed cells, cells transformed by Ad5 or Ad12 do not express the *sml* gene at detectable levels, suggesting that the Ad-transforming genes suppress the expression of this protease gene. The suppressing effect by Ad-transforming genes was confirmed by the observation that rat cells transformed by a combination of EJ*ras* plus Ad E1A show a greatly reduced expression of *sml* compared with its expression in cells transformed by EJ*ras* only (Fig. 2A). To investigate whether

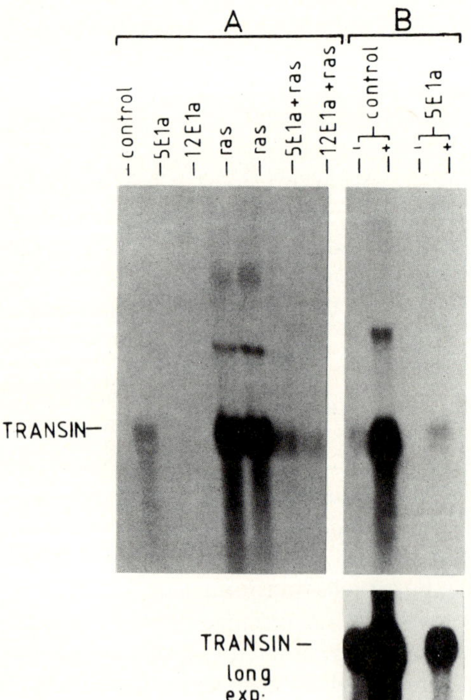


Fig. 2A, B. Adenovirus E1A suppresses transin (stromelysin) mRNA expression in NRK cells but does not prevent a low level of induction by EJ*ras* or TPA. Cytoplasmic RNA was isolated from subconfluent cell cultures grown on minimal essential medium with 8% fetal calf serum (A), or from confluent cell cultures subjected to serum starvation for 24 h and subsequently exposed to serum-free medium with (+) or without (—) TPA 100 ng/ml for 16 h (B). Then 10 μg of each RNA sample was analyzed for expression of transin mRNA by standard Northern blotting procedures using a radiolabeled transin cDNA clone as a probe

Ad E1A can also suppress the transient activation of *sml* gene expression after serum starvation, untransformed and Ad E1A-transformed NRK cells were serum-starved for 24 h, after which 8% FCS or TPA 100 ng/ml was added for 16 h. Figure 2 B shows that *sml* expression is strongly activated by serum or TPA in normal NRK cells but that a low level of expression is also observed in E1A-transformed NRK cells. This suggests that Ad E1A can reduce the basal expression level of *sml* but not the induction phenomenon itself. Nuclear run-on experiments show that the effects on cytoplasmic mRNA levels are caused by differences in the rate of transcription initiation, and hence reflect changes in promoter activity.

### 3.3 Inhibition of Expression of the 27K Heat Shock Gene Correlates with Oncogenicity in Immunocompetent Animals

We have recently isolated a monoclonal antibody, 2C8, which immunoprecipitates a 22K and a 27K protein from [³H] leucine-labeled BRK cells transformed by nononcogenic Ad5. The proteins could not be precipitated from cells transformed by oncogenic Ad12 (Fig. 3 A). Gel filtration studies indicate that 2C8 recognizes the two proteins only when they form a complex, which has an apparent molecular weight of 700K. Since the 27K protein shows properties characteristic for the 27K (28K) heat shock protein, we investigated whether the 27K polypeptide precipitated by 2C8 represents the heat shock protein. Comparison of the proteins by two-dimensional gel analysis and heat-shock experiments shows that the two are identical. The 27K protein recognized by 2C8 indeed increased in concentration in Ad5-transformed cells after heat shock, but remained undetectable under these conditions in lysates of Ad12-transformed cells. The inhibition of 27K/*hsp27* expression not only occurred at the protein level but also at the mRNA level since no *hsp27* mRNA could be detected in Ad12-transformed cells (Fig. 3 B), even not after a heat

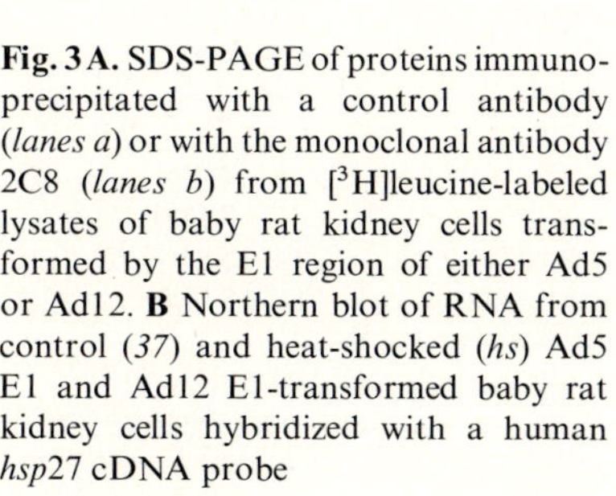


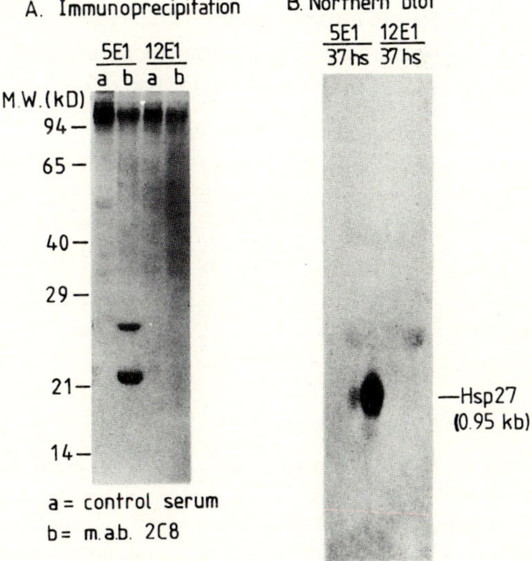


**Fig. 3 A.** SDS-PAGE of proteins immuno-precipitated with a control antibody (*lanes a*) or with the monoclonal antibody 2C8 (*lanes b*) from [³H]leucine-labeled lysates of baby rat kidney cells transformed by the E1 region of either Ad5 or Ad12. **B** Northern blot of RNA from control (*37*) and heat-shocked (*hs*) Ad5 E1 and Ad12 E1-transformed baby rat kidney cells hybridized with a human *hsp27* cDNA probe

shock when *hsp*70 mRNA was readily detectable. This phenomenon resembles the inhibition of class I MHC expression which occurs only in Ad12-transformed cells. The two phenomena differ, however, in that class I MHC is not suppressed in Ad12 E1A + EJ*ras*-transformed cells, whereas the 27K protein is undetectable in these cells. In fact, the absence of the 27K/*hsp*27 protein correlates more strictly with oncogenicity in immunocompetent animals than absence of class I MHC antigens. Further work must focus on the possible significance of the lack of expression of the 27K/*hsp*27 gene, and on the mechanism of suppression.

## 3.4 The Role of the p53-55K E1B Complex in Oncogenicity

In cells transformed by nononcogenic Ad5, part of the 55K E1B protein occurs in a complex with the cellular p53 T antigen (SARNOW et al. 1982; ZANTEMA et al. 1985a). Most of this complex is concentrated in a discrete cytoplasmic body localized close to the nuclear membrane. A similar complex does not occur in cells transformed by the oncogenic Ad12 (ZANTEMA et al. 1985a). The precise nature of the cytoplasmic body is unknown, but experimental results so far suggest that it consists of the p53-55K complex arranged in filamentlike structures (ZANTEMA et al. 1985a, b), although the presence of other proteins has not been excluded. Further work showed that the cytoplasmic body is missing in rat 3Y1 cells transformed by Ad5. However, by selecting transformed 3Y1 clones with a high E1B expression level we found cells which did contain these structures. Analysis of the concentrations of the p53 and the 55K E1B proteins in these transformed 3Y1 lines showed that the presence of cytoplasmic bodies was correlated with the concentration of the 55K protein: the bodies are only present when the concentration of the E1B protein exceeds a certain threshold level. To investigate whether the presence of the cytoplasmic bodies influences oncogenicity, $3 \times 10^6$ transformed cells were injected into nude mice. The results indicate that cells containing cytoplasmic bodies are less oncogenic than cells lacking the bodies, in the sense that the average latency periods for tumor formation are much longer. The tumors that finally arose in the animals injected with cytoplasmic body-containing cells did not carry the bodies any longer, suggesting that selection had taken place for a loss of these structures. Quantitative immuno-

Table 2. Tumorigenicity of Ad5-transformed 3Y1 cell lines in nude mice

| Cell line | E1B-55k level[a] | Cytoplasmic body | p53, Nuclear concentration[b] | Tumorigenicity latency period (weeks) |
|---|---|---|---|---|
| untransformed | — | — | + | — |
| 3 × 3 | + + | — | + + + | 6– 7 (4/4[c] |
| 3 × 4 | + + | — | + + + | 6– 7 (4/4) |
| 3 × 5 | + + + | ± | + + | 14–15 (3/4) |
| 3 × 6 | + + + + | + | ± | 14–15 (4/4) |

[a] E1B-55K protein levels were determined by Western blotting.
[b] Relative concentration of nuclear p53, as determined by quantitative immunofluorescence.
[c] No. of animals with tumors per total number injected.

fluorescence results indicate that the concentration of the p53 protein in the nucleus is much lower in cells carrying cytoplasmic bodies than in cells lacking them, suggesting that a correlation exists between oncogenicity and free p53. The results are summarized in Table 2.

# 4 Discussion

In the present paper we report that transformation by human adenoviruses (Ad) 5 and 12 is accompanied by an inhibition of expression of a number of cellular genes, including c-*myc*, JE, stromelysin (*sml*), and fibronection (FN). The observation that adenovirus transformation is associated with suppression of gene activity is in agreement with results of mutational studies of region E1A showing that the sequences of E1A that are responsible for cell transformation are also associated with suppression of transcriptional enhancers (BORRELLI et al. 1984; VELCICH and ZIFF 1985; HEN et al. 1985). Since only little is known about the functions of the genes that are suppressed in Ad-transformed cells, the significance of the phenomena is difficult to assess. The c-*myc* gene presumably plays a role in the regulation of cell proliferation, whereas the excreted protease stromelysin may have functions in changing the extracellular matrix and hence could also play a role in cell proliferation. Fibronectin is an extracellular matrix protein which influences cell shape and/or attachment to other cells or substrates. The function of JE is still unknown, but the amino acid sequence deduced from the primary structure of the cDNA predicts that it may be an excreted protein (C. Stiles, personal communication). Although expression of JE and *sml* is strongly suppressed or even undetectable in Ad-transformed cells, the genes are nevertheless still inducible by serum, after preincubation in serum-free medium. This suggests that E1A does not interfere with the GF-mediated induction phenomenon but only reduces the expression levels by a certain factor, irrespective of whether expression is low or high. The observation that the c-*fos* gene can be induced by serum to normal levels implies that the GF-mediated signal transduction pathway (at least the part that activates c-*fos*) is basically intact and that E1A does not interfere at this level.

Since the c-*myc* gene appears to have a function in the control of cell proliferation and *myc* mRNA is normally expressed in growing cells, the question can be asked how Ad-transformed cells can proliferate in the absence of c-*myc* expression. A plausible explanation could be that E1A products are able to bypass either directly or indirectly the c-*myc* protein and take over its function in the control of cell proliferation. If this hypothesis is correct, it also may explain why cells transformed by Ad E1 can grow in the absence of GFs (except for IGF-I or insulin, which is still required for Ad12-transformed cells). The level at which the E1A proteins act is not known. The fact that expression of c-*myc* is inhibited, particularly in Ad5-transformed cells, predicts that the site of interaction is "downstream" of *myc*.

So far, nothing is known about the mechanism of suppression of the gene of the 27K/*hsp*27 protein (and possibility the 22K protein). Expression of this heat shock gene is inhibited only in cells that are oncogenic in immunocompetent animals, and as such it resembles the class I MHC genes which are suppressed by Ad12 E1A (provided E1A is expressed at a sufficiently high level). However, the 27K/*hsp*27 gene is also inhibited in cells transformed by Ad12 E1A + EJ*ras* or Ad5 E1A + EJ*ras*, as opposed to the class I MHC genes which are not inhibited in these cells. This argues against an identical mechanism of suppression of the *hsp*27 gene and the class I MHC genes.

Apart from the question how E1A inhibits the expression of these divergent genes, a second aspect of interest is the biological significance of these phenomena for the transformed phenotype. The consequences of c-*myc* expression have already been discussed. Very little can be said about the effects of inhibition of expression of JE and the gene coding for the 27K/*hsp*27 protein, mainly because nothing is known about their normal function. Their possible role in transformation and oncogenicity will be investigated by reintroducing the genes into the transformed cells under conditions such that they are again expressed. Similar experiments are in progress with the *sml* gene and the FN gene. The effects of re-expression of the latter two genes on the metastasizing ability of the transformed cells will also be studied, since it has been suggested that excreted proteases may play a role in this phenomenon (GARBISA et al. 1987).

Finally, studies on the p53 cellular T antigen in Ad5-transformed cells have shown that the molecular complex of the p53 protein and 55K E1B antigen is concentrated in a discrete cytoplasmic body. We have shown that this body is not formed when the 55K-protein concentration drops below a certain level. Interestingly, Ad5-transformed 3Y1 cells lacking this cytoplasmic body are more strongly tumorigenic than cells containing this body, in the sense that the former type of cells induce tumors after a short latency period and the latter after a longer latency period. Quantitative immunofluorescence studies have shown that the concentration of the p53 antigen in the nucleus is low when the cytoplasmic bodies are present and high when these structures are absent. This result suggests that the 55K E1B protein level can influence the concentration of free nuclear p53 antigen, and thereby modulates the oncogenicity of the cells. The fact that the p53/55K E1B bodies do not occur in Ad12-transformed cells may possibly explain why the latter cells are always highly oncogenic in nude mice, with tumors appearing after short latency periods.

# References

Bernards R, Van der Eb AJ (1984) Adenovirus: transformation and oncogenicity. Biochim Biophys Acta 783:187–204

Bernards R, Schrier PI, Houweleling A, Bos JL, Van der Eb AJ, Zijlstra M, Melief CJ (1983) Tumorigenicity of cells transformed by adenovirus type 12 by evasion of T-cell immunity. Nature 305:776–779

Borrelli E, Hen R, Chambon P (1984) Adenovirus-2 E1 A products repress enhancer-induced stimulation of transcription. Nature 312:608–612

Branton PE, Bayley ST, Graham FL (1985) Transformation by human adenoviruses. Biochim Biophys Acta 780:67–94

Cochran BH, Reffel AC, Stiles CD (1983) Molecular cloning of gene sequences regulated by platelet-derived growth factor. Cell 33:939–947

Garbisa S, Pozzatti R, Muschel RJ, Saffiotti U, Ballin M, Goldfarb RH, Khoury G, Liotta LA (1987) Secretion of type IV collagenolytic protease and metastatic phenotype: induction by transfection with c-Ha*ras* but not c-Ha*ras* plus Ad2-E1A. Cancer Res 47: 1523–1528

Greenberg ME, Ziff EB (1984) Stimulation of 3T3 cells induces transcription of the c-*fos* proto-oncogene. Nature 311: 433–448

Hen R, Borrelli E, Chambon P (1985) Repression of immunoglobulin heavy chain enhancer by the adenovirus-2 E1A products. Science 230: 1391–1394

Jochemsen AG, Bernards R, Van Kranen HJ, Houweling A, Bos JL, Van der Eb AJ (1986) Different activities of the adenovirus types 5 and 12 E1A regions in transformation with the EJ Ha-*ras* oncogene. J Virol 59: 684–691

Kelly K, Cochran BH, Stiles CD, Leder P (1983) Cell-specific regulation of the c-myc gene by lymphocyte mitogens and platelet-derived factor. Cell 35: 603–610

Matrisian LM, Leroy P, Ruhlmann C, Gesnel MC, Breathnach R (1986) Isolation of the oncogene and epidermal growth factor-induced transin gene: complex control in rat fibroblasts. Mol Cell Biol 6: 1679–1686

McKay IA, Malone P, Marshall CJ, Hall A (1986) Malignant transformation of murine fibroblasts by a human c-Ha-*ras*-1 oncogene does not require a functional EGF-receptor. Mol Cel Biol 6: 3382–3387

Sarnow P, Shih HY, Williams J, Levine AJ (1982) Adenovirus E1B-58kD tumor antigen and SV40 large tumor antigen are physically associated with the same 54kD cellular protein in transformed cells. Cell 38: 387–394

Schrier PI, Bernards R, Vaessen RTMJ, Houweling A, Van der Eb AJ (1983) Expression of class I major histocompatibility antigens switched off by highly oncogenic adenovirus 12 in transformed rat cells. Nature 305: 771–775.

Timmers HTM, Van Zoelen EJJ, Bos JL, Van der Eb AJ (1988a) Cells transformed by adenovirus type 12 but not type 5 are dependent on insulin or IGF-1 for their proliferation. J Biol Chem 263: 1329–1335

Timmers HTM, De Wit D, Bos JL, Van der Eb AJ (1988b) E1A products of adenovirus reduce the expression of cellular proliferation-associated genes. Oncogene Res (in press)

Vaessen RTMJ, Houweling A, Van der Eb AJ (1987) Post-transcriptional control of class I MRC mRNA expression in adenovirus 12-transformed cells. Science 35: 1486–1488

Van der Eb AJ, Oostra BA (1985) Expression of class I major histocompatibility antigens in adenovirus-transformed cells. In: Doerfler W (ed) Adenovirus DNA, vol 10. pp 343–366

Velcich A, Ziff E (1985) Adenovirus E1A proteins repress transcription from the SV40 early promoter. Cell 40: 705–716

Zantema A, Schrier PI, Davis-Olivier A, Van Laar T, Vaessen RTMJ, Van der Eb AJ (1985a) Adenovirus serotype determines association and localization of the large E1B tumor antigen with cellular tumor antigen p53 in transformed cells. Mol Cell Biol 5: 3084–3091

Zantema A, Fransen JAM, Davis-Olivier A, Ramaekers FCS, Vocija GP, DeLeys B, Van der Eb AJ (1985b) Localization of the E1B proteins of adenovirus 5 in transformed cells as revealed by interaction with monoclonal antibodies. Virology 142: 44–58

# Recombination Between Adenovirus DNA and the Mammalian Genome

W. Doerfler, R. Jessberger, and U. Lichtenberg

## 1 Introduction

Eukaryotic cells, particularly mammalian cells, exhibit an amazingly high propensity to take up foreign DNA from the environment and to incorporate extraneously added genetic material into their own genomes. We are interested in the mechanism of foreign DNA integration and have used adenovirus DNA as a model for our investigations.

For the molecular biologist and geneticist, "viruses are packages of genes and genetic elements with millennia of biological experience" (Doerfler 1987). In keeping with the experimental promises of this definition, viral genomes have traditionally been used as models for investigating certain aspects of the molecular biology of eukaryotic systems. In contrast to the in general more conservative behavior of a cellular genome that is restrained not only in a cell but in a long-lived organism, viral genomes can be passed in rapid progression from cell to cell, from organism to organism, and — if the molecular properties permit — from species to species. In the course of such promiscuous perambulations viral genomes have constantly been subjected to varying demands and selective pressures. Moreover, these genomes have had the opportunity to incorporate into their own continuity foreign genetic information by legitimate and more frequently perhaps by illegitimate mechanisms. As a consequence, viral genomes today reflect a history of most interesting encounters which have been profitable to the survivors.

For the researcher, genomes with such sophistication of biological experience offer a challenge and a highly developed pool of different molecular mechanisms. In our own research on some of these mechanisms we have selected two formidable examples of viral genomes which have not only had a long-standing biological history but have also enjoyed several decades of acquaintance with molecular biologists. As a consequence, these genomes have already been studied in considerable detail. Our model genomes have been those of human adenovirus types 2 (Ad2) and 12 (Ad12).

Apparently, foreign DNA can be integrated at many different sites into the cellular genome. In our studies this impression was initially garnered from the results of restriction and blot hybridization analyses performed on Ad2- and Ad12-transformed cell lines or on Ad12-induced hamster tumors. The integration patterns

Institute of Genetics, University of Cologne, Weyertal 121, 5000 Cologne 41, FRG unravelled then suggested that adenovirus DNA had been integrated at many different sites in the cellular genome (SUTTER et al. 1978; STABEL et al. 1980; for review, see DOERFLER 1982). In order to substantiate this, at the time, tentative conclusion, the junction sequences at nine different sites of transition between adenovirus DNA and cellular DNA were cloned from six different Ad-transformed or Ad12-induced tumor lines and from a symmetric recombinant (SYREC2) in which Ad12 DNA was linked to cellular DNA. The nucleotide sequences at these junction sites were determined and compared (for reviews, see: DOERFLER et al. 1983, 1987; ROTH and WILSON 1988). No evidence was adduced for a specific cellular integration sequence for adenovirus DNA. All cellular sequences that had recombined with adenovirus DNA proved in fact different. However, it was conceivable that complex recognition signals for the cellular recombination machinery were present but not yet recognized in these cellular sequences. Patchy homologies between cellular and viral DNA sequences were observed in some but not in all instances.

## 2 A Glance at the Structure of Integration Sites

Frequently, but with the exception of the Ad12-transformed hamster cell line HA12/7, terminal viral nucleotides were deleted, 8–174 nucleotides at different junctions. In the Ad2-transformed hamster cell line HE5, not a single cellular nucleotide was lost at the site of insertion (GAHLMANN and DOERFLER 1983), whereas in the Ad12-induced mouse tumor CBA-12-1-T a major deletion of cellular DNA comprising about 1.5–1.6 kilobase pairs (kbp) was apparent at the site of integration (SCHULZ and DOERFLER 1984).

In the vicinity of the adenovirus DNA sequences from the Ad12-induced tumor cell line CLAC1 and the Ad12-induced tumor T1111(2) known cellular sequences were detected: the coding sequence for the 4.5 S RNA (SCHULZ et al. 1987) and for an intracisternal A particle (IAP), an endogenous retrovirus-like genome (LICHTEN-BERG et al. 1987), respectively. The arrangement of Ad12 DNA in hamster DNA from the tumor T1111(2) is complex in that nucleotide 65 at the left end of an unstable Ad12 DNA copy is linked to 127 nucleotides of cellular DNA which are followed to the left by a short inverted sequence of Ad12 DNA corresponding to nucleotides 1361 to 1290. This sequence abuts a stretch of 620 nucleotides of hamster cell DNA which is joined to the long terminal repeat of the hamster IAP sequence (LICHTENBERG et al. 1987). We have speculated that the presence of the transposable IAP sequence (KUFF et al. 1983) might be related to the instability of this integrated Ad12 genome.

It has also been shown that the cellular 5'-CCGG-3' (HpaII) and the 5'-GCGC-3' (HhaI) sites adjoining the integrated Ad12 DNA molecule in the tumor T1111(2) are not methylated, and neither are the neighboring Ad12 DNA sequences in the E1 region. In contrast, the same cellular DNA sequences on the second nonoccupied chromosome in the same tumor or on either chromosome in normal hamster cells or tissues are completely methylated (LICHTENBERG et al. 1988). Thus, the insertion of foreign, unmethylated (Ad12) DNA into the cellular genome at a methylated site can lead to local demethylations.

The cellular DNA sequences adjacent to adenovirus DNA in the symmetric recombinant SYREC2 (Deuring and Doerfler 1983), in the Ad12-induced hamster tumor line CLACI (Stabel and Doerfler 1982), in the Ad12-transformed hamster cell line HA12/7 (Jessberger et al. 1989a), in the Ad12-induced mouse tumor CBA-12-1-T (Schulz and Doerfler 1984), and in the Ad12-transformed hamster cell line HE5 (Gahlmann et al. 1984) have all been shown to be transcribed, not only in the transformed and tumor cells mentioned, but also in the normal host cells from which the transformed and tumor cells have been derived. This finding indicates that the chromatin structure associated with transcriptional activaties could facilitate the recombination of the host genome with foreign DNA (Gahlmann et al. 1984; Schulz et al. 1987). Of course, it cannot be claimed that only transcriptionally active cellular sites would be prone to recombination with newly introduced DNA such as adenovirus DNA after classic viral infection or after transfection.

We have briefly alluded to the finding that integrated adenovirus DNA can be lost again from the cellular genome. This observation with the Ad12-induced tumor T1111(2) has not been an isolated one. In the past, a number of revertant cell lines have been established from the Ad12-transformed cell line T637. The genomes of these revertants have lost all or part of the 20 copies of Ad12 DNA (Groneberg et al. 1978; Eick et al. 1980; Eick and Doerfler 1982). Similarly, Ad12 DNA has been shown to be excised from Ad12-induced tumor cells upon continuous passage in culture. Surprisingly, the cells devoid of Ad12 genomes have been shown to be still as oncogenic in hamsters as the original Ad12-induced tumor cells. Persistence of the Ad12 genome, at least in these cell lines, is not an absolute requirement for the maintenance of the oncogenic phenotype (Kuhlmann et al. 1982). These provocative observations must not, however, detract from the fact that in general integrated adeno-virus genomes are very stable when cell lines, like T637, are continuously passaged in culture over several years (Sutter et al. 1978; Stabel et al. 1980). Another remarkable example of integrational stability is that of cell line HE3/ATL which has been tumor-passaged through hamsters 36 times without loss of the viral genome or alterations in the viral integration patterns (Cook et al. 1988). On critical analysis, it is obvious that we do not understand the mechanisms that govern the stability or permit the loss of integrated foreign DNA in mammalian hosts. In view of the fact that viral genome integration is one of the indispensible parameters in viral persistence, research on the fascinating aspects of viral genome excision will have to assume a foremost position in work on viral pathogenicity.

Detailed analyses of sites of viral DNA integration have revealed interesting subtleties about the way in which foreign DNA can be built into an apparently stable host genome. Although we now understand that extensive sequence homologies or similarities between integration partners do not play a predominant role in foreign DNA insertion, comprehension of the enzymatic mechanisms involved in the re-structuring of the viral and host genomes at the sites of insertion is still lacking. In designing an overall concept of how foreign DNA insertion proceeds, the data have alerted us to the possibility that the transcriptional activity of host DNA may predispose certain of its sites to recombination with foreign DNA. While these genetically active sites may not be the only ones prone to undergo genetic exchange reactions, they may offer the incoming foreigner a ready chance to express itself through the regulatory instruments of the host genome.

## 3  Development of a Cell-Free System for the Study of Recombination/Integration in Mammalian Cells

We reasoned that investigations in a cell-free system, though initially fraught with many expected and probably even more unforeseen complications, might eventually help to elucidate some of the steps involved in the biochemistry of foreign DNA insertion in mammalian cells. Based on previous experience with viral-host DNA recombination events that led to the establishment of adenovirus-transformed cell lines or of Ad12-induced tumors, we employed nuclear extracts of BHK21 cells. These extracts have been prepared according to published procedures (DARBY and BLATTNER 1984; KUCHERLAPATI et al. 1985).

### 3.1  Design of the Experiment

The genetic signals that could direct viral-cellular DNA recombination in mammalian cells are not known. It is conceivable that the host cell's recombination enzymes recognize certain DNA sequences or specific structures of DNA-protein complexes. One way of confirming the presence of these presumptive signals in DNA, that is used in cell-free recombination experiments, is the selection of DNA sequences which have previously been involved in an integration reaction. Even without knowing the decisive signals for recombination in DNA, these signals would thus be included in the reaction.

A cell-free system of nuclear extracts from BHK21 cells was developed to catalyze recombination in vitro between the DNA of Ad12 and a pBR322-cloned 1768-bp BHK21 hamster DNA fragment (JESSBERGER et al. 1989b). This sequence had previously been identified as the preinsertion site corresponding to the junction between the left end of Ad12 DNA and hamster DNA in cell line CLAC1 (STABEL and DOERFLER 1982). A preinsertion sequence, which had recombined previously with foreign (Ad12) DNA, might again be recognized by the recombination system of the cell. *Pst*I-cleaved Ad12 DNA and the circular or the *Eco*RI-linearized preinsertion DNA were incubated with nuclear extracts. Recombinants were isolated by transfecting the DNA, which was reextracted from the in vitro recombination reaction, into recA$^-$ *E. coli* strains and by screening for Ad12 DNA-positive colonies. Without a selectable eukaryotic marker, all Ad12 DNA-positive recombinants were registered. Of a total of >90 recombinants, 21 were studied by restriction hybridization and 4 by partial nucleotide sequence analyses: The sites of linkage between Ad12 DNA and the preinsertion hamster DNA were all different and distinct from the original CLAC1 junction site between Ad12 and the preinsertion hamster DNA. The in vitro recombinants were not generated by simple end-to-end joining of DNA fragments but by an exchange reaction. Thirteen recombinants were derived from the 61–71 map unit fragment of Ad12 DNA. Control experiments as well as comparisons of nucleotide sequences and of restriction maps between the 25 recombinant DNAs and previously cloned junction sequences of Ad12 DNA and hamster DNA identified the recombinants as authentic and newly generated during in vitro incubation. Work currently in progress in our laboratory aims at determining whether randomly selected hamster DNA sequences can undergo similar recombination reactions with Ad12 DNA.

The recombination reaction could not be elicited when uncut Ad12 DNA was used. Possibly, the intact viral genome was too long to be effectively inserted into and maintained in the plasmid which carried the preinsertion sequence. Recombination was observed both with the uncut circular and the *Eco*RI-linearized plasmid containing the preinsertion sequence. The in vitro recombination reaction might depend on the availability of DNA termini; their importance in certain recombination reactions was demonstrated earlier (WILSON et al. 1982; ROTH et al. 1985; ROTH and WILSON 1986). Nuclear extracts similar to those utilized in the present experiments were previously used by other investigators (DARBY and BLATTNER 1984; KUCHER-LAPATI et al. 1985) who successfully rejoined DNA molecules carrying homologous sequences. The structure of the recombinants generated between Ad12 DNA and the hamster preinsertion sequence precluded the possibility of simple end-to-end joining. Ad12 DNA sequences were inserted into the preinsertion hamster DNA as evidenced by the nucleotide sequence data obtained for several of the recombinants (JESSBERGER et al. 1989b). The plasmid containing the preinsertion sequence and Ad12 DNA, which were used as recombination partners, was not extensively degraded by incubation with nuclear extracts. It is unknown to what extent the described recombination reactions resemble the actual integration of foreign DNA into the mammalian genome. Some of the elements of the integration reaction might have been mimicked.

It is interesting to recall in the context of the in vitro recombination results that both Ad12 and Ad2 DNA have frequently been found to integrate via their terminal sequences and that viral nucleotides at these termini can be deleted in the process of insertion (DOERFLER et al. 1983). When the nucleotide sequences of the *Pst*I fragments of Ad12 DNA were compared with the Ad12 sequences actually linked in vitro to the hamster preinsertion sequence, terminal deletions of viral DNA were apparent. The in vitro recombinants were generated with linearized or with circular preinsertion DNA and with linear fragments of Ad12 DNA. Possibly, these fragments were of a more suitable size for the recombination reaction. Between the time point of the incubation with the nuclear extract and the isolation of Ad12 DNA-positive recombinants, the previously linearized plasmid DNA had apparently been resealed.

Striking sequence repeats of CTG, GCCC, CCTC, and CCTT were observed in the preinsertion hamster sequence (JESSBERGER et al. 1989b). Although their role in serving as landmarks for the recombination reaction remains uncertain, it should be mentioned that the CCTT repeat has also been observed, e.g., in the mouse kappa-immuno-globulin gene (HOECHTL and ZACHAU 1983), in the mouse MHC class II H2-Ia-$\beta$ gene (LARHAMMAR et al. 1983), in the hamster alpha A crystalline gene (VAN DEN HEUVEL et al. 1985), and in the human HLA-DP-$\beta$1 and -$\alpha$1 genes (KELLY and TROWSDALE 1985).

A large number of different control experiments were carried out to confirm that recombination had taken place during incubation with nuclear extracts. Details of these control experiments are described elsewhere (JESSBERGER et al. 1989b). The *E. coli* strain recA$^-$ was transfected with (i) a mixture of the plasmid-pre-insertion sequence and Ad12 DNA, (ii) a mixture of these DNAs which had been incubated with proteinase K-inactivated nuclear extracts, or (iii) the pBR322-cloned preinsertion DNA and Ad12 DNA which had been incubated separately with nuclear extracts from BHK21 cells under conditions of recombination (JESSBERGER et al. 1989b). The negative outcome of these experiments and the fact that the sites of recombination between Ad12 DNA and the hamster preinsertion sequence were different among each other and from the CLAC1 and other known viral-cellular junction sites excluded the possibility that previously isolated junction plasmids might have been accidentally reisolated as contaminants. Similarly, the possibility of the recombinants having been generated in the intermediary *E. coli* host could be excluded. The finding that the separate incubation of the preinsertion DNA and of Ad12 DNA and the subsequent transfection of these molecules into the recA$^-$ *E. coli* strain did not give rise to recombinants also ruled out the possibility that only part of the recombination reaction had occurred in the *E. coli* host.

*Acknowledgements.* We thank Petra Böhm for excellent editorial work. This research was supported by grants from the Bundesministerium für Forschung und Technologie, Bonn (BCT 0390/2), and the Ministerium für Wissenschaft und Forschung, Düsseldorf (IV B 5-50002786). R. J. was supported by a stipend from Fonds der Chemischen Industrie, Frankfurt, during part of this work.

*Note Added in Proof*
Five randomly selected hamster DNA fragments (total lengths 21 kilo basepairs), which were unrelated to the preinsertion sequence, were cloned in pBR322 DNA and used in similar recombination experiments with Ad12 DNA in cell-free systems. Recombinants were not observed. These data supported the notion of selective recombination of Ad12 DNA *in vitro* with the preinsertion site DNA (JESSBERGER et al. 1989b).

# References


Cook JL, Lewis AM Jr, Klimkait T, Knust B, Doerfler W, Walker TA (1988) In vivo evolution of adenovirus 2-transformed cell virulence associated with altered E1A gene function. Virology 163: 374–390

Darby V, Blattner F (1984) Homologous recombination catalyzed by mammalian cell extracts in vitro. Science 226: 1213–1215

Deuring R, Doerfler W (1983) Proof of recombination between viral and cellular genomes in human KB cells productively infected by adenovirus type 12: structure of the junction site in a symmetric recombinant (SYREC). Gene 26: 283–289

Doerfler W (1982) Uptake, fixation, and expression of foreign DNA in mammalian cells: the organization of integrated adenovirus DNA sequences. Curr Top Microbiol Immunol 101: 127–194

Doerfler W (1987) Genetic principles and viral oncogenesis. In: Grundmann E (ed) Experimental neurooncology, brain tumor and pain therapy. Fischer, Stuttgart, pp 47–55 (Cancer campaign, vol 10)

Doerfler W, Gahlmann R, Stabel S, Deuring R, Lichtenberg U, Schulz M, Eick D, Leisten R (1983) On the mechanism of recombination between adenoviral and cellular DNAs: the structure of junction sites. Curr Top Microbiol Immunol 109: 193–228

Doerfler W, Spies A, Jessberger R, Lichtenberg U, Zock C, Rosahl T (1987) Recombination of foreign (viral) DNA with the host genome. In: Rott R, Goebel W (eds) Molecular basis of viral and microbial pathogenesis. Springer, Berlin Heidelberg New York, pp 60–72 (Colloquium der Gesellschaft für Biologische Chemie in Mosbach (Baden). vol 38)

Eick D, Doerfler W (1982) Integrated adenovirus type 12 DNA in the transformed hamster cell line T637: sequence arrangements at the termini of viral DNA and mode of amplification. J Virol 42: 317–321

Eick D, Stabel S, Doerfler W (1980) Revertants of adenovirus type 12-transformed hamster cell line T637 as tools in the analysis of integration patterns. J Virol 36: 41–49

Gahlmann R, Doerfler W (1983) Integration of viral DNA into the genome of the adenovirus type 2-transformed hamster cell line HE5 without loss or alteration of cellular nucleotides. Nucleic Acids Res 11: 7347–7361

Gahlmann R, Schulz M, Doerfler W (1984) Low molecular weight RNAs with homologies to cellular DNA at sites of adenovirus DNA insertion in hamster or mouse cells. EMBO J 3: 3263–3269

Groneberg J, Sutter D, Soboll H, Doerfler W (1978) Morphological revertants of adenovirus type 12-transformed hamster cells. J Gen Virol 40: 635–645

Hoechtl J, Zachau HG (1983) A novel type of aberrant recombination in immunoglobulin genes and its implications for V-J joining mechanism. Nature 302: 260–263

Jessberger R, Weisshaar B, Stabel S, Doerfler W (1989a) Arrangement and expression of integrated adenovirus type 12 DNA in the transformed hamster cell line HA 12/7: amplification of Ad12 and c-myc DNAs and evidence for hybrid viral-cellular transcripts. Virus Res. 13: 87–102

Jessberger R, Heuss D, Doerfler W (1989b) Recombination in hamster cell nuclear extracts between adenovirus type 12 DNA and two hamster preinsertion sequences. EMBO J 8: 869–878

Kelly A, Trowsdale J (1985) Complete nucleotide sequence of a functional HLA DP beta gene and the region between the DP beta 1 and DP alpha 1 genes: comparison of the 5′ ends of HLA class II genes. Nucleic Acids Res 13: 1607–1621

Kucherlapati RS, Spencer J, Moore PD (1985) Homologous recombination catalyzed by human cell extracts. Mol Cell Biol 5: 714–720

Kuff EL, Feenstra A, Lueders K, Smith L, Hawley R, Hozumi N, Shulman M (1983) Intracisternal A-particle genes as movable elements in the mouse genome. Proc Natl Acad Sci USA 80: 1992–1996

Kuhlmann I, Achten S, Rudolph R, Doerfler W (1982) Tumor induction by human adenovirus type 12 in hamsters: loss of the viral genome from adenovirus type 12-induced tumor cells is compatible with tumor formation. EMBO J 1: 79–86

Larhammar D, Hammerling U, Denaro M, Lund T, Flavell RA, Rask L, Peterson PA (1983) Structure of the murine immune response I-A-beta locus: sequence of the I-A-beta gene and an adjacent beta chain second domain exon. Cell 34: 179–188

Lichtenberg U, Zock C, Doerfler W (1987) Insertion of adenovirus type 12 DNA in the vicinity of an intracisternal A particle genome in Syrian hamster tumor cells. J Virol 61: 2719–2726

Lichtenberg U, Zock C, Doerfler W (1988) Integration of foreign DNA into mammalian genome can be associated with hypomethylationn at site of insertion. Virus Res 11: 335–342

Roth DB, Wilson JH (1986) Nonhomologous recombination in mammalian cells: role for short sequence homologies in the joining reaction. Mol Cell Biol 6: 4295–4303

Roth D, Wilson J (1988) Illegitimate recombination in mammalian cells. In: Kucherlapati R, Smith GR (eds) Genetic recombination. ASM, Washington pp 621–653, DC

Roth DB, Porter TN, Wilson JH (1985) Mechanism of nonhomologous recombination in mammalian cells. Mol Cell Biol 5: 2599–2607

Schulz M, Doerfler W (1984) Deletion of cellular DNA at site of viral DNA insertion in the adenovirus type 12-induced mouse tumor CBA-12-1-T. Nucleic Acids Res 12: 4959–4976

Schulz M, Freisem-Rabien U, Jessberger R, Doerfler W (1987) Transcriptional activities of mammalian genomes at sites of recombination with foreign DNA. J Virol 61: 344–353

Stabel S, Doerfler W (1982) Nucleotide sequence at the site of junction between adenovirus type 12 DNA and repetitive hamster cell DNA in transformed cell line CLAC1. Nucleic Acids Res 10: 8007–8023

Stabel S, Doerfler W, Friis RR (1980) Integration sites of adenovirus type 12 DNA in transformed hamster cells and hamster tumor cells. J Virol 36: 22–40
Sutter D, Westphal M, Doerfler W (1978) Patterns of integration of viral DNA sequences in the genomes of adenovirus type 12-transformed hamster cells. Cell 14: 569–585
Van den Heuvel R, Hendriks W, Quax WJ, Bloemendal H (1985) Complete structure of the hamster alpha A crystalline gene — reflection of an evolutionary history by means of exon shuffling. J Mol Biol 185: 273–284
Wilson JH, Berget PB, Pipas JM (1982) Somatic cells efficiently join unrelated DNA segments end-to-end. Mol Cell Biol 2: 1258–1269

# Part V: Herpesviruses:
## The Cellular and Molecular Biology of Epstein-Barr Virus

# Introduction

The herpesviruses all contain large, linear, double-stranded DNA genomes that vary between 150–230 kilobase pairs in length. In the virus particle, this DNA is contained within a toroidal core, made of proteins about 75 nm in diameter. The DNA-protein core is packaged in an icosahedral capsid composed of 162 capsomers, which in turn surrounds a protein tegument that appears as a granular zone of globular protein in electron micrographs. The icosahedron is wrapped in a lipid envelope containing several virus encoded glycoproteins that appear as short projections from the envelope in photomicrographs (ROIZMAN and LOPEZ 1985). The neurotropic herpesviruses, herpes simplex types 1 and 2 and varicella zoster virus (sometimes called the alpha group of herpesviruses), and the lymphotropic cytomegalovirus (a beta-group herpesvirus) have not convincingly been shown by epidemiological studies to be involved in oncogenic processes. Nor has it been possible to demonstrate convincingly that any of these viruses encodes a gene or induces an activity which results in the transformation of cells in culture. On the other hand, the gamma-group herpesviruses, including Epstein-Barr virus (EBV), herpesvirus saimiri, herpesvirus ateles, and Marek's disease virus, have all been shown to be causally associated with B- and T-cell lymphomas and, in the case of EBV, nasopharyngeal carcinoma (ROIZMAN and LOPEZ 1985). Part V contains a set of articles reviewing the gamma herpesviruses genes and gene products involved in viral episome replication and transformation of cells in culture. This introduction reviews the cellular and molecular biology of EBV infections, and it is meant to place the articles in this part into a context that permits the reader to appreciate the contributions in relation to the entire field of study.

EBV is most commonly transmitted between individuals via oral secretions, and the virus then replicates in the epithelial cells of the nasopharynx, tongue, and other organs of the oral cavity (ROIZMAN and LOPEZ 1985). The virus also infects B cells, where it replicates poorly, if at all. In those instances in which acute disease results (infectious mononucleosis), there is an initial leukopenia, but this is followed by an increased white cell count with the appearance of abnormal large lymphocytes (of the CD-8 T-cell class), which is apparently a reaction to the expression of foreign viral proteins on the surface of B cells (PATTENGALE et al. 1974). This antigen, termed LYDMA or lymphocyte-determined membrane antigen, appears to be identical to the latent membrane protein-1 (LMP-1) or the BNLF-1 membrane protein (two names for the same gene product) (THORLEY-LAWSON and ISRAELSOHN 1987). LMP-1 is one of the oncogenes of EBV (WANG et al. 1985). The B cells isolated from individuals infected with EBV contain the viral genome in a circular, nonintegrated form (LINDAHL et al. 1976). This genome often expresses eight EBV-encoded proteins, and at least some of these proteins appear to be involved in stimulating B cells to grow in culture indefinitely (in the absence of added antigens or special growth factors). EBV infection of B cells in culture also results in the immortalization of these cells for growth (Nonoyama and Pagano 1971; Pope et al. 1968). This is the most common assay for biological activity of the virus.

Burkitt's lymphoma, a B-cell lymphoma found predominantly in central Africa, is a childhood disease in which every tumor cell contains the EBV circular episomal genome (zur Hausen et al. 1970). The viral DNA expresses a similar set of gene products (about 10% of the genome coding capacity) to that observed in B cells derived from patients with infectious mononucleosis or in B cells immortalized in vitro with this virus (Wang et al. 1985; Hearing et al. 1984; Hennessy and Kieff 1985; Petti and Kieff 1988; Dillner et al. 1983). These human B-cell lymphoblastoid cell lines rarely produce virus, but some cell lines can be induced by several types of chemical agents to make small levels of infectious virus (Gerber 1972). EBV-associated nasopharyngeal carcinoma occurs predominantly in southern China where some individuals have chronic recurring EBV infections of the nasopharynx. The cells of the carcinoma all contain the EBV DNA plasmid (zur Hausen et al. 1970), expressing a similar subset of the viral proteins to that observed in latently infected B cells and B-cell lymphomas (Wang et al. 1985; Hearing et al. 1984; Hennessy and Kieff 1985; Petti and Kieff 1988; Dillner et al. 1983). That EBV genomes are always present in these cancer cells and that several viral gene products are commonly expressed in such cells (Wang et al. 1985; Hearing et al. 1984; Hennessy and Kieff 1985; Petti and Kieff 1988; Dillner et al. 1983) are reasons to suspect a causal relationship between EBV and cancer. The ability of the virus to immortalize B cells in culture (Pope et al. 1968) (i.e., a transformation assay) is another reason to postulate the involvement of EBV in these cancers. Finally, the identification of EBV-encoded genes and gene products (Wang et al. 1985; Baichwal and Sugden 1988) (LMP-1 or BNLF-1) that can transform cells in culture is sufficient to propose this causal relationship. Over the past 5 years, the complete nucleotide sequence of the EBV genome has been obtained (Baer et al. 1984), and some of the cDNAs (Sample et al. 1986) and protein products expressed in tumor cells and latently infected cells (Wang et al. 1985; Hearing et al. 1984; Hennessy and Kieff 1985; Petti and Kieff 1988; Dillner et al. 1983) have been elucidated. The cis- and trans-acting signals and gene products involved in viral plasmid replication and maintenance are being identified (Yates et al. 1984; Lupton and Levine 1985). The transforming activities of several viral genes have been tested, and onco-genes have now been recognized in this virus (Wang et al. 1985; Baichwal et al. 1988). It should be pointed out that an understanding of these events at the molecular level is at an early stage, but a firm foundation is in place.

The EBV genome (Baer et al. 1984) contains about 186 kilobase pairs of DNA and has terminal repeated sequences (TR; Fig. 1) of about 500 base pairs reiterated a variable number of times. About 18 kilobase pairs from one end of the genome (the left end by convention) is a variable number of a different set of internal repeated nucleotide sequences (IR1) composed of 3 kilobase pairs reiterated many times (*Bam* W fragments in Fig. 1). The terminal repeat and

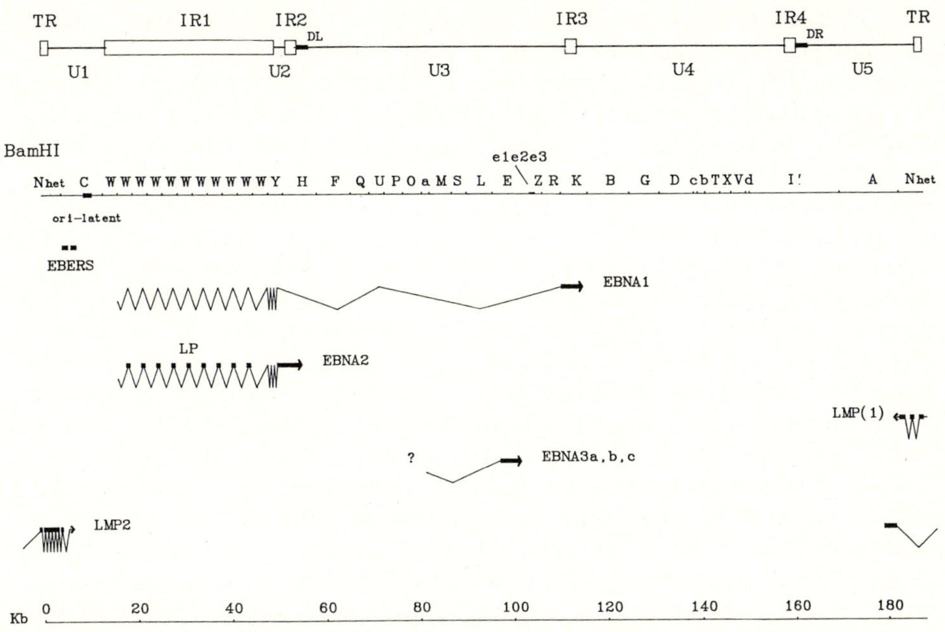

Fig. 1. Transcription of the Epstein-Barr virus genome in latently infected cells. *IR*, internal repeat sequence; *TR*, terminal repeated sequence; *U*, unique sequence; *LP*, leader protein

internal repeat have no sequence homology and do not recombine so there is (in contrast to some other herpesviruses) one unique orientation of the viral genes on the chromosome. The TR-IR segments a unique segment of viral sequences (U1) in the genome. There are several other repeated nucleotide sequences scattered throughout the viral DNA (IR2, 3, 4) which further serve to separate unique copy number genes (U2, 3, 4, 5). Figure 1 presents a schematic representation of the EBV genome and a *Bam*HI restriction enzyme cleavage map of this DNA. Open reading frames (ORFs) have been identified in this genome and named for their location in the *Bam*HI map (B), the direction of their transcript (L for leftward, R for rightward), and a number (1, 2, 3, etc.) for the position in the *Bam* fragment of the ORF. Thus, BNLF-1 is in the *Bam*HI-N fragment, and it is the first ORF in the leftward direction (see Fig. 1 where the BNLF-1 ORF is labelled LMP-1) (BAER et al. 1984).

Based upon the mapping of EBV mRNA to the *Bam*HI DNA fragments of the viral genome (VAN SANTEN et al. 1981), the isolation of cDNAs from these transcripts (SAMPLE et al. 1986), and the expression of EBV genomic and cDNA clones in cells transfected with these vectors (HEARING et al. 1984), several EBV transcripts and gene products have been identified and shown to be expressed in tumor-derived cells or latently infected cells. Of these, six are nuclear proteins (EBNA or Epstein-Barr virus nuclear antigen), and two are membrane proteins

(LMP, latent membrane proteins). It will be helpful to describe briefly each of these proteins and review what is known about their function.

The transcript for EBNA-1 appears to be initiated either in the IR region (*Bam* W, Fig. 1) or upstream of this region (*Bam* C) (SAMPLE et al. 1986). The entire coding region of EBNA-1, however, is in *Bam*HI K (HEARING et al. 1984). The protein has an unusual structure in that it is composed of an 89-amino acid *N*-terminal domain and a 313-amino acid *C*-terminal domain linked by a variable number (differs in different virus strains) of around 200 glycine-alanine repeated residues (GGA-GCA repeats are the IR3 repeat in Fig. 1) (BAER et al. 1984). Some of *C*-terminal domain (191 amino acids) binds specifically to a sequence (TAGCATATGCTA) (ROWLINS et al. 1985) found in *Bam* C which is a part of the latent origin of replication of the episome (ori-latent in Fig. 1) (YATES et al. 1984; LUPTON and LEVINE 1985). The EBNA-1 protein is required for long-term persistence of the episome and replication of this DNA (YATES et al. 1984; LUPTON and LEVINE 1985). The EBNA-1 binding site in *Bam*HI is part of a 30-bp repeated element that can act as an EBNA-1-dependent enhancer of transcription or gene expression (REISMAN et al. 1985). This brings up the possibility that EBNA-1 can regulate (in a positive sense) its own and other EBNA transcripts. The origin of replication of the episome has been identified by deletion analysis as being composed of two elements: a 30-bp repeat (20 times) and about 1 kilobase pairs away (in *Bam*HI C), a 65-bp dyad symmetry sequence which also contains four EBNA-1 binding sites (YATES et al. 1984; LUPTON and LEVINE 1985; REISMAN et al. 1985). Both of these *cis*-acting elements are required for long-term persistence of the viral DNA in these cells.

The mRNA for EBNA-2 also begins in *Bam*HI C or W fragments of DNA and is a bicistronic mRNA. Like EBNA-1, the 5'-end of this mRNA is composed of multiple splices of mRNA from the repeated IR1 elements in *Bam* W. These multiple splices join two different ORFs in the IR1 repeat, a 66-bp (W1 exon) and a 132-bp (W2 exon) element (SAMPLE et al. 1986). These exons are present in a variable number of repeated copies producing a leader protein (LP, Fig. 1) of variable length that has been observed in latently infected cells using antisera generated against a synthetic peptide with this sequence (DILLNER et al. 1986). Some W1-W2 leader proteins also contain an ORF sequence from *Bam*HI Y (33 base pairs from Y1, 122 base pairs from Y2, and 59 base pairs from Y3) or *Bam*HI H (144 base pairs from an H exon) as deduced from several diverse cDNA clones. The functions of the LP or LP-*Bam* Y or H fusion proteins are not clear. The second protein encoded in this mRNA is most likely derived entirely from *Bam*HI H and has been termed EBNA-2 (HENNESSY and KIEFF 1985). The EBNA-2 protein, like EBNA-1, is composed of an *N*- and *C*-terminal domain separated by a variable length of repeated sequence (CCC-CCA-CCA) which produces a polyproline tract (BAER et al. 1984). It is curious, but the U2 region (Fig. 1) encoding EBNA-2 differs extensively between two strains of EBV (AG876 and Jijoye versus B95-8 and Raji) and these 1.7-kilobase pair fragments are at best only partially homologous to each other (two alleles of EBNA-2). In addition, the P3HR1 virus carries a deletion (6.8 kilobase pairs) in the IR1-U2 region and fails to make EBNA-2. The progenitor of the P3HR1 virus (the Jijoye strain) retains the EBNA-2 coding sequences and is transformation competent while the P3HR1 strain (and the Daude strain) fail to immortalize B cells in culture. Expression of the EBNA-2 gene in Rat-1 cells rendered these cells able to grow in lower levels (0.5%–1%) of fetal calf serum, which is a property associated with transformation in cell culture (DAMBAUGH et al. 1986). These observations suggest a role for EBNA-2 in the transformation processes associated with this virus.

Very little is known about the functions of EBNA-3a, b, or c (sometimes termed EBNA-3, 4, and 5) except for the isolation of their cDNAs and the detection of these proteins in latently infected cells (PETTI and KIEFF 1988). These have been mapped to the BERF-1, BERF-2b, and BERF-4 ORFs encoding proteins of 145K (EBNA-3a), 155K (EBNA-3c), and 165K (EBNA-3b), all detected in latently infected cells by antibody from human patients (PETTI and KIEFF 1988). Two different membrane proteins (LMP-1 and 2, see Fig. 1) have been detected in latently infected cells. LMP-1 is produced from 3 exons localized in the *Bam* N fragment (WANG et al. 1985). When this protein is expressed in Rat-1 cells (WANG et al. 1985) or BALB/3T3 cells (BAICHWAL and SUGDEN 1988), it is able to induce these cells to form colonies in agarose (anchorage independence) and produce tumors in nude mice. Based upon the amino acid sequence of LMP-1 (BAER et al. 1984), it appears that the protein is composed of six membrane-spanning (transmembrane) regions which separate a 25-amino acid *N*-terminal and a 200-amino acid *C*-terminal domain, both of which appear to be in the cytoplasm. The mechanism by which this protein transforms cells in culture is unclear, but LMP-1 does represent the best example of an EBV-encoded oncogene demonstrated to date (WANG et al. 1985; BAER et al. 1984).

LMP-2 is transcribed from the opposite DNA strand (rightward) and is produced from multiple exons spliced together (BAER et al. 1984). The first exon (5′-end) is at the right hand end of the linear genome, and the next exons are at the left hand end of the viral DNA. Only the formation of a circular episome in an infected cell attaches these two ends and permits transcription across the terminal repeats, producing the primary transcript for this gene. The TR sequence and the LMP-1 exons (antisense) are spliced out of this mRNA to produce LMP-2. While the formation of a circular episome is probably required for latency and LMP-2 production, representing an interesting possible regulatory mechanism, no known function of LMP-2 has been found to date.

Finally, two RNA polymerase III transcripts, termed EBER-1 and -2 RNAs, are expressed in latently infected cells (in *Bam*HI C). Their functions remain to be elucidated.

The papers included in this part study the *cis*-acting elements required for the maintenance of the viral genome in the latent state as well as the protein responsible for transformation of cells in culture. BAICHWAL, HAMMERSCHMIDT, and SUGDEN characterize several mutants of the LMP-1 (BNLF-1) gene for their ability to transform cells in culture. They demonstrate that overproduction of this protein is toxic to cells and that BNLF-1 mutants which fail to transform cells are not toxic when expressed at high levels. Thus, it appears that high levels of the transforming gene product are detrimental to cells. FREY, CHITTENDEN, and LEVINE have carried out a detailed deletion analysis of the latent origin *cis*-acting element containing 20 of the 30-bp repeats. It appears that 2–6 repeats are unable to sustain plasmid maintenance (in D98-Raji cells) and that revertants of these deletion mutants contain at least 8 copies of this repeat as tandem duplications of the mutants. This paper also describes a new property of this EBV latent origin region, i.e., the ability to downregulate the SV40 origin of replication driven by the SV40 large T antigen. BIESINGER et al. point out that at least a portion of the transforming activity in herpesvirus saimiri and its tumorigenic potential are encoded in a region of the genome containing a gene product that is closely related in sequence to the human dihydrofolate reductase gene. The contribution of enzymes required to produce nucleic acid precursors is briefly discussed.

# References


Baer R, Bankier AT, Biggen MD, Deininger PL, Farrell PJ, Gibson TJ, Hatfull G, Hudson GS, Satchwell SC, Sequin C, Tuffnell PS, Barrell BG (1984) DNA sequence and expression of the B95-8 Epstein-Barr virus genome. Nature 310: 207–211

Baichwal VR, Sugden B (1988) Transformation of BALB-3T3 cells by the BNLF-1 gene of Epstein-Barr virus. Oncogene 2: 461–467

Dambaugh T, Wang F, Hennessy K, Woodlan E, Rickinson A, Kieff E (1986) Expression of the EBNA-2 protein in rodent cells J Virol 59: 453–462

Dillner J, Kadlin B, Alexander H, Ernberg I, Uno M, Ono Y, Klein G, Lerner RA (1986) An Epstein-Barr virus determined nuclear antigen partly encoded by the transformation associated Bam WYH region of EBV DNA. Proc Natl Acad Sci USA 83: 6641–6645

Gerber P (1972) Activation of Epstein-Barr virus by 5-BUDR in virus free human cells. Proc Natl Acad Sci USA 69: 83–85

Hearing JC, Nicolas JC, Levine AJ (1984) Identification of Epstein-Barr virus sequences that encode a nuclear antigen expressed in latently infected lymphocytes. Proc Natl Acad Sci USA 81: 4373–4377

Hennessy K, Kieff E (1985) A second nuclear protein is encoded by Epstein-Barr virus in latent infection. Science 227: 1238–1240

Lindahl T, Adams A, Bjursell G, Bornkamm GW, Kaschka-Dierich C, Jehn U (1976) Covalently closed circular duplex DNA of Epstein-Barr virus in a human lymphoid cell line. J Mol. Biol 102: 511–530

Lupton S, Levine AJ (1984) Mapping genetic elements of Epstein-Barr virus that facilitate extrachomosomal persistance of EBV derived plasmids in human cells. Mol Cell Biol 5: 2533–2542

Nonoyama M, Pagano J (1971) Complementary RNA specific to DNA of Epstein-Barr virus: detection of EB viral genomes in non-productive cells. Nature [New Biol] 233: 103–106

Pattengale PK, Smith RW, Perlin E (1974) Atypical lymphocytes in acute infectious mononucleosis. Identification by multiple T and B lymphocyte markers. N Engl J Med 291: 1145–1148

Petti L, Kieff E (1988) A sixth Epstein-Barr virus nuclear protein (EBNA-3B) is expressed in latently infected growth transformed lymphocytes. J Virol 62: 2173–2178

Pope JH, Horne MK, Scott W (1968) Transformation of foctal human leukocytes in vitro by filtrates of a human leukaemic cell line containing herpes-like viruses. Int J Cancer 3: 857–866

Reisman D, Yates J, Sugden B (1985) A putative origin of replication of plasmids derived from EBV is composed of two cis-acting components. Mol Cell Biol 5: 1822–1832

Roizman B, Lopez C (eds) (1985) The herpesviruses, vol 4. Plenum, New York

Rowlins DR, Milman G, Hayward SD, Hayward GS (1985) Sequence-specific DNA binding of the Epstein-Barr virus nuclear antigen to clustered sites in the plasmid maintenance region. Cell 42: 859–868

Sample J, Hummel M, Braun D, Birkenbach M, Kieff E (1986) Nucleotide sequences of mRNAs encoding EBV nuclear proteins: a probable transcriptional initiation site. Proc Natl Acad Sci USA 83: 5096–5100

Thorley-Lawson DA, Israelsohn E (1987) Generation of specific cytotoxic T cells with a fragment of the Epstein-Barr virus encoded p63-latent membrane protein. Proc Natl Acad Sci USA 84: 5384–5388

Van Santen V, Cheung A, Kieff E (1981) Epstein-Barr Virus RNA VII, size and direction of transcription of virus specific cytoplasmic RNA species in a transformed cell line. Proc Natl Acad Sci USA 78: 1930–1934

Wang D, Liebowitz D, Kieff E (1985) An EBV membrane protein expressed in immortalized lymphocytes transforms established rodent cells. Cell 43: 831–840

Yates J, Warren N, Reisman D, Sugden B (1984) A cis-acting element from Epstein-Barr viral genome that permits stable replication of recombinant plasmids in latently infected cells. Proc Natl Acad Sci USA 81: 3606–3610

zur Hausen H, Schulte-Holthausen H, Klein G, Henle W, Henle G, Clifford P, Sanesson L (1970) EBV DNA in biopsies of Burkitt tumors and aplastic carcinomas of the nasopharynx. Nature 228: 1056–1058

# Epstein-Barr Virus DNA Replication

A. Frey, T. Chittenden, and A. J. Levine

Epstein-Barr virus (EBV), a prevalent human herpesvirus, is maintained as an extrachromosomal, multicopy plasmid in latently infected B cells. Latent infection is characterized by limited viral gene expression (Epstein and Achong 1979; van Santon et al. 1981) and faithful replication of the EBV episome once (and only once) per cell cycle in parallel with cellular DNA (Adams 1987; Hamper et al. 1974). The EBV replicon, in contrast to the "runaway" replication characteristic of polyomaviruses, provides a model for the study of a replication origin regulated by and within the cell cycle. EBV plasmid replication is particularly attractive for this because only two viral genetic elements are required for plasmid maintenance: a *cis*-acting replication origin within the EBV *Bam*HI C region (Lupton and Levine 1985; Reisman et al. 1985; Yates et al. 1984) and the *trans*-acting EBV nuclear antigen 1 (EBNA-1) (Lupton and Levine 1985; Reisman et al. 1985). It seems likely, therefore, that EBV plasmid maintenance will be closely regulated by cellular mechanisms that initiate replication of host cell chromosomes and permit only one duplication cycle per generation.

Previous studies identified a putative EBV replication origin used in latent infection by the capacity of plasmids containing a selectable marker and the EBV *Bam*HI C region to be maintained extrachromosomally in EBV-infected cell lines (Yates et al. 1984). Deletion analysis defined two *cis*-acting elements within the *Bam*HI C region that are both required for plasmid maintenance (Lupton and Levine 1985; Reisman et al. 1985). One essential element was localized to a region of 65-bp dyad symmetry, a structural feature common to other viral replication origins (Tooze 1981). The second required site is about 1 kilobase pairs upstream of this dyad and is composed of a 30-bp sequence repeated tandemly, but imperfectly, twenty times (20 × 30-bp repeat motif).

EBNA-1 is the only viral protein required for plasmid maintenance (Lupton and Levine 1985; Reisman et al. 1985) and has been shown to bind in vitro to a core sequence present in each copy of the 30-bp repeats, as well as to four sites in the dyad symmetry region (Rawlins et al. 1985). The 20 × 30-bp motif thus can be considered to provide a sink of 20 tandem, evenly spaced EBNA-1 binding sites. In addition, the 20 × 30-bp motif can function as an EBNA-1-dependent enhancer element (Reisman and Sugden 1986), although it is not clear whether enhancer activity is important for EBV DNA replication or plasmid maintenance in transfected cells.

Princeton University, Department of Molecular Biology, Princeton, NJ 08544, USA

Current Topics in Microbiology and Immunology, Vol. 144
© Springer-Verlag Berlin · Heidelberg 1989

Given the activities of the 20+30-bp motif, experiments were designed to define at higher resolution the critical *cis*-acting requirements for plasmid maintenance which are present in this region of the genome. To this end, a series of *Bal*31 deletion mutants were constructed so that 15, 12, 6, or just 2 intact copies of the 30-bp repeats (p15, p12, p6, and p2) remained in the plasmid while the region of dyad symmetry was unaffected. The plasmids also contained the hygromycin phosphotransferase gene (GRITZ and DAVIES 1983) which confers resistance to hygromycin B in both animal cells and *Escherichia coli*. Previous experiments have indicated that extrachromosomal plasmid maintenance is associated with a substantial increase in the number of drug-resistant colonies relative to a vector plasmid that must integrate into cellular DNA (LUPTON and LEVINE 1985; REISMAN et al. 1985). Therefore, to assay the ability of a plasmid to be maintained, the deletion mutants were transfected into D98-Raji cells, an adherent cell line harboring multiple latent EBV genomes that supply EBNA-1 in *trans* (GLASER and NONOYAMA 1974). Drug-resistant colonies arose after 2–3 weeks of hygromycin selection. Typically, deletion mutants with 15 or 12 copies of the 30-bp repeat give rise to near wild-type numbers of drug-resistant colonies (Table 1). However, plasmids with just 6 or 2 copies produced fewer (100-fold) drug-resistant colonies (Table 1), indicating that these deletions may have abolished plasmid maintenance function.

To determine whether these plasmids were maintained extrachromosomally, drug-resistant colonies were expanded in cell culture, and low molecular weight DNA was prepared by the method of HIRT (1967). The DNA preparations were then analyzed by Southern blotting (SOUTHERN 1975) to determine whether the original plasmid had persisted. Extrachromosomal plasmid DNA of the appropriate size was detected for the p15 and p12 mutants. However, the plasmid DNA present in the low molecular weight DNA fractions of p6 and p2 transfections appeared to have undergone substantial rearrangements, as demonstrated by numerous bands of anomalous size on the Southern blots (Table 1).

**Table 1.** The ability of mutant Epstein-Barr virus plasmids to persist as episomes

| Plasmid | Relative number of drug-resistant colonies | Episomal | Estimated copy number |
|---|---|---|---|
| Hm vector | 1 | — | — |
| p20 | 1000 | + | 1– 5 |
| p15 | 700 | + | 1– 5 |
| p12 | 600 | + | 1– 5 |
| p6 | 10 | + (rearranged) | 10– 20 |
| p2 | 10 | + (rearranged) | 50–100 |
| p2R1 | 2000 | + | 10– 20 |
| p6R1 | 3000 | + | 10– 30 |

A wild-type EBV plasmid containing the origin of replication (20 repeated elements, p20) and mutant plasmids with lower numbers of repeated elements (p15, p12, p6, p2) were transfected into D98-Raji cells; the plasmid was selected by a drug-resistant marker (Hm). The vector alone, with no EBV origin (Hm vector) was employed as a control. The number of drug-resistant colonies, episomal state of the plasmid, and copy number is presented. The p2 and p6 plasmids failed to establish an episomal state. Rare colonies contained rearranged plasmids (p2R1, p6R1) which acted as revertants.

One interpretation of these observations is that 6 or 2 copies of the 30-bp repeats are not sufficient for plasmid maintenance. The anomalously sized extra-chromosomal DNA present in the rare drug-resistant colonies may then represent plasmid structural rearrangements that have somehow restored plasmid maintenance function.

To test this possibility, the rearranged plasmids from Hirt extracts of p6 and p2 transfections were rescued by transforming competent *E. coli*. Of more than 50 plasmids rescued from p6 or p2 Hirt extracts, none are structurally equivalent to the transfected parent plasmid. Two rescued plasmids which had rearranged structures were chosen for further analysis. Plasmid p2R1, isolated from p2 Hirt extracts, consists of four copies of the transfected plasmid, p2, linked head to tail as a tetramer. Plasmid p6R1, rescued from p6 Hirt extracts, is essentially a head to head dimer of p6 with the inversion points within the plasmid maintenance regions. Additional rearrangements that remain to be characterized have occurred in the vicinity of the 30-bp repeats.

The possibility that these rearrangements have restored plasmid maintenance ability was then tested directly by transfecting the rescued plasmids back into D98-Raji cells. It is clear (Table 1, p2R1, p6R1) that the rearranged plasmids have a restored capacity to produce drug-resistant colonies, compared with the parent plasmids from which they were derived. Southern blot analysis of Hirt extracts indicated that p2R1 and p6R1 are maintained extrachromosomally and do not undergo further rearrangements.

These experiments demonstrate that more than six copies of the 30-bp repeat are required for extrachromosomal maintenance of the EBV episome (in D98-Raji cells). The p2R1 revertant provides a total of eight copies and is more efficient than the wild-type in producing drug-resistant colonies (two-fold) and maintaining a high copy number (2–20-fold increase) (Table 1). Furthermore, the eight copies of the 30-bp repeats need not be contiguous in the plasmid since they are widely separated in the p2R1 tetramer structure. It is not yet clear if the altered spatial arrangements of these repeated elements permits a lower number to be effective.

In order for the EBV genome to be maintained as a stable episome replicated once (and only once) per cell cycle, it requires mechanisms to regulate the number of rounds of duplication per generation. In the bovine papillomavirus (BPV), viral genetic elements have been shown to regulate negatively an adjacent SV40 origin of replication in the same plasmid, thereby preventing multiple rounds of replication within one cell cycle (ROBERTS and WEINTRAUB 1986, 1988). Having identified two *cis*-acting elements in the *Bam*HI C restriction fragment required for each round of replication, they were tested for their impact upon the ability of a plasmid containing the SV40 origin of DNA replication to duplicate itself multiply in a single cell cycle. When plasmids containing the SV40 origin of DNA replication linked to a selectable drug resistance gene are transfected into monkey cells expressing the SV40 large T antigen (COS 7 cells) (GLUZMAN 1981), the plasmid DNA replicates multiple times and eventually kills the host cell, preventing it from forming a clone of cells harboring the plasmid. A plasmid such as pSLneo (Fig. 1) (LUPTON and LEVINE 1985), which contains the SV40 origin of replication and a positive selectable marker for neomycin resistance (or G418 resistance), when transfected into COS 7 cells produces very few or no

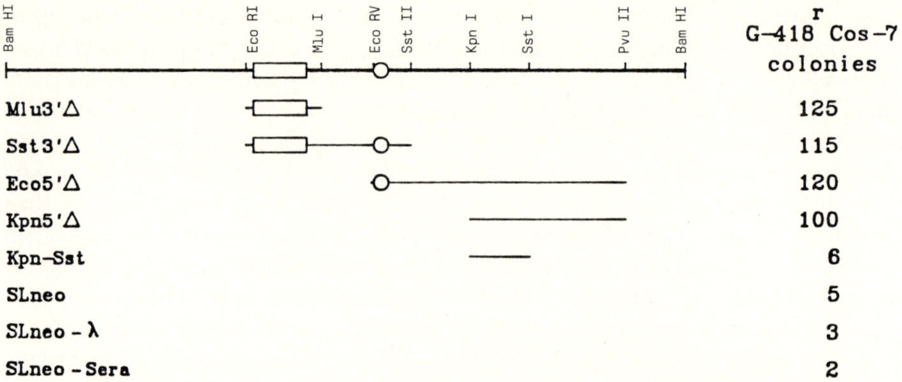

Fig. 1. The ability of Epstein-Barr virus (EBV) origin fragments to modulate down SV40 "runaway" DNA replication. The *Bam*HI C fragment of EBV DNA is presented with two *cis*-acting origin elements (*rectangle* is 20 × 30-bp repeats; *circle* is the 65-bp dyad). Deletion mutants of this fragment were employed to map its ability to permit plasmids containing the SV40 origin of replication to persist as an episome in monkey cells expressing SV40 large T antigen (colony numbers above 100)

G418-resistant colonies (Fig. 1). This is due to the "runaway" replication of the plasmid DNA, which is lethal to the cells. If the EBV latent origin region could negatively regulate the function of the SV40 origin, one would expect an increase in the number of G418-resistant COS 7 cells that produce colonies as has been reported for SV40-BPV hybrid plasmids (ROBERTS and WEINTRAUB 1986). As shown in Fig. 1, plasmids that contain the 20 × 30-bp repeat element (Mlu3'Δ), the 65-bp dyad element without the 30-bp repeats (Eco5'Δ), or both of these elements (Sst3'Δ) inserted into the pSLneo plasmid result in a 20–25-fold increase in the number of G418-resistant COS 7 colonies in this assay. These three constructions demonstrate that at least two different EBV sequences from the *Bam*HI C restriction fragment can regulate the SV40 origin in this assay. Indeed, a fragment of DNA between the *Kpn*I-*Pvu*II sites, which does not contain any known required elements for EBV plasmid replication but is adjacent to these two *cis*-acting positive elements, also has the ability to stimulate (20-fold) the number of G418-resistant colonies made by COS 7 cells transfected with this plasmid (Fig. 1, Kpn5'Δ). Not all DNA fragments cloned into pSLneo are able to prevent SV40 DNA runaway replication and cell death. A subfragment of Kpn5'Δ (the Kpn-Sst) failed to function in this assay as did a 5.6-kb *Eco*RI fragment of lambda DNA (SLneo-λ) and a mouse cellular DNA 0.6-kb *Eco*RI fragment (SLneo-Sera) (Fig. 1). These experiments demonstrate that at least two distinct segments of the *Bam*HI C DNA fragment from the EBV genome, localized at and around the EBV latent origin of DNA replication, can act to prevent SV40 runaway replication and consequent cell death in COS 7 cells. In contrast to BPV, which requires one *trans*-acting viral function in order to regulate negatively the SV40 origin (ROBERTS and WEINTRAUB 1986), the *Bam*HI C sequences downregulate the SV40 origin in the absence of any known EBV *trans*-acting factors, suggesting that the EBV DNA sequences themselves are involved.

The G418-resistant colonies of COS 7 cells containing the EBV DNA (Mlu3′Δ, Sst3′Δ, or Kpn5′Δ) were cloned and expanded in culture. The low molecular weight DNA was extracted from these cells by the Hirt procedure and analyzed by Southern blotting. Based upon this analysis, it was concluded that the DNA was maintained in an episomal state (in the small molecular weight fraction) at a level of 500–1000 copies per cell. The DNA was predominantly in form I configuration, however, as was demonstrated in the example of BPV-SV40 hybrid plasmids (ROBERTS and WEINTRAUB 1986), a variable proportion of the hybridizable material was present in an apparently multimeric state.

It remains possible that the cloned cell lines of G418-resistant COS 7 cells containing the Sst3′Δ plasmid are really a mixture of cells, some lacking episomal Sst3′Δ plasmid and a small number exhibiting runaway replication at the SV40 origin (socalled jackpot cells) which provide G418 resistance to the cell population. An analysis of the entire culture would find episomal copies of Sst3′Δ (from a few jackpot cells) in this mixed culture of cells. If that were the case, incubating these cells with bromodeoxyuridine (BrdU) in a density transfer experiment (ROBERTS and WEINTRAUB 1986) should yield heavy-heavy density DNA (with both strands substituted with BrdU) in one cell cycle since plasmids in jackpot cells undergo multiple rounds of replication. On the other hand, if all of the cells replicated their episomes only once per cell cycle, heavy-heavy plasmid DNA would be absent and the majority of the plasmid DNA would appear in the heavy-light (or hybrid density) portion of the CsCl density gradient. When a COS 7 Sst3′Δ clone was incubated in BrdU-containing medium for 24 h (less than one cell cycle), only light-light and heavy-light plasmid DNA was observed in the CsCl density gradient. These results strongly suggest that replication of the EBV-SV40 hybrid plasmid DNAs is tightly regulated within the cell cycle.

These experiments demonstrate that the inserted EBV DNA prevents runaway replication at the SV40 origin and cell death. The EBV DNA also exerts a negative control upon the plasmid so that it replicates only once per generation. Interestingly, there are at least two distinct DNA elements in the EBV *Bam*HI C fragment that can do this, one of which does not overlap with the positive regulatory elements that require EBNA-1 in *trans* to initiate DNA replication. Additional experiments are now required to elucidate the mechanisms regulating EBV episomes in a cell cycle.

*Acknowledgements.* We thank A. K. Teresky for technical help and K. James for typing the manuscript. This work was supported by grant MV47H from the American Cancer Society.

# References

Adams A (1987) Replication of latent Epstein-Barr virus genomes in Raji cells. J Virol 61: 1743–1746

Epstein M, Achong B (eds) (1979) The Epstein Barr virus. Springer, Berlin Heidelberg New York

Glaser R, Nonoyama M (1974) Host cell regulation of induction of Epstein-Barr virus. J Virol 14: 174–176

Gluzman Y (1981) SV40-transformed simian cells support the replication of early SV40 mutants. Cell 23: 175–182

Critz L, Davies J (1983) Plasmid-encoded hygromycin B resistance: the sequence of hygromycin B phosphotransferase gene and expression in *Escherichia coli* and *Saccharomyces cerevisia*. Gene 25: 179–188

Hamper B, Tanaka A, Nonoyama M, Derge J (1974) Replication of the resident repressed Epstein-Barr virus genome during the early S phase (S-1 period) of non-producer Raji cells. Proc Natl Acad Sci USA 71: 631–633

Hirt B (1967) Selective extraction of polyoma DNA from infected mouse cell cultures. J Mol Biol 26: 365–369

Lupton S, Levine A (1985) Mapping genetic elements of Epstein-Barr virus that facilitate extrachromosomal persistence of Epstein-Barr virus derived plasmids in human cells. Mol Cell Biol 5: 2533–2542

Rawlins D, Milman G, Hayward S, Hayward G (1985) Sequence-specific DNA binding of the Epstein-Barr virus nuclear antigen to clustered sites in the plasmid maintenance region. Cell 42: 859–868

Reisman D, Yates J, Sugden B (1985) A putative origin of replication of plasmids derived from Epstein-Barr virus is composed of two *cis*-acting componentes. Mol Cell Biol 5: 1822–1832

Reisman D, Sugden B (1986) *Trans*-activation of an Epstein-Barr viral transcriptional enhancer by the Epstein-Barr viral nuclear antigen 1. Mol Cell Biol 6: 3838–3846

Roberts J, Weintraub H (1986) Negative control of DNA replication in composite SV40-bovine papilloma virus plasmids. Cell 46: 741–752

Roberts J, Weintraub H (1988) *Cis*-acting negative control of DNA replication in eukaryotic cells. Cell 52: 397–404

Southern E (1975) Detection of specific sequences among DNA fragments separated by gel electrophoresis. J Mol Biol 98: 503–517

Tooze J (ed) (1981) Molecular biology of the tumor viruses. Part 2: DNA tumor viruses, 2nd edn. Cold Spring Harbor Laboratory, New York

Van Santen V, Cheung A, Kieff E (1981) Epstein-Barr virus RNA. VII. Size and direction of virus specified cytoplasmic RNA species in a transformed cell line. Proc Natl Acad Sci USA 78: 1930 –1934

Yates J, Warren N, Reisman D, Sugden B (1984) A *cis*-acting element from the Epstein-Barr viral genome that permits stable replication of recombinant plasmids in latently infected cells. Proc Natl Acad Sci USA 31: 3306–3310

# Characterization of the BNLF-1 Oncogene of Epstein-Barr Virus

V. R. Baichwal, W. Hammerschmidt, and B. Sugden

## 1 Introduction

Epstein-Barr virus (EBV) is a human herpesvirus that is causally associated with several malignancies including nasopharyngeal carcinoma, African Burkitt's lymphoma, and lymphoproliferative disorders in immunosuppressed people (reviewed in Miller 1985). In vitro EBV infects human B cells and immortalizes between 10%–100% of them (Sugden and Mark 1977; reviewed in zur Hausen 1981). It seems likely that EBV's capacity to immortalize cells in culture is related to its tumorigenicity in vivo. This likelihood has led researchers to study the mechanism of immortalization of B cells by this virus. The EBV genes required to initiate and/or to maintain the immortalized state in vitro have not been identified. However, at least eight viral genes are found to be transcribed consistently in B cells immortalized in vitro, and circumstantial evidence exists that at least several of these gene products will be required for this process (reviewed in Knutson and Sugden 1988). Until recently, there was no evidence that causally linked any one of these eight viral genes with changes in the growth control of EBV-infected cells. It has now been shown that one viral gene, BNLF-1 (or LMP), when expressed from either of two heterologous promoters and introduced into either of two established rodent cell lines, Rat-1 (Wang et al. 1985) or BALB/3T3 cells (Baichwal and Sugden 1988), induces these cells to grow in an anchorage-independent fashion in agarose. Most of these anchorage-independent clones (9 out of 9 for Rat-1 and 2 out of 5 for BALB/3T3) also grow as tumors in nude mice. Thus, the BNLF-1 gene of EBV scores as an oncogene both in its capacity to induce anchorage-independent growth in established cell lines and in its ability to induce a tumorigenic phenotype in these cells.

The product of the BNLF-1 gene has been characterized in a variety of laboratories with a variety of approaches. A summary of the information obtained by these studies is given below.

First, the amino acid sequence of the BNLF-1 open reading frame as deduced from the sequence of the viral genome reveals that the protein derived has six hydrophobic domains which alternate with short hydrophilic stretches, indicating that these hydrophobic domains are likely to be inserted into a membrane (Bankier et al. 1983; Fennewald et al. 1984). These six potential membrane-

McArdle Laboratory for Cancer Research, 450 N. Randall Avenue, Madison, WI 53706, USA

Current Topics in Microbiology and Immunology, Vol. 144
© Springer-Verlag Berlin · Heidelberg 1989

spanning domains are flanked by approximately 25 amino acids at the *N*-terminus of the protein and 200 amino acids at its *C*-terminus.

Second, a series of experimental findings are consistent with the six membrane-spanning domains of the BNLF-1 protein being embedded in the plasma membrane, with the *N*- and *C*-termini being cytoplasmic. When viable cells are exposed to proteases, the extracellular domains of approximately 30% of the BNLF-1 in a cell are cleaved (LIEBOWITZ et al. 1986). Antibodies directed towards the *N*- or the *C*-terminus fail to stain viable cells as measured by immunofluorescence assays (HENNESSY et al. 1984; MANN et al. 1985). These same reagents do stain cells which are permeabilized, indicating that the *N*- and *C*-termini of BNLF-1 are within the cytoplasm. When cells are fractionated biochemically into particulate or soluble fractions, BNLF-1 is found to be associated with the particulate fraction (HENNESSY et al. 1984; MANN et al. 1985). This association is also consistent with BNLF-1 being localized in a cellular membrane. The BNLF-1 protein has been found in patches in permeabilized cells, again indicating that at least some of it is located in the plasma membrane (LIEBOWITZ et al. 1986).

Third, the BNLF-1 protein has been shown to be associated with cimentin, which is a component of the cytoskeletal framework (LIEBOWITZ et al. 1987).

Fourth, the BNLF-1 protein is phosphorylated on serine residues (BAICHWAL and SUGDEN 1987), and to a lesser extent on threonine residues (MANN and THORLEY-LAWSON 1987), but apparently lacks any protein kinase activity of its own (BAICHWAL and SUGDEN, unpublished observations).

Fifth, the BNLF-1 protein is turned over in cells considerably more rapidly than the average protein. For example, its half-life measured in one lymphoblastoid cell line is approximately 5 h as compared with 16 h for total cellular protein (BAICHWAL and SUGDEN 1987). All of these findings apply equally to the BNLF-1 protein in EBV-immortalized lymphoblastoid cell lines and in rodent cell lines rendered anchorage-independent by the BNLF-1 gene (BAICHWAL and SUGDEN 1988, and unpublished observations). These findings, however, do not reveal what functions the protein has that cause it to confer the phenotype of anchorage-independent growth on established rodent cell lines or potentially participate in the immortalization of human B cells.

We have dissected the BNLF-1 gene in order to determine which domains, when expressed in the BALB/3T3 cell line, induce anchorage-independent growth. Our goal is to gain some insight into the functions of the BNLF-1 protein that contribute to this phenotype. In particular, we have asked whether the *C*-terminal half of the protein is sufficient or even necessary for inducing anchorage-independent growth. We have found that the *C*-terminal 200 amino acids of the BNLF-1 protein can be deleted and the protein will still induce anchorage-independent growth in BALB/3T3 cells (BAICHWAL and SUGDEN, submitted for publication). The *N*-terminal 25 amino acids and the six membrane-spanning domains must be intact for expression of the full transforming activity. Also, the wild-type gene and all mutants that yield anchorage-independent growth when expressed in BALB/3T3 cells are toxic for a variety of cells when expressed at high levels (HAMMER-SCHMIDT et al., submitted for publication). All of the variants that fail to induce anchorage-independent growth in BALB/3T3 cells are not toxic when expressed at high levels in any cell line tested. These findings indicate that it is likely that the function of the BNLF-1 protein responsible for inducing the anchorage-independent growth phenotype in BALB/3T3 cells is also toxic to these and other cells when expressed at higher levels.

## 2 Results

Mutants of the BNLF-1 gene have been constructed and inserted into vectors in which the mutant genes are expressed either from the SV40 early promoter or from the CMV immediate early promoter (STINSKI and ROEHR 1985). In these plasmids expression from the CMV immediate early promoter is substantially higher (30- to 40-fold) than from the SV40 early promoter. The probable structure of each of the eight mutant proteins, as predicted from their amino acid sequence, is shown in Fig. 1. There is no strong experimental evidence that the mutants assume the indicated structure within the plasma membrane as shown in this figure. However, seven of these eight mutants have been found to be expressed in cells using an antibody against the *C*-terminal portion of the protein (BAICHWAL and SUGDEN, manuscript submitted; HAMMERSCHMIDT et al., manuscript submitted). The eighth (Cdel-199) lacks all of the epitopes detected by this antibody. All the mutants that have five or more putative membrane-spanning domains have been found to be localized in the particulate fraction by biochemical fractionation, or in the plasma membrane as monitored by immunofluorescence. Ndel-128 and Del-25-132, which have only two membrane-spanning domains, are found either in the plasma membrane and in the cytoplasm (Ndel-128) or only in the plasma membrane (Del-25-132). Del-27-212, which lacks all six membrane-spanning domains, fractionates with the soluble portion of the cell and accordingly yields only cytoplasmic staining when detected by immunofluorescence. When these mutants are expressed from the SV40 early promoter in BALB/3T3 cells, only those that have all six membrane-spanning domains induce anchorage-independent growth in recipient cells. The mutant Cdel-155 induces anchorage-independent growth as efficiently as the wild-type. This mutant lacks the *C*-terminal 155 amino acids of BNLF-1. Mutants lacking the *C*-terminal 174, the *C*-terminal 199, or the *N*-terminal 9 amino acids also induce anchorage-independent growth but not as efficiently as wild-type. In contrast, mutants that contain no membrane-spanning segments (Del-27-212) or contain only two membrane-spanning segments (Del-25-132) fail to induce anchorage-independent growth, even though the *N*-terminal and *C*-terminal cytoplasmic domains are intact. Taken together, these findings indicate that the six membrane-spanning domains of the BNLF-1 protein are required for this protein to induce anchorage-independent growth in BALB/3T3 cells.

The capacity of the wild-type BNLF-1 gene or of its mutants to induce anchorage-independent growth in BALB/3T3 cells in culture has been difficult to measure. This difficulty is manifested by a wide variation in the number of anchorage-independent colonies induced by the wild-type gene in consecutive experiments. It seems possible that this variation might result from a limited expression of the BNLF-1 gene or in its mutant derivatives, from the SV40 early promoter in the recipient BALB/3T3 cells. To circumvent this potential difficulty,

| PROTEIN NAME | SCHEMATIC STRUCTURE | SUBCELLULAR LOCATION | TRANSFORMING ACTIVITY | TOXICITY |
| --- | --- | --- | --- | --- |
| BNLF-1 (1-386) | 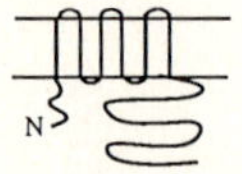 | MEMBRANE | ++ | ++ |
| C del-155 (1-231) | 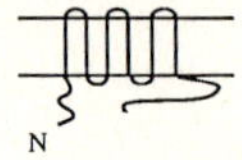 | MEMBRANE | ++ | ++ |
| C del-174 (1-212) | 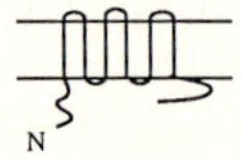 | MEMBRANE | + | ++ |
| C del-199 (1-187) | 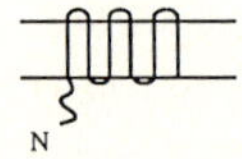 | ? | + | ++ |
| N del-9 (10-386) | 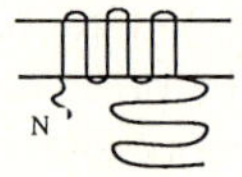 | ? | ± | ± |
| Ndel-43 (44-386) | 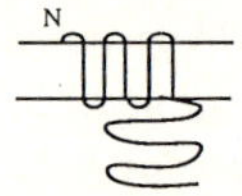 | MEMBRANE | − | − |
| Ndel-128 (129-386) | 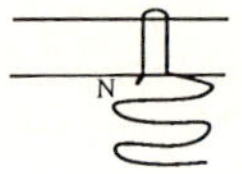 | MEMBRANE & CYTOSOL | − | − |
| Del-25-132 (1-24 & 133-386) | 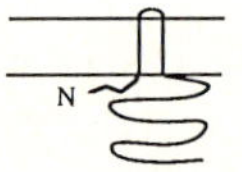 | MEMBRANE | − | − |
| Del -27-212 (1-26 & 213-386) |  | CYTOSOL | − | − |

**Fig. 1.** A summary of the subcellular location, transforming activity, and toxicity of mutant BNLF-1 proteins. The *first column* gives the name of the protein. The *numbers in parentheses* are the amino acids present in the mutant proteins relative to the wild-type protein. The *second column* gives a schematic structure of the proteins and indicates the regions of the wild-type protein retained in the mutants. The *N*-terminus of the protein is indicated by *N*. The *horizontal lines* indicate the membrane. The exact arrangement of the polypeptide chain in the plasma membrane is not known for either the wild-type or the mutant proteins. The *third column* gives the subcellular location of the protein. The location was determined either by biochemical fractionation of cell lysates or by immunofluorescence analysis, as in the case of Del-25-132. The presence of a protein in the particulate fraction of cells up on fractionation of cell lysates is assumed to reflect its location in the cellular membranes. The *fourth column* gives the transforming activity of the protein, as judged by its capacity to induce anchorage-independent growth in BALB/3T3 cells, when expressed from an SV40 enhancer and early promoter. The *fifth column* gives the toxicity of the mutant proteins, when expressed from the CMV immediate-early enhancer and promoter. The range of transforming and toxic activity is + + (activity similar to wild-type) to − (no activity), based on the information given in the text. The transforming activity of Ndel-9 and Cdel-174 or Cdel-199 has been determined with different assays and cannot be compared with one another

the BNLF-1 open reading frame and that of all its mutant derivatives were placed downstream of the immediate-early promoter of CMV, and these constructions were tested for anchorage-independent growth in BALB/3T3 cells. We found that the wild-type CMV BNLF-1 gene does not induce anchorage-independent growth in BALB/3T3 cells. Rather, when CMV BNLF-1 is introduced along with pSV$_2$neo into BALB/3T3 cells and G418-resistant colonies are selected, the presence of CMV BNLF-1 leads to a decreased number of G418-resistant colonies. More impressively, the BNLF-1 protein is not detected in these G418-resistant colonies even though the cells express high levels of BNLF-1 transiently 48 h after introduction of the DNA into the cells. This finding that a cytotoxic activity is associated with expression of BNLF-1 at high levels helps to explain the observed variation in our anchorage-independent growth assays. To examine this phenomenon of toxicity in greater detail and to extend it to EBV-positive B-cell lines, we have constructed a vector that expresses BNLF-1 from the CMV immediate-early promoter, contains *ori*P, which EBV's origin of plasmid replication, and expresses resistance to hygromycin. Similar vectors have been made which express seven of the mutants of BNLF-1 (all mutants listed in Fig. 1 except Cdel-199). These vectors, or a vector lacking BNLF-1, were then introduced into six lymphoblastoid cell lines immortalized by EBV and into one adherent cell line which expresses an integrated copy of EBNA-1 constitutively (EBNA-1 is the EBV gene required in *trans* for *ori*P-mediated plasmid replication). The number of hygromycin-resistant colonies has been measured and compared with the number of hygromycin-resistant colonies obtained by introduction of the vector that does not express BNLF-1. The wild-type gene and its mutants, which when expressed from the SV40 early promoter induce anchorage-independent growth in BALB/3T3 cells, lead to a 10–1000-fold decrease in the number of hygromycin-resistant colonies relative to the negative control. All of the mutants which fail to induce anchorage-independent growth in BALB/3T3 fail to induce toxicity in any of these seven cell lines. Also, in contrast to the wild-type gene, mutants that fail to induce toxicity are expressed at high levels in the hygromycin-resistant colonies. The wild-type gene expressed from the CMV immediate-early promoter also kills HEp-2 cells in the absence of any selection for hygromycin resistance. The HEp-2 cell line takes up and expresses DNA introduced by electroporation efficiently. Fifty percent of Hep-2 cells electroporated with pCMV BNLF-1 die within 4 days after electroporation, while no cells electroporated with the negative control die. These observations indicate that mutants containing the six membrane-spanning domains plus the short *N*-terminal domain of this protein kill cells when expressed at high levels. Mutants which lack one, four, or all of the membrane-spanning domains do not kill cells when expressed at high levels. Although this cytotoxicity can be monitored conveniently in experiments measuring survival after selection for drug-resistant cells, the toxicity is manifest in cells in the absence of such selection.


# 3 Discussion

The phenotypes induced in cells by expression of the BNLF-1 gene from the SV40 early promoter or the CMV immediate-early promoter differ. In the former case, two established rodent cell lines are induced to grow in an anchorage-independent fashion. In the latter case, all cell lines tested are killed. Eight mutants of the BNLF-1 gene expressed from the SV40 early promoter and seven from the CMV immediate-early promoter were tested for their capacity to induce these two phenotypes. Those mutants that induce one phenotype induce the second. Perhaps most striking is the finding that the mutant, Ndel-9, is intermediate both in its capacity to induce anchorage-independent growth in BALB/3T3 cells and in its capacity to kill cells. This correlation indicates that the function of the BNLF-1 protein, which renders BALB/3T3 cells anchorage-independent, is likely to be toxic when expressed at higher levels. What can this function be?

We speculate that the BNLF-1 protein is related to cell-surface receptors. Its location at the cell surface, its rapid turnover, its presence in patches in the membrane, and its being phosphorylated are consistent with this notion. Its six membrane-spanning domains are reminiscent of the seven membrane-spanning segments common to the rhodopsin family of receptors (HANLEY and JACKSON 1987). The observation that BNLF-1 requires its six membrane-spanning domains to transform BALB/3T3 cells underscores the importance of this portion of the molecule. Perhaps this portion binds its ligand. The corresponding region of the $\beta_2$-adrenergic receptor, a member of the rhodopsin family, does bind its ligand (STRADER et al. 1987; reviewed in LEFKOWITZ and CARON 1988). If BNLF-1 is related to a receptor, then it seems likely that either it will function as a receptor for some growth factor or that it will not bind a ligand but rather be an activated receptor akin to v-*erbB* (reviewed in BISHOP 1985). In either case, it could transduce a signal to the cytoplasm that affects the growth control of the cell. This proposed function for BNLF-1 consistent with its capacity to induce anchorage-independent growth in some cells, with its participation in the immortalization of EBV-infected B cells, and with its killing capability when expressed at high levels.

It is now important to test our speculation that the BNLF-1 protein of EBV is functionally related to a receptor. In order to carry out this test technically it is desirable to have functional assays that are rapid and unequivocal. It appears that measuring the cytotoxicity resulting from a high level of expression of BNLF-1 is such an assay. In particular it is now possible to determine with this assay whether removing putative ligands from tissue culture medium or modulating known signal transduction pathways in cells can abrogate the toxicity associated with a high level of BNLF-1 expression in these cells. These approaches are feasible and should aid in identifying whether or not BNLF-1 is a virally encoded receptor.

*Acknowledgements.* We thank our laboratory colleagues for helpful discussions. The work was supported by the Public Health Service grants CA-22443 and CA-07175 from the National Institutes of Health. V.R.B. was supported in part by the Wisconsin Power and Light Foundation Fellowship in Cancer Research. W. H. was supported by the Deutsche Forschungsgemeinschaft.

# References

Baichwal VR, Sugden B (1987) Posttranslational processing of an Epstein-Barr virus-encoded membrane protein expressed in cells transformed by Epstein-Barr virus. J Virol 61: 866–875

Baichwal VR, Sugden B (1988) Transformation of BALB/3T3 cells by the BNLF-1 gene of Epstein-Barr virus. Oncogene 2: 461–467

Bankier AT, Deininger PL, Satchwell SC, Baer R, Farrell PJ, Barrell BG (1983) DNA sequence analysis of the EcoRI $D_{HET}$ fragment of B958 Epstein-Barr virus containing the terminal repeat sequences. Mol Biol Med 1: 425–446

Bishop JM (1985) Viral oncogenes. Cell 42: 23–38

Fennewald S, Van Santen V, Kieff E (1984) Nucleotide sequence of an mRNA transcribed in latent growth-transforming virus infection that indicates that it may encode a membrane protein. J Virol 51: 411–419

Hanley MR, Jackson T (1987) News and views. Substance K receptor: return of the magnificent seven. Nature 329: 766–767

Hennessy K, Fennewald S, Hummel M, Cole T, Kieff E (1984) A membrane protein encoded by Epstein-Barr virus in latent growth transforming infection. Proc Natl Acad Sci USA 81: 7207–7211

Knutson JC, Sugden B (1988) In: Klein G (ed) Advances in viral oncology, vol 8. Raven, New York

Lefkowitz RJ, Caron MG (1988) Adrenergic receptors: models for the study of receptors coupled to guanine nucleotide regulatory proteins. J Biol Chem 263: 4993–4996

Liebowitz D, Wang D, Kieff E (1986) Orientation and patching of the latent infection membrane protein encoded by Epstein-Barr virus. J Virol 58: 233–237

Liebowitz D, Kopan R, Fuchs E, Sample J, Kieff E (1987) An Epstein-Barr virus transforming protein associates with vimentin in lymphocytes. Mol Cell Biol 7: 2299–2308

Mann KP, Thorley-Lawson D (1987) Posttranslational processing of the Epstein-Barr virus encoded p63/LMP protein. J Virol 61: 2100–2108

Mann KP, Staunton D, Thorley-Lawson D (1985) Epstein-Barr virus encoded protein in the plasma membranes of transformed cells. J. Virol. 55: 710–720.

Miller G (1985) Epstein-Barr virus. In: Fields BN (ed) Virology. Raven, New York, pp 563–589

Stinski MF, Roehr TJ (1985) Activation of the major immediate early gene of human cytomegalo-virus by *cis*-acting elements in the promoter-regulatory sequence and by virus-specific *trans*-acting components. J. Virol. 55: 431–441.

Strader CD, Sigal IS, Register RB, Candelore MR, Rands E, Dixon RAF (1987) Identification of residues required for ligand binding to the β-adrenergic receptor. Proc. Natl. Acad. Sci. USA 84: 4384–4388.

Sugden B, Mark W (1977) Clonal transformation of adult human leukocytes by Epstein-Barr virus. J. Virol. 23: 503–508.

Wang D, Liebowitz D, Kieff E (1985) An EBV membrane protein expressed in immortalized lymphocytes transforms established rodent cells. Cell 43: 831–840

zur Hausen H (1981) In: Tooze T (ed) DNA Tumor Viruses. Cold Spring Harbor Laboratory, New York, pp 747–795

# Genes for the Synthesis of Deoxythymidylate Monophosphate in T-Cell Lymphoma-Inducing Herpesviruses of Nonhuman Primates

B. Biesinger[1], R. Grassmann[1], B. Fleckenstein[1], S. C. S. Murthy[2], J. Trimble[2], and R. C. Desrosiers[2]

Herpesvirus saimiri (HVS) and herpesvirus ateles (HVA) are T-lymphotropic viruses common in certain New World primates. Both viruses appear not to be pathogenic in their natural host species, squirrel monkeys (*Saimiri sciureus*) and spider monkeys (*Ateles* spp.), respectively. However, they can induce rapidly progressing T-cell lymphomas in a wide spectrum of South American monkeys (reviewed in Fleckenstein and Desrosiers 1982), and both viruses are capable of transforming T cells of marmoset monkeys in vitro (Falk et al. 1978; Schirm et al. 1984; Desrosiers et al. 1986). The herpesvirus group is in general highly heterogeneous with respect to the molecular structure of genomic DNA and pathogenic properties. Herpes simplex virus and varizella zoster virus, members of the alpha herpesvirus subgroup, and human cytomegalovirus, the beta herpesvirus prototype, have never been shown to be oncogenic or to possess transforming genes that are expressed in

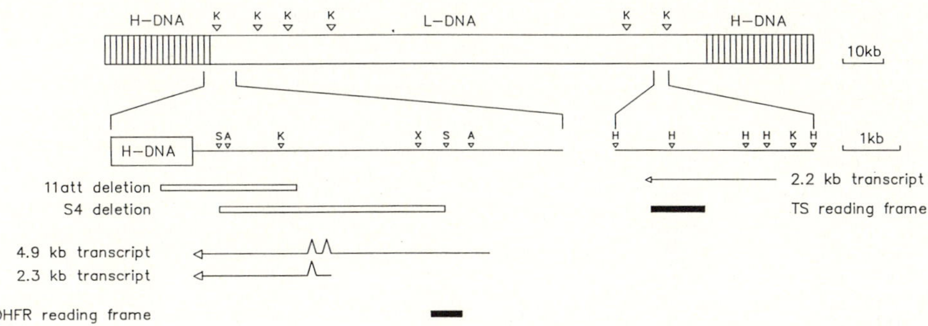


**Fig. 1.** The herpesvirus saimiri M-genome with the unique L-region flanked by repetitive H-DNA. The left terminal L-DNA including one H-DNA repeat is expanded on the *left*. Restriction sites for *Ava*II (*A*), *Kpn*I (*K*), *Sac*I (*S*), and *Xmn*I (*X*) are indicated. Deletions of the non-oncogenic mutants 11att and S4 are represented by *open boxes*; mRNAs transcribed from left terminal L-DNA in lytically infected OMK cells are given by *arrows*; the DHFR reading frame is shown by a *filled box*. The TS region is expanded on the *right* with sites for the restriction endonucleases *Hin*dIII (*H*) and *Kpn*I (*K*). Positions and direction of the TS mRNA (size without polyadenylation) and the open reading frame are shown

[1] Institut für Klinische und Molekulare Virologie der Universität Erlangen—Nürnberg, 8520 Erlangen, FRG
[2] New England Regional Primate Research Center, Harvard Medical School, Southborough, MA 01772, USA

Current Topics in Microbiology and Immunology, Vol. 144
© Springer-Verlag Berlin · Heidelberg 1989

tumors or cell lines (reviewed by Galloway and McDougall 1983). This is in contrast to the gamma herpesviruses, such as the human Epstein-Barr virus, HVS, and HVA, that are associated with tumors of the lymphatic system and whose DNAs are capable of persisting as large episomes in transformed cells (Kaschka-Dierich et al. 1977, 1982; Werner et al. 1977).

The genomes of HVS and HVA are linear, double-stranded DNA molecules of about 160 kilobase pairs; each contains a unique AT-rich region (L-DNA) of some 110 kilobase pairs that is flanked at both ends by noncoding, GC-rich,

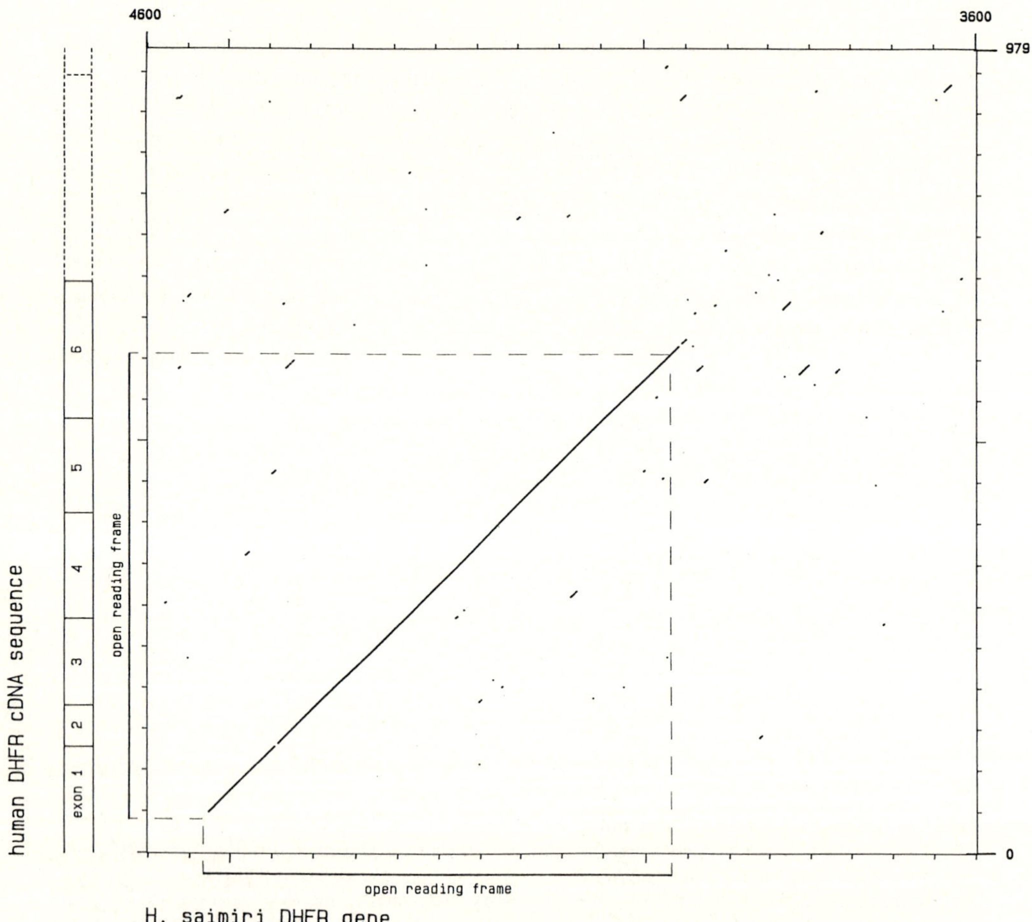


Fig. 2. Dot matrix comparison of the herpesvirus saimiri dihydrofolate reductase (*DHFR*) gene and a human DHFR cDNA sequence (stringency 14 out of 21). Homology is restricted to the open reading frames; adjacent untranslated sequences do not share significant homologies. The exons of the human gene are indicated

Fig. 3. Dot matrix comparison of the herpesvirus saimiri TS transcription unit and a human ▶ thymidylate synthase (*TS*) cDNA sequence (stringency 14 out of 21). Only the open reading frames show homologies and colinear organization

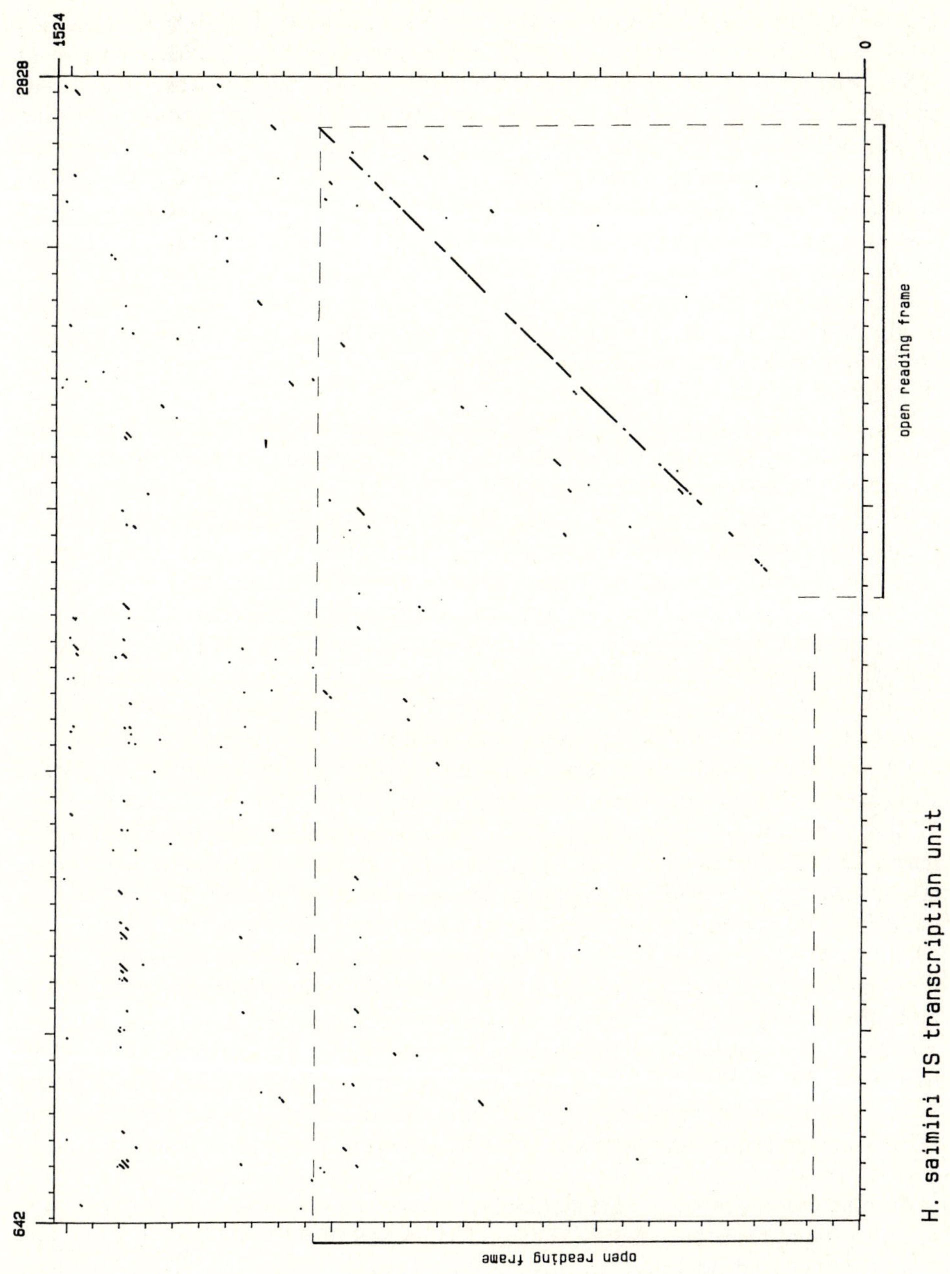
1524
2828
642
0
open reading frame
open reading frame
H. saimiri TS transcription unit
human TS cDNA sequence

tandem repetitions (H-DNA) in variable numbers (BORNKAMM et al. 1976; FLECKEN-STEIN et al. 1978; BANKIER et al. 1985; STAMMINGER et al. 1987; RICHTER et al. 1988) (Fig. 1). The left terminal L-DNA sequence was shown to be linked with the oncogenic and T-cell-transforming phenotype of HVS. Spontaneous deletion of a 2.3-kbp segment spanning the left H/L junction resulted in a replication competent non-oncogenic variant of HSV 11 (11att) (KOOMEY et al. 1984). A deletion mutant artificially deprived of a 4-kbp segment from left terminal L-DNA (Fig. 2) also did not induce tumors in owl monkeys (DESROSIERS et al. 1985) and lost its in vitro-transforming capability (DESROSIERS et al. 1986). Restoration of the deleted sequences restored the transforming and tumor-inducing capacity (DESROSIERS et al. 1985, 1986). The left terminal L-DNA of HVS is transcribed during lytic replication into two, 3′-coterminal, spliced mRNAs of 4.9 and 2.3 kilobases (KAMINE et al. 1984).

The nucleotide sequence revealed two open reading frames, one of which predicted a polypeptide of 187 amino acids which was found by computer analysis to be highly related to dihydrofolate reductase (DHFR, EC 1.5.1.3) of human cells (TRIMBLE et al. 1988; MASTERS and ATTARDI 1983) (Fig. 2). The putative DHFR polypeptides of HVS strain 11 and human DHFR are equal in size and identical in 83.3% of their amino acids. Nucleotide sequence homologies between the HVS gene and the human DHFR cDNA sequence (MASTERS and ATTARDI 1983) were found strictly confined to the open reading frames, while 5′- and 3′-untranslated sequences appeared entirely unrelated (TRIMBLE et al. 1988) (Fig. 2). The human DHFR gene contains five introns with sizes up to 14 kilobase pairs (CHEN et al. 1984), while the DHFR gene of HVS has no introns (TRIMBLE et al. 1988).

Various HVS strains are diverse in their oncogenic potential and the nucleotide sequences of the left terminal 7 kilobase pairs of L-DNA (MEDVECZKY et al. 1984). Group A virus strains (including strain 11) efficiently induce lymphomas in common marmosets (*Callithrix jacchus*) and transform peripheral blood lymphocytes in vitro (DESROSIERS et al. 1986; SZOMOLANYI et al. 1987). Group B viruses, such as strain SMHI, are much less effective in transforming in vitro (DESROIERS et al. 1986; SZOMOLANYI et al. 1987), and they have been reported to be non-oncogenic for common marmosets (LAUFS and FLECKENSTEIN 1973). Two further strains, 484-77 and 488-77 (non A — non B) do readily transform in vitro and in vivo. Southern blots employing a fragment of the HVS strain 11 DHFR gene indicated that strains of all three subgroups contain a DHFR-homologous sequence (TRIMBLE et al. 1988). Preliminary experiments have indicated that HVA also contains DHFR-related sequences; however, the precise nature of an equivalent gene awaits to be investigated.

Another enzyme of nucleotide metabolism, thymidylate synthase (TS, EC 2.1.1.45), was also found to be encoded by HVS and HVA. TS of HVS is highly homologous (70% amino acid identity) to the human enzyme. The viral gene has been mapped to the right end of the HVS genome (Fig. 1). The TS gene is transcribed into a 2.5-kb mRNA (BODEMER et al. 1986). Sequence comparison of the human TS cDNA (TAKEISHI et al. 1985) and the viral TS transcription unit is shown in Fig. 3. The open reading frames are strongly conserved. However, the flanking regions, including the very long 5′ untranslated mRNA sequence (1207 nucleotides), do not reveal significant homology. The protein encoded by the viral gene is

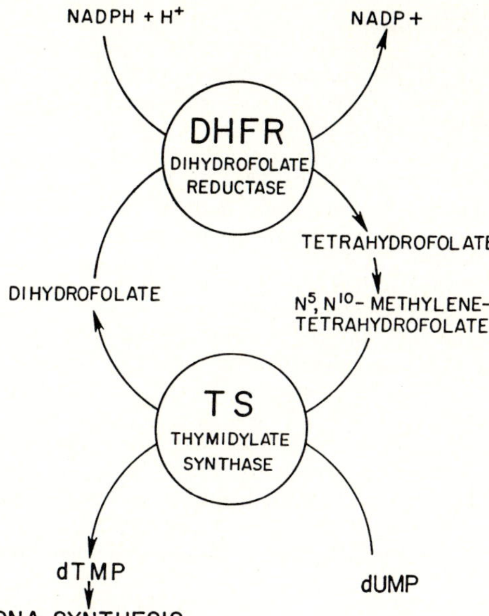


**Fig. 4.** Cooperation of dihydrofolate reductase (*DHFR*) with thymidylate synthase (*TS*) in deoxythymidylate monophosphate (*dTMP*) synthesis. TS catalyzes the methylation of deoxyuridine monophosphate (*dUMP*) to dTMP. The cofactor $N^5,N^{10}$-*methylene-tetrahydrofolate* serves both as methyl and as hydrogen donor and is converted to *dihydrofolate*. DHFR reduces dihydrofolate again to *tetrahydrofolate*, regenerating the cofactor needed for continued dTMP synthesis. After conversion to deoxythymidylate triphosphate (*dTTP*), dTMP can be used for DNA synthesis

a functional thymidylate synthase (HONESS et al. 1986). Though very large amounts of TS transcripts are detectable in the later phases of infection, rather little protein seems to be synthesized. This suggests a complex regulation of TS gene expression during virus growth, possibly related to the presence of multiple minicistrons in the 5′ untranslated sequence (BODEMER et al. 1986). The TS gene of HVA was mapped to the equivalent position near the right end of L-DNA (RICHTER et al. 1988).

Synthesis of deoxythymidylate monophosphate requires coordinated regulation of DHFR and TS genes (Fig. 4). In bacteriophage T4 and *Leishmania*, this coordination has been achieved by molecular linkage of the genes for these two enzymes (PUROHIT and MATHEWS 1984; BEVERLY et al. 1986). The virus-encoded TS and DHFR may possibly provide a growth advantage for HVS and HVA with their AT-rich genomes. However, DHFR is dispensable at least in vitro since DHFR-defective virus is still capable of replicating (DESROSIERS et al. 1985, 1986). The intron-less structure of the viral DHFR gene indicates that an RNA intermediate was involved in the viral acquisition of this cellular gene.

*Acknowledgements.* The work was supported by the Deutsche Forschungsgemeinschaft and the United States Public Health Service.

# References


Bankier AT, Dietrich W, Baer R, Barrell BG, Colbère-Garapin F, Fleckenstein B, Bodemer W (1985) Terminal repetitive sequences in herpesvirus saimiri virion DNA. J Virol 55: 133–139

Beverly SM, Ellenberger TE, Cordingley JS (1986) Primary structure of the gene encoding the bifunctional dihydrofolate reductase-thymidylate synthase of Leishmania major. Proc Natl Acad Sci USA 83: 2584–2588

Bodemer W, Niller HH, Nitsche N, Scholz B, Fleckenstein B (1986) Organization of the thymidylate synthase gene of herpesvirus saimiri. J Virol 60: 114–123

Bornkamm GW, Delius H, Fleckenstein B, Werner F-J, Mulder C (1976) Structure of herpesvirus saimiri genomes: arrangement of heavy and light sequences in the M genome. J Virol 19: 154–161

Chen M-J, Shimada T, Moulton AD, Cline A, Humphries RK, Maizel J, Nienhuis AW (1984) The functional human dihydrofolate reductase gene. J Biol Chem 259: 3933–3943

Desrosiers RC, Bakker A, Kamine J, Falk LA, Hunt RD, King NW (1985) A region of the herpesvirus saimiri genome required for oncogenicity. Science 228: 184–187

Desrosiers RC, Silva DP, Waldron LM, Letvin NL (1986) Nononcogenic deletion mutants of herpesvirus saimiri are defective for in vitro immortalization. J Virol 57: 701–705

Falk L, Johnson D, Deinhardt F (1978) Transformation of marmoset lymphocytes in vitro with herpesvirus ateles. Int J Cancer 21: 652

Fleckenstein B, Desrosiers RC (1982) Herpesvirus saimiri and herpesvirus ateles. In: Roiznan B (ed) The Herpesviruses Vol 1. Plenum, New York, pp 253–332

Fleckenstein B, Bornkamm GW, Mulder C, Werner F-J, Daniel MD, Falk LA, Delius H (1978) Herpesvirus ateles DNA and its homology with herpesvirus saimiri nucleic acid J Virol 25: 361–373

Galloway DA, McDougall JK (1983) The oncogenic potential of herpes simplex viruses: evidence for a "hit-and-run" mechanism. Nature 302: 21–24

Honess RW, Bodemer W, Cameron KR, Niller H-H, Fleckenstein B, Randall RE (1986) The A + T-rich genome of herpesvirus saimiri contains a highly conserved gene for thymidylate synthase. Proc Natl Acad Sci USA 83: 3604–3608

Kamine J, Bakker A, Desrosiers RC (1984) Mapping of RNA transcribed from a region of the herpesvirus saimiri genome required for oncogenicity. J Virol 52: 532–540

Kaschka-Dierich C, Falk L, Bjursell G, Adams A, Lindahl T (1977) Human lymphoblastoid cell lines derived from individuals without lymphoproliferative disease contain the same latent forms of Epstein-Barr virus DNA as those found in tumor cells. Int J Cancer 20: 173–180

Kaschka-Dierich C, Werner FJ, Bauer I, Fleckenstein B (1982) Structure of nonintegrated, circular herpesvirus saimiri and herpesvirus ateles genomes in tumor cell lines and in vitro-transformed cells. J Virol 44: 295–310

Koomey JM, Mulder C, Burghoff RL, Fleckenstein B, Desrosiers RC (1984) Deletion of DNA sequences in a nononcogenic variant of herpesvirus saimiri 11. J Virol 50: 662–665

Laufs R, Fleckenstein B (1973) Susceptibility to Herpesvirus saimiri and antibody development in Old and New World monkeys. Med Microbiol Immunol (Berl) 158: 227–236

Masters JN, Attardi G (1983) The nucleotide sequence of the cDNA coding for the human dihydrofolic acid reductase. Gene 21: 59–63

Medveczky P, Szomolanyi E, Desrosiers RC, Mulder C (1984) Classification of herpesvirus saimiri into three groups based on extreme variation in a DNA region required for oncogenicity. J Virol 52: 938–944

Purohit S, Mathews CK (1984) Nucleotide sequence reveals overlap between T4 phase genes encoding dihydrofolate reductase and thymidylate synthase. J Biol Chem 259: 6261–6266

Richter J, Puchtler I, Fleckenstein B (1988) The thymidylate synthase gene of herpesvirus ateles. J Virol 62: 3530–3535

Schirm S, Müller I, Desrosiers RC, Fleckenstein B (1984) Herpesvirus saimiri DNA in a lymphoid cell line established by in vitro transformation. J Virol 49: 938–946

Stamminger T, Honess RW, Young DF, Bodemer W, Blair ED, Fleckenstein B (1987) Organization of terminal reiterations in the virion DNA of herpesvirus saimiri. J Gen Virol 68: 1049–1066

Szomolanyi E, Medveczky P, Mulder C (1987) In vitro immortalization of marmoset cells with three subgroups of herpesvirus saimiri. J Virol 61: 3485–3490

Takeishi K, Kaneda S, Ayusawa D, Shimizu K, Gotoh O, Seno T (1985) Nucleotide sequence of a functional cDNA for human thymidylate synthase. Nucleic Acid Res 13: 2035–2043

Thompson R, Honess RW, Taylor L, Morran J, Davison AJ (1987) Varicella-zoster virus specifies a thymidylate synthetase. J Gen Virol 68: 1449–1455

Trimble JJ, Murthy SCS, Bakker A, Grassmann R, Desrosiers RC (1988) A gene for dihydrofolate reductase in a herpesvirus. Science 239: 1145–1147

Werner F-J, Bornkamm GW, Fleckenstein B (1977) Episomal viral DNA in a herpesvirus saimiri-transformed lymphoid cell line. J Virol 22: 794

# Part VI: Hepatitis B Virus and Liver Cancer

# Introduction

Today it is widely accepted that hepatitis B virus (HBV) is involved in the development of hepatocellular carcinoma (HCC). Consequently, besides smoking, papillomaviruses, and some chemicals, HBV might be considered as one of the most potent carcinogens worldwide. These assumptions are mainly based on two observations. (a) There is a strong positive correlation between chronic infection with HBV and liver cancer. In chronically infected people the risk factor for HCC goes up at least 100-fold and in some areas the life-time risk of death from HCC for infected male patients is as high as 40% (Beaseley 1982). (b) In woodchucks, experimental infections with woodchuck hepatitis virus, a related hepadnavirus, leads to HCC in 90% of animals within 1–2 years (POPPER et al. 1987).

However, it must be considered that in the human situation it may take up to 3 decades from primary infection with HBV to detection of HCC. This situation, which is, for example, very similar to human papillomavirus-associated cancers, makes it difficult to envisage a one-step mechanism. A multistep mechanism in which factors additional to HBV may play a role appears to be more likely.

So far, integrated HBV DNA has been detected in most of the tumors studied, whereas HBV replication ceases in most cases. The functional role of the integrated DNA is not yet clear. There are no oncogenes encoded in the viral genome which could become activated upon integration. Integration is not site-specific, and with one exception insertional mutagenesis within a known oncogene or related genes could not be detected (DEJEAN et al. 1986; see SCHLICHT and SCHALLER).

Only more recently in model experiments using different promoters with an attached CAT gene as indicator could a *trans*-activating function be demonstrated for the viral X gene as well as for integrated HBV sequences (for details see KOSHY and HOFSCHNEIDER). As shown by NAGAYA et al. (1987) HBx sequences are present and expressed in most of the approx. 50 studied tumor samples. Therefore, as a new hypothesis one may consider HBV-specific *trans*-activation of cellular genes as part of a more complex mechanism leading to HCC.

# References

Beaseley RP (1982) Hepatitis B virus as the etiological agent in hepatocellular carcinoma–epidemiologic considerations. Hepatology 2: 21–26

Dejean A, Bougueleret L, Grzeschik KH, Tiollais P (1986) Hepatitis B virus DNA integration in a sequence homologous to v-*erb* B and steroid receptor genes in hepatocellular carcinoma. Nature (London) 322: 70–73

Miyaki M, Sato C, Gotanda T, Matsui T, Mishiro S, Imai M, Mayumi M (1986) Integration of region X of hepatitis B virus genome in human primary hepatocellular carcinomas propagated in nude mice. J. Gen. Virol. 67: 1449–1454

Popper H, Roth L, Purcell PH, Tennant BC, Gerin JL (1987) Hepatocarcinogenicity of the woodchuck hepatitis virus. Proc Natl Acad Sci USA 84: 866–870

Nagaya T, Nakamura T, Tokino T, Tsurimoto T, Imai M, Mayumi T, Kamino K, Yamamura K, Matsubara K (1987) The mode of hepatitis B virus integration in chromosomes of human hepatocellular carcinoma. Genes & Development 1: 773–782

# Analysis of Hepatitis B Virus Gene Functions in Tissue Culture and In Vivo

H.-J. Schlicht and H. Schaller

## 1 Introduction

The hepatitis B viruses, also called hepadnaviruses, are enveloped DNA viruses which primarily infect liver cells (reviewed by Ganem and Varmus 1987). The most important member of this virus family is the human hepatitis B virus (HBV; Dane et al. 1970), which is the causative agent of a severe form of hepatitis in humans. Since the virus is preferentially transmitted by contact with infected blood or blood products, this form of hepatitis was called serum hepatitis or hepatitis B in contrast to hepatitis A, which is caused by a picornavirus typically transmitted by contaminated food.

Although the course of the disease can be quite severe, HBV infections are usually self-limited, and most patients recover completely. However, a certain proportion of those infected enter a chronic course which is characterized by the continuous presence of virus or viral proteins in the serum (Hoofnagle et al. 1987). Apparently, in these patients at least some of the HBV-infected cells are not eliminated by the immune system and therefore, since HBV is not cytopathic, become permanent virus producers. The factors which determine whether the infection becomes chronic or not are unknown. However, the immune status appears to be of major importance since up to $100\%$ of infected newborns, but only about $5\%$ of infected adults, become chronic virus carriers.

The consequences of chronic HBV infection are highly variable. Whereas many virus carriers are in good health, others show signs of a chronic active hepatitis with sporadic jaundice. However, all chronically infected people are at high risk of developing liver cirrhosis or liver cancer (Popper et al. 1987a). Therefore, besides Epstein-Barr virus, HTLV-I, and certain papillomaviruses, HBV belongs to the small group of viruses pathogenic in humans which have been shown to cause or to promote tumor development. Since worldwide several hundred million people, especially in the Far East, are chronically infected with HBV, the HBVs are currently under intense investigation.

Zentrum für Molekulare Biologie, Universität Heidelberg, Im Neuenheimer Feld 282, 6900 Heidelberg 1 FRG

Current Topics in Microbiology and Immunology, Vol. 144
© Springer-Verlag Berlin · Heidelberg 1989

## 2 Virus Structure and Replication

In this chapter we will describe the organization of the virus particle, the genome structure, and the replication strategy, as it is essential for the understanding of the viral life cycle. For a comprehensive discussion of HBV molecular biology, the reader is referred to a recent review article (GANEM and VARMUS 1987).

The hepadnaviruses are enveloped viruses with a simple morphology (Fig. 1a). In the case of HBV, the complete virus particle is also referred to as the Dane particle (DANE et al. 1970). The envelope contains several partially glycosylated transmembrane proteins which all possess the same *C*-terminus, but differ in the *N*-terminal sequences (referred to as S-, preS1-, and preS2-antigens or HBsAg and HBpreSAg). Enclosed by this envelope is the viral core particle which consists of about 180 copies of a single protein species (referred to as core protein or HBcAg). The core particle contains the viral genome and the viral DNA polymerase which also has reverse transcriptase activity. The genome consists of a partially double-stranded DNA molecule with a length of slightly more than 3000 base pairs (Fig. 1b). At the 5′-end of the minus strand, a protein is covalently bound (referred to as DNA-linked terminal protein) which is believed to function as a primer for genome replication. Due to the bound protein, the DNA molecule is not covalently closed. During plus strand synthesis, which starts several hundred nucleotides upstream from the 3′-end of the minus strand in a region called DR2, the circular structure is achieved due to a template switch of the nascent plus strand. Usually at the time when the virus particle is exported from the cell, the plus strand has not been completed. Thus, a large gap remains within this strand which is repaired only after entry of the virus into a new host cell.

Cloning and sequence analysis of the HBV genome revealed four overlapping open reading frames which are exclusively located on one strand (Fig. 1b). The C-gene encodes the core protein. At its 3′-end it overlaps with the P-gene which encodes the polymerase. Completely overlapping with the P-gene is the preS/S-gene which encodes the surface proteins. The fourth open reading frame, which overlaps with the P- and the C-genes, is called the X-gene because its significance is unclear. Very recently, evidence has been obtained that the X-gene product might function as a *trans*-activator of gene expression (SPANDAU and LEE 1988 KOSHY and HOFSCHNEIDER, this volume).

Basically, two types of mRNAs are transcribed from the genome (Fig. 1b). The large transcripts of about 3.5 kb start near the 5′-end of the C-gene and have more than genome length. Because of the heterogenous 5′-ends, translation of these mRNAs gives rise to at least two different core gene products. Whereas one of these proteins (the HBcAg) forms the viral capsid, the other (the HBeAg) is secreted from the infected cells. A possible function of this secretory core gene product will be discussed below. Furthermore, recent evidence obtained in our laboratory strongly suggests that the viral polymerase is also encoded by one of the long mRNAs.

The short transcripts of about 2.1–2.4 kb start within or 5′ from the preS/S-gene and encode the surface proteins. Since these mRNAs only differ with respect to their 5′ sequences, they give rise to proteins with different *N*-terminal but identical

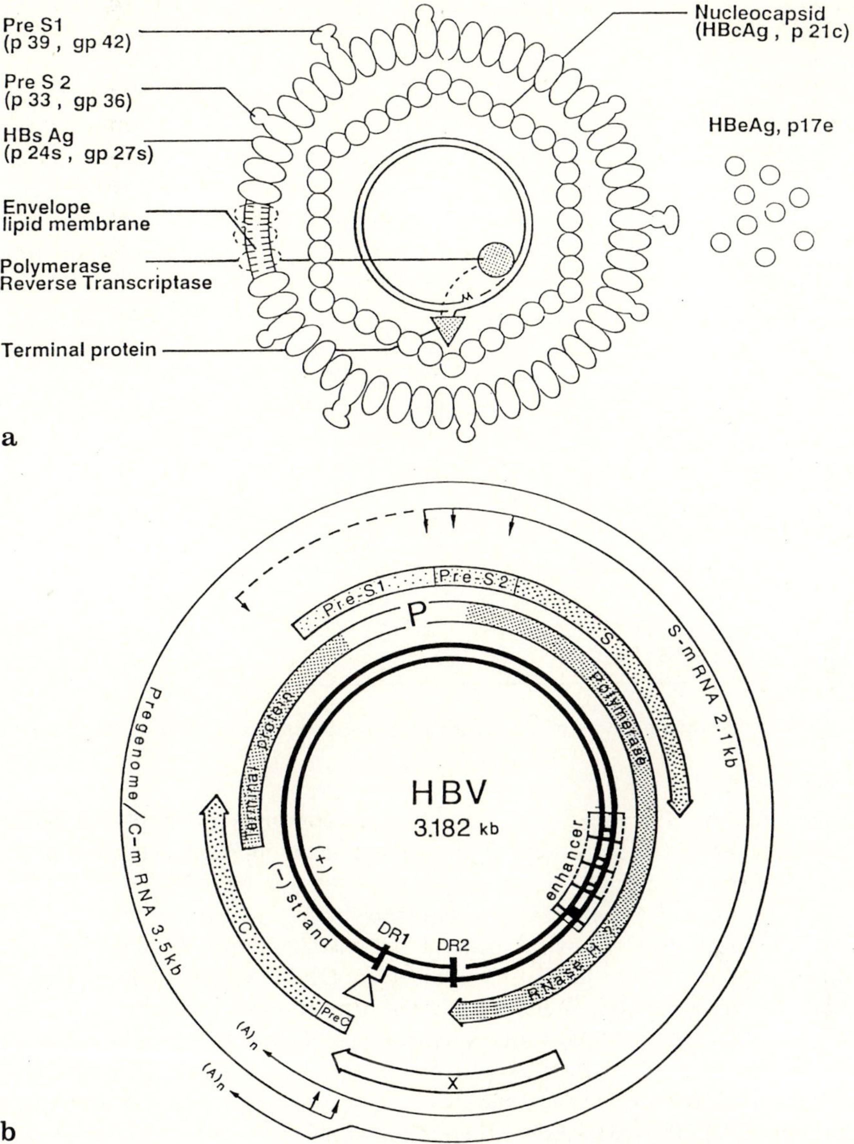


**Fig. 1a, b.** Hepatitis B virus (HBV) structure and genome organization. **a** Schematic representation of the HBV particle. **b** Genetic, physical, and transcriptional map of the HBV genome. The *inner circles* represent the viral DNA. The nicked minus strand and the incomplete plus strand are depicted as *heavy lines.* The four open reading frames as defined by their respective start and stop codons are depicted as *arrows.* The *outer lines* represent the major viral transcripts

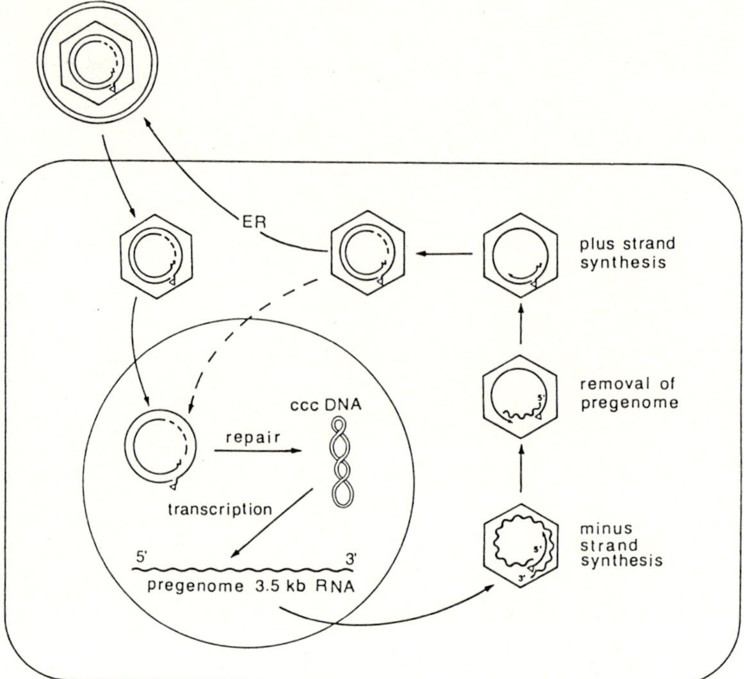


**Fig. 2.** Proposed replication pathway of the hepatitis B viruses. Details are given in the text

*C*-terminal domains. For unknown reasons, large amounts of empty envelopes which contain viral surface proteins are secreted from infected cells during the course of infection.

The replication strategy of HBV is outlined in Fig. 2. After entry of the virus into the host cell, the viral genome is uncoated and repaired to form a covalently closed circular DNA molecule (ccc DNA) which then is used as a template for transcription. Whereas the short transcripts function as normal mRNAs, one of the large 3.5 kb transcripts (the pregenome) is not only translated into viral protein but also serves as a template for the synthesis of a new viral genome. This process is believed to start with the encapsidation of this RNA into a viral core particle. DNA synthesis is then initiated, possibly primed by the protein which remains covalently bound to the 5'- end of the newly synthesized DNA minus strand. During reverse transcription, the RNA template is degraded. A small fragment of its 5'-end, however, remains intact and serves as a primer for plus strand synthesis. As has been noted above, virus budding, which appears to take place at the endoplasmic reticulum (ER), usually occurs before plus strand synthesis has been completed.

One characteristic feature of HBV infection is that replication takes place almost exclusively in differentiated hepatocytes. This is at least in part due to the fact that expression of the pregenomic RNA, which serves as a template for reverse transcription, is controlled by a tissue-specific promoter which requires activation by cellular proteins which are only produced in a few differentiated tissues. In fact, experiments undertaken with one of the animal viruses have conclusively shown that primary hepatocytes obtained by liver perfusion lose their competence for virus production within a few days in culture (TUTTLEMAN et al. 1986; GALLE et al. 1988). As will be discussed below, only recently have permanent cell lines been found which are differentiated enough to support virus replication.

## 3 Possible Mechanisms of Carcinogenesis

In this section we will summarize some of the ideas as to how HBV infection can result in the development of liver cancer. In the following sections we will then show how the in vivo and in vitro systems which are now available can be used to test some of these hypotheses experimentally.

Firm evidence that HBV infection can result in tumor formation has been obtained by epidemiological studies as well as by clinical observations (POPPER et al. 1987a). These investigations show that people chronically infected with HBV are at high risk of developing liver cirrhosis or liver cancer whereas people who suffered only from acute HBV infection are not. Consequently, the major risk factor appears to be the continuous presence of virus or viral components which, by unknown mechanisms, lead to a malignant transformation and tumor development usually 20–40 years after primary HBV infection. Since infection becomes chronic in only a few percent of adults, but in up to 100% of newborns, the immune status of the affected individuals must play a crucial role in this determination. Therefore, it appears likely that the hepatitis B-viruses have developed mechanisms by which elimination of infected cells is at least partially prevented. Probably the secretion of large amounts of surface proteins and core gene products (HBeAg) by the infected cells leads to an irritation of the immune system which is then unable to eliminate these cells efficiently.

One consequence of sustained HBV infection is the integration of viral DNA sequences into the host cell chromosome. Since there are numerous examples of transforming retroviruses which, like HBV, also replicate by reverse transcription and integrate into the cellular genome, much effort has been undertaken to test whether there are any similarities between retroviruses and hepadnaviruses with respect to the mechanism of cell transformation. However, since virtually all HBV integrates are rearranged and therefore can no longer give rise to infectious virus, integration is not an essential step during the HBV life cycle but rather appears to be a dead end street. Interestingly, with rare exceptions, these rearrangements interrupt the core gene, whereas the S-gene usually remains intact. This finding suggests that there is a counter-selection against cells which express core protein, possibly because of the host's immune response. On the other hand, these findings also show that the continuous presence of infectious virus is not necessary for tumor induction.

To date, it is still unclear as to how integration of viral sequences is related to cell transformation. Since analysis of many different tumors did not reveal any common integration site, integration of HBV sequences appears to be random (NAGAYA et al. 1987). Therefore, it seems unlikely that the malignancies are in general due to the activation of cellular oncogenes by viral promoters or enhancers. Moreover, no oncogenes have been demonstrated within the HBV genome. However, the recent finding that the HBV X-gene product might function as a *trans*-activator of gene expression has again stimulated the idea that cell transformation might be due to an alteration of cell metabolism caused by a viral gene product (see KOSHY and HOFSCHNEIDER, this volume). The significance of the X-gene product for tumor induction, however, remains open, since the X-gene is interrupted in most of the integrated HBV genomes isolated from liver tumors (NAGAYA et al. 1987).

Another viral protein whose action could unbalance the host cell metabolism is the viral polymerase. As this enzyme has the characteristics of a reverse transcriptase, it has been speculated that transformation might be caused by gene amplification due to reverse transcription of cellular RNAs and integration of the generated DNA copies. A prerequisite for such a mechanism would be the availability of active reverse transcriptase in the cytoplasm of infected cells. Although it is assumed that the HBV reverse transcriptase is first synthesized as an enzymatically inactive precursor which becomes activated only after incorporation into a viral core particle, it cannot be ruled out that small amounts of the precursor can also be activated outside core particles. Furthermore, it has been shown for retroviruses that cellular mRNAs can be packaged into virus particles and, after reverse transcription, can be transduced as DNA copies into a new host cell (LINIAL 1987). In principal, such a mechanism appears possible also for HBV.

Probably the most coherent theory to explain how HBV infection could induce hepatocellular carcinoma is based on the observation that the most frequent precursor lesion of HBV-induced liver cancer, at least in Western countries, is liver cirrhosis (POPPER et al. 1987a). This cirrhosis appears to be due to a chronic, cytotoxic, immune response directed against hepatocytes expressing viral proteins. If so, this process, referred to as necroinflammation, should result in a continuous destruction of liver tissue which, in the long run, is only partially replaced. Furthermore, the need to substitute for the destroyed hepatocytes provides a permanent mitogenic stimulus for the hepatic tissue, which therefore is in a continuous state of proliferation. It can well be imagined that this status, which can last several decades, eventually gives rise to a transformed hepatocyte which then develops into a tumor.

Though very attractive, this theory still awaits confirmation. One major problem is that, to date, no cytotoxic immune cells have been unambiguously demonstrated in HBV-infected individuals. However, examination of the HBV-specific cellular immune response is still very preliminary. Thus, future studies, which will employ the whole machinery of modern immunological techniques, must show whether tumor development is due to a long-lasting destruction of the liver tissue.


## 4 Animal Model Systems

A major obstacle for the analysis of HBV infection was, and still is, the very narrow host range of this virus which, besides man, is only infectious for chimpanzees. Though the limited availability of chimpanzees prevents systematic investigations, a major contribution of this system during the early phase of HBV research was the unambiguous demonstration that one of the HBV genomes which had been cloned and sequenced was indeed infectious (WILL et al. 1982, 1985).

The need for an HBV model system more amenable to experimental studies then led to the discovery of several animal pathogenic hepadnaviruses. Furthermore, as a completely different approach to assess the consequences of the permanent production of HBV proteins, transgenic mice have been generated (CHISARI et al. 1985; BABINET et al. 1985). In order to mimic a common state of chronic HBV infection, the gene encoding one of the HBV surface proteins was injected into fertilized mouse eggs, and offspring were generated which permanently produced the major HBV surface antigen. In the reports published to date, no disease symptoms have been described with respect to these animals.

In the following we will concentrate on two well-characterized hepadnaviruses which infect animals, the duck (DHBV) and the woodchuck (WHV) forms, since they have proven to be especially useful for the detailed analysis of many aspects of hepadnavirus infection. For all hepadnaviruses it has been shown that infection can be achieved by intrahepatic injection of cloned viral DNA. Thus, with these systems it is possible to assess the infectivity and pathogenicity of virus mutants in vivo.

The major advantage of the woodchuck system is that it represents an excellent model system for the study of tumor induction. The first evidence that WHV infection can cause liver tumor in woodchucks was a very high incidence of hepatocellular carcinoma in a woodchuck colony held in captivity in the Philadelphia Zoo. Indeed, it was this accumulation of liver cancer that resulted in the isolation and characterization of the virus (SUMMERS et al. 1978). Recently, it could be convincingly shown that chronic WHV infection can result in hepatocellular carcinoma in the absence of other risk factors (POPPER et al. 1987 b). In this study, woodchucks were infected with WHV either immediately after birth or as adults. Within 17–36 months, all animals (8 out of 8) which had become chronic WHV carriers developed hepatocellular carcinoma. In the future with this system it should be possible to examine in detail the pathological changes which are connected to chronic WHV infection and to correlate these changes to cancer development. Particularly, this system should allow the definition of the mechanism by which chronic infection is established and to assess the contribution of the host immune response with respect to liver cell damage.

A major advantage of the duck system compared with WHV is the better availability of the experimental animals. Moreover, since there is now also a tissue culture system for DHBV (see below), the duck virus is very well suited for the analysis of gene functions common to all hapadnaviruses, especially by mutational analysis. However, since development of liver cancer appears to be rather rare in ducks chronically infected with DHBV, this system is not suited for the study of hepadnavirus-induced tumors. Whether this finding is due to differences in the immune reactions between mammals and birds or reflects a different hepato-carcinogenicity of the viruses remains to be seen. One major difference between the mammalian and the duck virus is that there is no X-gene in DHBV. While it has been shown that the X-gene product is not essential for the production of HBV in tissue culture (YAGINUMA et al. 1987; unpublished data from our laboratory), to date there has been no conclusive study which assesses the significance of the putative X-gene product in vivo.

## 5 Tissue Culture Systems

A major breakthrough in HBV research was the recent discovery that certain permanent human hepatoma cell lines (HepG2 und HUH-7) can support HBV replication in tissue culture following transfection with cloned viral DNA (SUREAU et al. 1986; CHANG et al. 1987; TSURIMOTO et al., 1987 YAGINUMA et al. 1987). Though the cells cannot be infected and therefore may lack the virus receptor, they are differentiated enough to give rise to HBV particles which are infectious in vivo (ACS et al. 1987). Thus, the HepG2 and the HUH-7 cell lines are currently intensively used for the elucidation of the molecular details of HBV gene expression and replication.

A major drawbeck to HBV study is that the infectivity of the virus particles obtained cannot be routinely tested in vivo since such experiments have to be carried out with chimpanzees. Therefore, we examined whether the respective human cell lines could also support replication of DHBV. These experiments showed that large amounts of infectious DHBV particles are produced by HepG2 cells upon transfection with cloned DHBV DNA (GALLE et al. 1988). Since it has also been shown by us and others that primary liver cells obtained by perfusion can be infected in vitro (TUTTLEMAN et al. 1986; SCHLICHT et al. 1987a), the duck system now allows the investigation of virus mutants in tissue culture and in vivo. Mutated genomes can be transiently expressed in HepG2 cells, and the biosynthesis of virus and virus components can be analyzed biochemically. If virus particles are produced, they can be subsequently analyzed for infectivity or pathogenicity either in vitro by using primary liver cell cultures or in vivo. Moreover, the fact that human cell lines can give rise to infectious DHBV shows that the narrow host range which is characteristic for hepdnaviruses is not due to species-specific factors regulating gene expression but, most likely, to species-specific host cell receptors.

A first application of this new system was the analysis of the different proteins which are encoded by the viral core gene (SCHLICHT et al. 1987b; JUNKER et al. 1987). It has been known for a long time that usually two core gene products which react differently with certain antisera could be detected in HBV-infected individuals. One of these proteins, referred to as HBcAg, was predominantly found in hepatocytes and was later shown to form the viral capsid. The other protein, referred to as HBeAg, was predominantly detected in the serum. Since treatment with proteases or detergents converted the HBcAg into a protein which reacted with HBeAg-specific antisera, it was assumed for a long time that HBeAg represents a degraded form of the HBcAg. However, the molecular analysis of HBV and DHBV core gene expression shows that production of these two proteins is due to the selective usage of a signal sequence which is located in front of the C-gene. If the core protein is synthesized without the signal sequence, the protein remains intracellular and aggregates to form core particles. However, if the protein is synthesized with the signal sequence, it enters the secretory pathway and is secreted after proteolytic processing.

What is the significance of the secretory core protein? Since such proteins are produced not only during HBV but also during DHBV infection, generation of such proteins appears to be beneficial for these viruses. However, as could be shown in the duck system, production of the secretory core protein is not essential for the development of an acute infection (SCHLICHT et al. 1987b). Thus, it is tempting to speculate that these proteins play a role in the establishment of the chronic carrier state, probably by modulation of the HBcAg-specific immune response.

This new experimental system also allowed us to analyze the mode of polymerase synthesis. Interestingly, the data obtained to date strongly suggest that the strategy used for polymerase synthesis and initiation of reverse transcription differs fundamentally between retroviruses and HBVs (SCHLICHT et al., 1989; outlined in Fig. 3). According to these data, hapdnaviral reverse transcriptase is not synthesized via a core/polymerase fusion protein, as is the case for retroviruses, but rather as an authentic P-gene product. Synthesis of this P-protein starts by internal translation initiation within the long 3.5-kb mRNA which also encodes the core protein and serves as the template for reverse transcription. Since this mRNA apparently gives rise to two different proteins, it is here referred to as the C/P mRNA. The P-protein thus generated consists of at least two domains (Fig. 3, black and hatched boxes; SPRENGEL et al. 1985) which are separated by a spacer domain that at least in part can be deleted without influencing polymerase function (Fig. 3, open

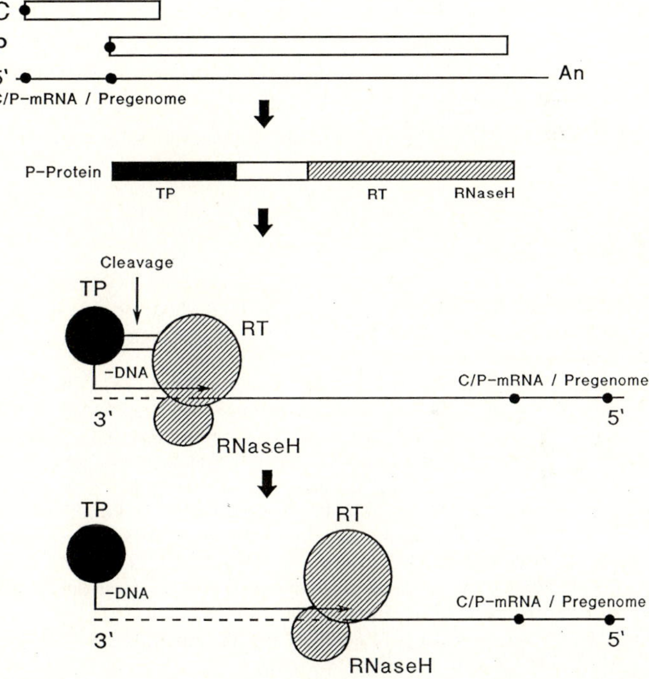



**Fig. 3.** Expression strategy of the hepadnaviral P-gene and a model for initiation of reverse transcription. The *open bars* in the upper part of the figure represent the viral C- and P-genes. The respective translation initiation codons are shown as *dark dots*. Internal initiation within the long C/P-mRNA gives rise to a P-protein which appears to consist of two domains. Upon binding of the P-protein to the pregenomic RNA, which probably is accompanied by a conformational change of the primary P-protein, reverse transcription is initiated. The P-protein may have to be cleaved within the spacer region before reverse transcription can proceed. Further details are given in the text. *TP*, terminal protein; *RT*, reverse transcriptase

box; unpublished data from our laboratory). The N-terminal domain most likely represents the terminal protein (TP) which functions as the primer for reverse transcription (BARTENSCHLAGER et al., 1988). The C-terminal domain contains the reverse transcriptase (RT) as well as the RNaseH activity (RADZIWILL et al., manuscript in preparation). Upon binding of this multifunctional protein to the pregenomic RNA, reverse transcription is initiated. Elongation of the DNA strand may require cleavage of the P-protein within the spacer region. Reverse transcription then proceeds with the primer protein remaining covalently linked to the newly synthesized DNA strand.

As demonstrated by the examples described above, there are now powerful experimental systems available which allow the analysis of all aspects of viral gene functions on the molecular level. Thus, it can be expected that it will be possible in the near future to gain a much deeper insight into the mechanisms by which liver disease can be caused by hepatitis B viruses. With this knowledge it might then be possible to develop specific drugs which interfere with the viral life cycle. With several hundred million, chronically HBV-infected individuals worldwide, the need for such a therapy is evident.

*Acknowledgement*. We thank R. Bartenschlager, V. Bosch, and G. Radziwill for the communication of unpublished results. We also thank W. Tucker and R. Rigg for critical reading of the manuscript. This work was supported by the Deutsche Forschungsgemeinschaft, SFB 229, the Bundesministerium für Forschung und Technologie, BCT 0381-5, and the Fonds der Chemischen Industrie.

# References

Acs G, Sells M, Purcell R, Price P, Engle R, Shapiro M, Popper H (1987) Hepatitis B virus produced by transfected Hep G2 cells causes hepatitis in chimpanzees. Proc Natl Acad Sci USA 84: 4641–4644

Babinet C, Farza H, Morello D, Hadchouel M, Pourcel C (1985) Specific expression of hepatitis B surface antigen (HBsAg) in transgenic mice. Science 230: 1160–1163

Bartenschlager R, Schaller H (1988) The amino terminal domain of the hepadnaviral P-gene encodes the terminal protein (genome linked protein) believed to prime reverse transcription The EMBO Journal 7: 4185–4192.

Chisari F, Pinkert C, Milich D, Filipi P, McLachlan A, Palmiter R, Brinster R (1985) A transgenic mouse model of the chronic hepatitis B surface antigen carrier state. Science 230: 1157–1160

Chang C, Jeng K, Hu C, Lo S, Su T, Ting L, Chou C, Han S, Pfaff E, Salfeld J, Schaller H (1987) Production of hepatitis B virus in vitro by transient expression of cloned HBV DNA in a hepatoma cell line. EMBO J. 6: 675–680

Dane D, Cameron C, Biggs M (1970) Virus like particles in serum of patients with Australia antigen associated hepatitis. Lancet 1: 695–698

Galle P, Schlicht HJ, Fischer M, Schaller H (1988) Production of infectious duck hepatitis B virus in a human hepatoma cell line. J Virol 62: 1736–1740

Ganem D, Varmus H (1987) The molecular biology of the hepatitis-B viruses. Ann. Rev. Biochem. 56: 651–693

Hoofnagle J, Shafritz D, Popper H (1987) Chronic type B hepatitis and the "healthy" HBsAg carrier state. Hepatology 7: 758–763

Junker M, Galle P, Schaller H (1987) Expression and replication of the hepatitis-B virus genome under foreign promoter control Nucleic Acids Res 15: 10117–10132

Linial M (1987) Creation of a processed pseudogene by retroviral infection. Cell 49: 93–102

Nagaya T, Nakamura T, Tokino T, Tsurimoto T, Imai M, Mayumi T, Kamino K, Yamamura K, Matsubara K (1987) The mode of hepatitis B virus DNA integration in chromosomes of human hepatocellular carcinoma. Genes and Development 1: 773–782

Popper H, Shafritz D, Hoofnagle J (1987a) Relation of the hepatitis B virus carrier state to hepatocellular carcinoma. Hepatology 7: 764–772

Popper H, Roth L, Purcell R, Tennant B, Gerin J (1987b) Hepatocarcinogenicity of the woodchuck hepatitis virus. Proc Natl Acad Sci USA 84: 866–870

Schlicht HJ, Galle P, Schaller H (1987a) The hepatitis B viruses: molecular biology and recent tissue culture systems J Cell Sci [Suppl] 7: 197–212

Schlicht HJ, Salfeld J, Schaller H (1987b) The pre-C region of the duck hepatitis-B virus is essential for synthesis and secretion of processed core proteins but not for virus formation. J Virol 61: 3701–3709

Schlicht HJ, Radziwill G, Schaller H (1989) Synthesis and encapsidation of duck hepatitis B virus reverse transcriptase do not require formation of core polymerase fusion proteins. Cell 56: 85–92

Spandau D, Lee C (1988) *trans*-Activation of viral enhancers by the hepatitis B virus X protein. J Virol 62: 427–434

Sprengel R, Kuhn C, Will H, Schaller H (1985) Comparative sequence analysis of duck and human hepatitis-B virus genomes. J Med Virol 15: 323–333

Summers J, Smolec J, Snyder R (1978) A virus similar to human hepatitis B virus associated with hepatitis and hepatoma in woodchucks. Proc Natl Acad Sci USA 75: 4533–4537

Sureau C, Roup-Lemonne J, Mullins J, Essex M (1986) Production of hepatitis B virus by a differentiated human hepatoma cell line after transfection with cloned circular HBV DNA. Cell 47: 37–47

Tsurimoto T, Fujiyama A, Matsubara K (1987) Stable expression and replication of hepatitis B virus genome in an integrated state in a human hepatoma cell line transfected with the cloned viral DNA. Proc Natl Acad Sci USA 84: 444–448

Tuttleman J, Pugh J, Summers J (1986) In vitro experimental infection of primary duck hepatocyte cultures with duck hepatitis-B virus. J Virol 58: 17–25

Will H, Cattaneo R, Koch H, Darai G, Schaller H, Schellekens P, van Eerd C, Deinhardt F (1982) Cloned HBV DNA causes hepatitis in chimpanzees. Nature (London) 299: 740–741

Will H, Cattaneo R, Darai G, Deinhardt F, Schellekens H, Schaller H (1985) Infectious hepatitis B virus from cloned DNA of known nucleotide sequence. Proc Natl Acad Sci USA 82: 891–895

Yaginuma K, Shirakata Y, Kobayashi M, Koike K (1987) Hepatitis B virus (HBV) particles are produced in a cell culture system by transient expression of transfected HBV DNA. Proc Natl Acad Sci USA 84: 2678–2682

# Transactivation by Hepatitis B Virus May Contribute to Hepatocarcinogenesis

R. KOSHY and P. H. HOFSCHNEIDER

## 1 Introduction

Hepatitis B virus (HBV) is a major factor in the development of primary liver cancer in many areas of the world (BEASLEY 1982). The mechanisms by which HBV causes cell transformation are not understood (see SCHLICHT and SCHALLER, this volume). Viral DNA is integrated into hepatocyte chromosomes early after infection, but the cellular sites of integration are varied, and no common sites have been observed in tumors thus far (KOSHY 1987). Furthermore, there is no evidence of integration at a genetic locus known to be concerned with transformation or cell proliferation. A single exception to this is a recently reported case in which HBV DNA was inserted into host DNA sequences homologous to the cellular v-*erb*-A protooncogene and certain steroid hormone receptors (DEJEAN et al. 1986; DETHE et al. 1987) and which has since been shown to encode a novel retinoic acid receptor (BRAND et al. 1988; BENBROOK et al. 1988). It is also known that the integration of HBV DNA promotes genetic instability by chromosome rearrangements including deletion of cellular DNA and translocations (KOCH et al. 1984; ROGLER et al. 1987). The significance of these findings is at present unclear. Thus, as a rule, there is little evidence to provide a basis for a common *cis*-acting virus function in tumorigenesis.

The elucidation (SUMMERS and MASON 1982) of the replicative strategy of hepadnaviruses involving a viral reverse transcriptase (hitherto a distinctive feature of retroviruses) has stimulated further comparisons between HBV and retroviruses. Such studies have revealed analogies in the organization of genes on their respective genomes (MILLER and ROBINSON 1986). The X open reading frame (ORF) of HBV is one of four viral genes (see Fig. 1 in SCHLICHT and SCHALLER, this volume) and the only one without a known function. This gene occupies a position corresponding to the *tat* gene of human T lymphotropic viruses (HTLV) which encodes a *trans-activating* function necessary for virus replication (WONG-STAAL and GALLO 1985) and which has also been suggested to be a factor in T-lymphocyte transformation (CROSS et al. 1987). These reasons encouraged us to investigate the X ORF of HBV for *trans*-activational activity. This chapter describes these investigations and discusses the implication of the properties of the X-gene in the development of HBV-associated hepatoma.

Max-Planck Institut für Biochemie, 8033 Martinsried bei München, FRG

Current Topics in Microbiology and Immunology, Vol. 144
© Springer-Verlag Berlin · Heidelberg 1989

## 2 Stimulation of Gene Expression by the X-Gene

In order to test whether the HBV X-gene has the ability to stimulate gene expression, plasmid constructs consisting of the X ORF and 5′-sequences were constructed. These plasmids were cotransfected into cells of human liver origin (CCL13) along with pSV2cat plasmid (GORMAN et al. 1982) containing the chloramphenicol acetyl transferase gene (CAT) under control of the SV40 early promoter. The measure of stimulation of CAT expression was provided by the amount of acetylation of [$^{14}$C]-labelled chloramphenicol in lysates of transfected cells.

The results of these experiments are depicted in Fig. 1. In the constructs used in these experiments the preS- and S-genes were retained for the expression of the X-gene. The plasmids used were as follows: Plasmid pHBV 2836 consisted of a 2.3-kb *Bgl*II-*Bgl*II fragment of HBV DNA bearing the preS- and S-genes and their promoters as well as the X ORF and the downstream termination sequences. This contiguous fragment of DNA has the X ORF which can encode a polypeptide of 154 amino acids having a molecular wight of 16560 daltons (for a review, see TIOLLAIS et al. 1985) and also the HBV enhancer element (SHAUL et al 1985; TOGNONI et al. 1985) at map position 1000–1200. This plasmid very efficiently stimulates CAT expression (ca. 50-fold). In order to investigate whether the observed effect is due to the presence of the enhancer, plasmid pHBV 2836ΔNco/Xho was constructed in which the enhancer was deleted. This plasmid stimulates CAT expression to a similar degree as pHBV 2836, indicating that the enhancer is not the reason for the activity. Furthermore, the enhancer appeared not to influence the level of expression in these experiments in which the X transcripts arose from the S- or preS-promoters. When the promoter region (5′ to the X ORF) was removed, as in pHBV 824 and pHBV 1371 there was no stimulation of CAT expression in transfected cells. The expression of transcripts was investigated by Northern blot analyses of RNA isolated from cells transfected with the plasmids described. X-specific transcripts were made in cells transfected with pHBV 2836 and pHBV 2836 ΔNco/Xho, whereas no RNA was seen in cells transfected with pHBV 1371, which is devoid of a promoter (Fig. 2). Thus it was clear from these results that pSV2cat stimulation was a function of X-gene expression.

**Fig. 1.** HBV plasmid constructs and CAT expression in CCL13 cells. Levels of *trans*-activation in CCL13 cells by cloned subgenomic HBV fragments containing the X-gene. The different HBV plasmids are depicted as well as a physical map of the relevant genome sequences as reference. CAT expression from pSV2cat in the presence of the respective test plasmids is shown to the *right* of the figure. A 10-fold molar excess of the cotransfected test plasmids over pSV2cat was used in experiments with HBV promoters. For pU3R-IXΔ cat and pRSVX the molar ratios of test and indicator plasmids were 1:1 and 4:1, respectively. Optimum molar ratios were experimentally determined. Negative control CAT assays were performed with lysates of cells transfected with pSV2cat plasmid alone (for pU3R-IXΔ cat and pRSVX) and in other experiments with cotransfected vector plasmid pMLΔBS. Drawings are not made to scale

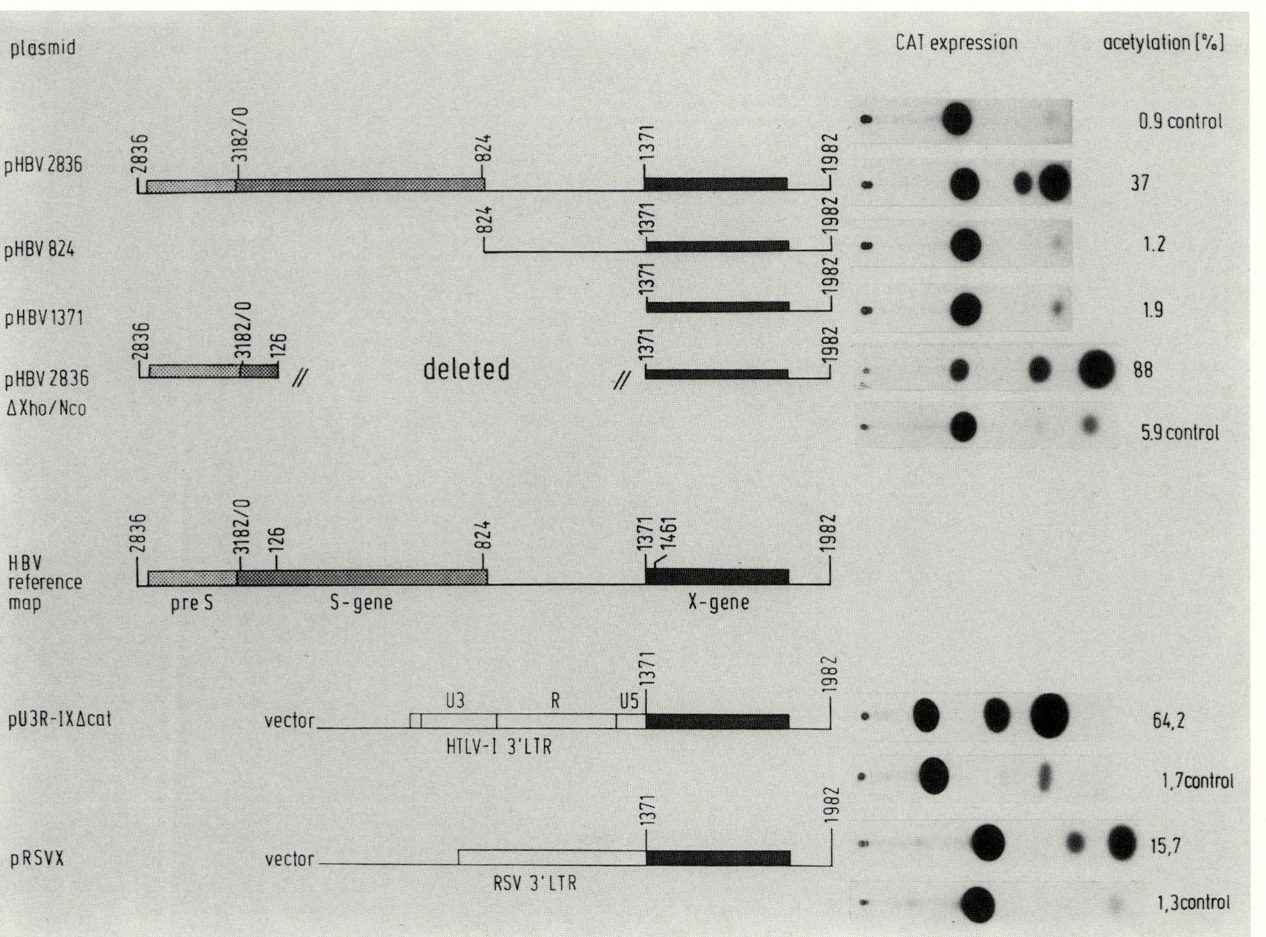
plasmid
CAT expression
acetylation [%]
pHBV 2836
2836
3182/0
824
1371
1982
0.9 control
37
pHBV 824
824
1371
1982
1.2
pHBV 1371
1371
1982
1.9
pHBV 2836 ΔXho/Nco
2836
3182/0
126
deleted
1371
1982
88
5.9 control
HBV reference map
2836
3182/0
126
824
1371
1461
1982
pre S
S-gene
X-gene
pU3R-IXΔcat
vector
U3
R
U5
1371
1982
HTLV-I 3'LTR
64,2
1,7control
pRSVX
vector
1371
1982
RSV 3'LTR
15,7
1,3control

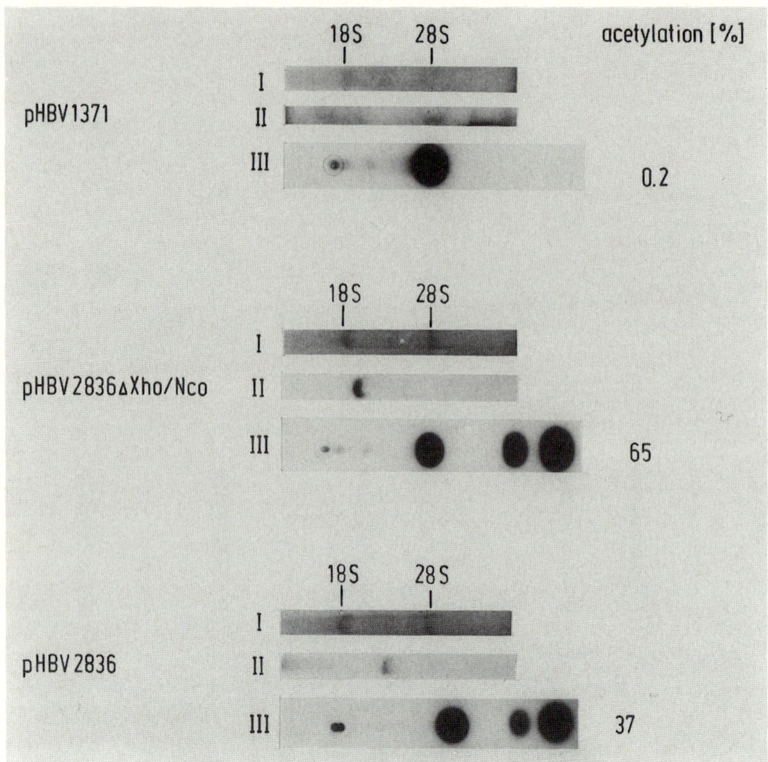



**Fig. 2.** *trans*-Activation of pSV2cat is dependent on the presence of an X-specific mRNA. Northern analyses of total RNA (*II*) and CAT activity (*III*) of extracts from CCL13 cells transfected with pSV2cat and test plasmids pHBV1371 (molar ratio 1:10), pHBV2836Δ Xho/Nco, and pHBV2836 are shown. Methylene blue staining of the filters (*I*) indicates comparable amounts of total RNA in each case. The percentage of [$^{14}$C]-labelled chloramphenicol acetylated is also given. The value for pHBV2836Δ Xho/Nco is taken from an independent experiment to that depicted in Fig. 1

# 3 Identification of the Gene Promoter

In hepatocytes naturally infected with HBV the major viral RNA transcripts are 2.3 kb and 3.5 kb in size arising from the S-promoter and the C-promoter, respectively (for a review, see TIOLLAIS et al. 1985), either of which may give rise to an X protein. However, in cells transfected with viral DNA a smaller transcript has been seen (GOUGH 1983), which suggests the presence of another promoter immediately upstream of the X ORF. We investigated this by means of a plasmid construct consisting of an *Acc*I-*Nco*I fragment of HBV (nucleotides 824–1374) containing the putative X promoter and the HBV enhancer cloned upstream of the CAT gene in a promoterless vector (pCAT3M). Clones were obtained with the HBV fragment in both orientations (pHBVXcat+ and pHBVXcat—) with respect to the CAT gene (Fig. 3). Plasmids with the HBV DNA in the same orientation as the CAT gene

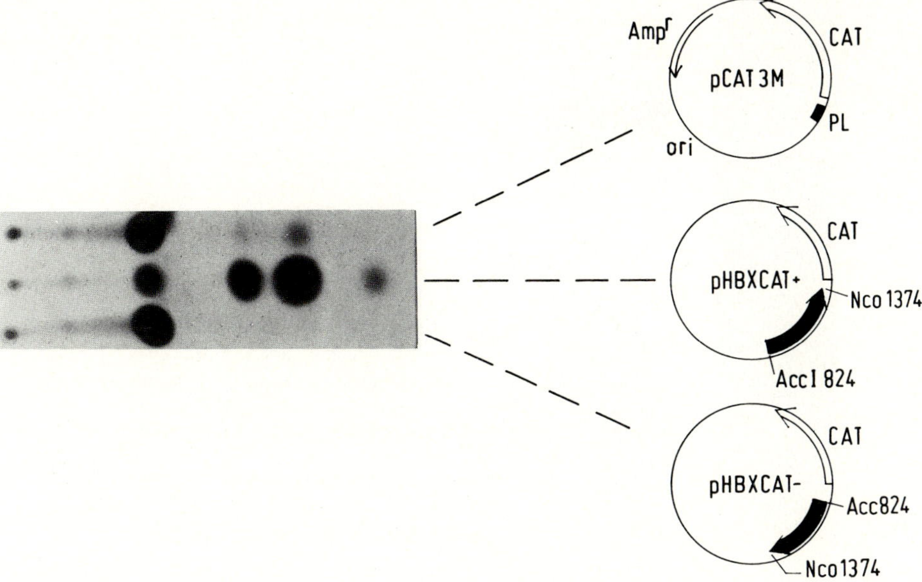


**Fig. 3.** Evidence for a promoter upstream of the X-gene. A fragment of HBV DNA (*AccI-NcoI*, nucleotides 824-1376) was inserted into the *Bgl* II site of vector *pCAT3M* after Klenow I treatment of vector and insert DNA: *pHBXcat*(+) has the HBV DNA in the correct orientation with respect to the CAT gene and *pHBXcat*(−) has the HBV segment in the wrong orientation. In transfected CCL13 cells the CAT gene is expressed only when the HBV fragment is present in the correct orientation

efficiently expressed the CAT gene in transfected cells, whereas no expression was seen when the HBV DNA was in an inverted orientation (Fig. 3). This strongly suggests the presence of a promoter within the fragment of HBV DNA used. X-transcripts of about 0.8 kb are abundantly expressed in transfected cells (see Fig. 6C). Assuming poly A residues of about 0.2 kb, the promoter should lie within 100 bases of the X start site. These results corroborate other recently presented evidence (Treinin and Laub 1988) on the existence of this promoter.

## 4 A Protein is Encoded by the X Protein

Two additional plasmid constructs, pRSVX and pU3R-IXΔcat, in which the X ORF was placed under the control of heterologous promoters, namely, the human T lymphotropic virus I (HTLV-I) long terminal repeat (LTR) and the Rous sarcoma virus (RSV) LTR, respectively, were also able to stimulate pSV2cat expression very efficiently (Fig. 1). In order to prove unequivocally that a protein product of the X ORF is required for the stimulatory activity, frame-shift mutations were introduced into plasmids pRSVX and pU3R-IXΔcat by *Bam*HI cleavage of the plasmid in the unique site within the X-gene, filling in the cohesive ends so generated with a T$_4$ DNA

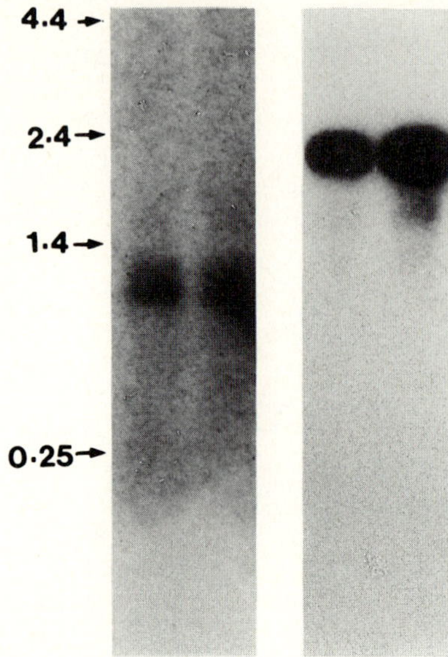



**Fig. 4.** Transcription of X-specific RNA in cells transfected with plasmid DNA bearing a frameshift mutation in the X coding sequences. Polyadenylated selected RNA from CCL13 cells transfected respectively with equal amounts of the frame-shifted plasmid, pU3R-IXfsΔ cat (*left lane*) and with the wild-type plasmid pU3R-IXΔ cat (*right lane*) were analyzed. The *panel on the left* shows hybridization to an X probe, and the *panel to the right* shows the same filter strip-washed and rehybridized to a β-actin probe

polymerase and religating the newly created blunt ends using $T_4$ DNA ligase. This procedure introduced 4 bases into the sequences, thus shifting the reading frame by 1 base. Neither of the frame-shifted mutant plasmids pRSVXfs and pU3R-IXΔcatfs was able to provide stimulatory activity in CAT assays following transfection (Fig. 1), even though Northern analyses showed that the RNA expression in each case was comparable to that obtained with the wild-type plasmids. The RNA analyses using pU3R-IXΔcat and its frame-shifted version, pU3R-IxΔcatfs, are presented in Fig. 4. These experiments confirm that the stimulatory activity provided by the HBV X-gene is dependent on the expression of a functional protein.

## 5 HBV X Protein Stimulates Transcription

Northern analyses of RNA isolated from cells transfected with the various plasmids showed that the levels of CAT RNA were elevated in cells transfected with plasmids that stimulated CAT activity, i.e., pGEMHBV2 (which consists of two head-to-tail copies of the HBV genome in the vector pGEM), pHBV 2836, and pHBV2836ΔNco/Xho, as compared with the very minimal expression of the same transcripts in cells transfected with vector plasmid alone (Fig. 5). Similar amounts of total RNA were present in different lanes as confirmed by rehybridization of the filters with a β-actin probe (not shown) and methylene blue staining (Fig. 5) compare lanes a, b, and d). The results of this experiment as well as the frame-shift experiment described above demonstrate that the stimulation of CAT expression by the X product occurs by stimulation of transcription.

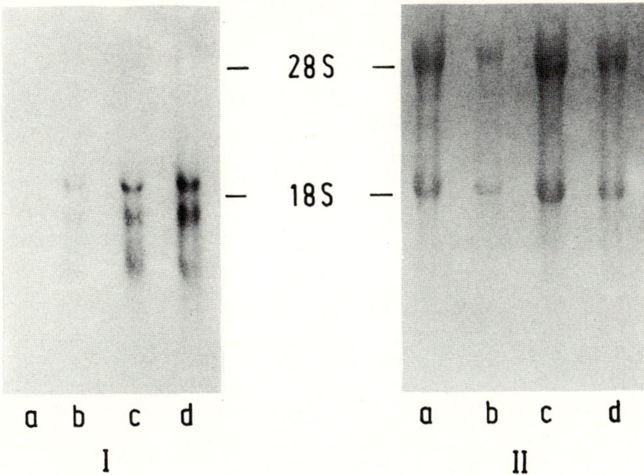


**Fig. 5.** Transcriptional activation of pSV2cat by different HBV plasmids. *I*, Northern-blot hybridization of CAT RNA in CCL13 cells transfected with pSV2cat and a 10-fold molar excess of the test plasmids (*a*) pMLΔBS, (*b*) pGEMHBV2, (*c*) pHBV2836, and (*d*) pHBV2836-ΔXho/Nco. The probe was nick-translated *Hin*dIII-*Bam*HI CAT fragment from pSV2cat. *II*, Methylene blue stain of the same filter as in I

# 6 HBV X Protein is a TRANS-Activator

The observations discussed above provide strong evidence that the stimulation of gene expression by the X product is by *trans* activation. In order to demonstrate this conclusively, we obtained clones of cells in which the X-gene is stably integrated into the cellular chromosomes. HepG2 hepatoblastoma cells were cotransfected with plasmids pAG2HBV, containing two head-to-tail copies of the HBV genome, and pAG60, which encodes resistance to the drug G418. Resistant clones of cells were isolated after selection of the transfected cultures with G418. Clones containing HBV DNA were identified by hybridization of cellular DNA with HBV probe and radioimmunoassay of culture medium for secreted hepatitis B surface antigen. X protein-mediated *trans*-activation was studied in several such cell clones after transfection of pSV2cat DNA. Parental HepG2 cells lacking HBV DNA were used as a control in these experiments. The results of such experiments with a representative clone 15/1 are shown in Fig. 6: Southern blot analyses show that intact viral DNA is integrated in the cellular DNA (A), and X mRNA is abundantly expressed in in these cells (C). The size of the RNA is consistent wich expression from the X promoter. As seen in Fig. 6B, CAT expression is stimulated 15-fold in 15/1 cells but is absent in HepG2 cells. These experiments establish that the stimulatory potential of the X protein is in *trans*.

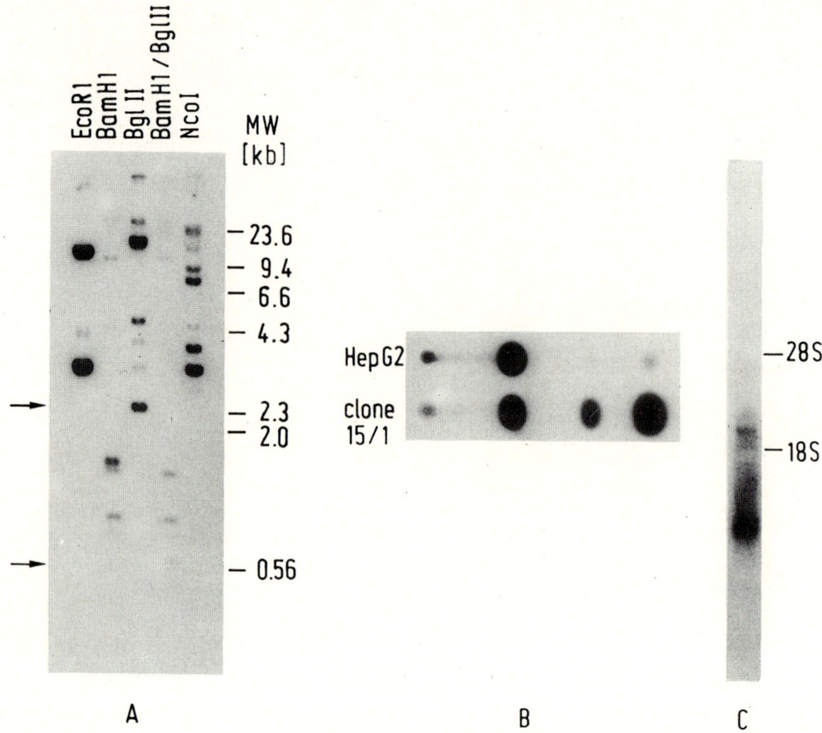

**Fig. 6A–C.** Analyses of HepG2 clone 15/1 cells. **A** Southern blot of 15/1 cellular DNA digested with the indicated enzymes and hybridized with an HBV probe. The 2.3-kb *Bgl* II fragment and the 0.58-kb *Bam*HI/*Bgl* II fragment of HBV (*arrows*) indicate intact X sequences integrated in the cellular DNA. **B** Transient CAT expression in parental HepG2 and 15/1 cells transfected with pSV2cat DNA. In HepG2 the acetylation was 3.4%, and in clone 15/1 it was 45.4%. **C** Northern blot of polyadenylated selected RNA of 15/1 cells hybridized with an HBV probe showing abundant expression of X transcripts of about 800 nucleotides, derived from integrated HBV DNA

# 7 HBV Protein Stimulates Various Viral Enhancer-Promoters

The ability of the X protein to *trans*-activate different promoters was tested by using various plasmids constructed as previously described (ZAHM et al. 1988) in which the expression of the CAT gene was controlled by various eukaryotic viral promoters, viz, the herpes simplex virus I thymidine kinase promoter (pBLcat2), the Rous sarcoma virus LTR (pRSV-LTRcat), the mouse mammary tumor virus LTR (pMMTV-LTRcat), and the human T lymphotropic virus I LTR (pHTLV-LTRcat). The results demonstrate a general ability of the X product to stimulate CAT expression (Fig. 7). However, the MMTV LTR was not significantly stimulated. In that case, the experiment was performed in the absence of dexamethasone in order to dissociate the possible stimulatory effect of the X protein from that which would be conferred by hormones. Thus, even though the effect of the X protein is not a very specific one, there seems to be some degree of preference. Only

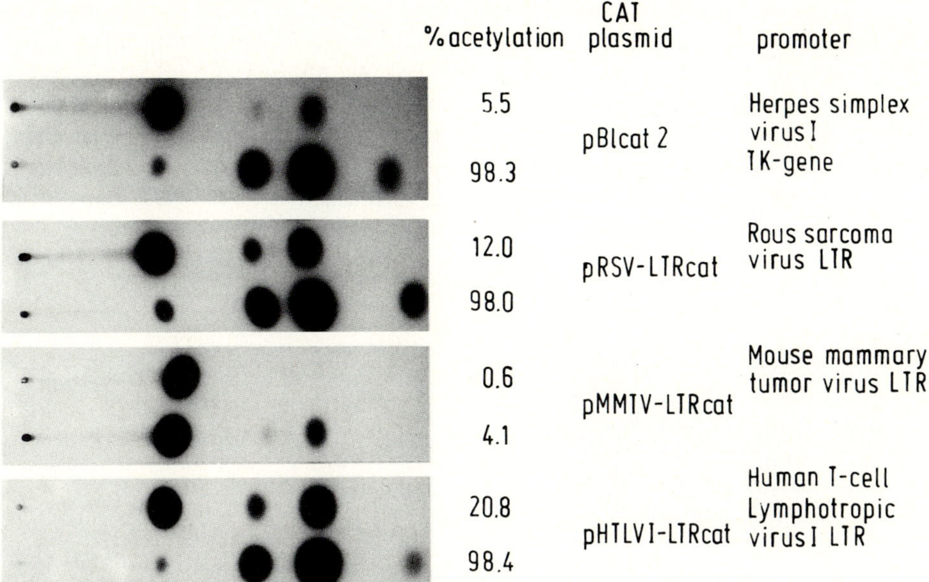



**Fig. 7.** *trans*-Activation of different eukaryotic promoters by pHBV2836 or pHBV1982. Stimulation of transient CAT expression from the indicated promoters by HBV DNA is presented. The control assays were done with cells transfected with the same molar amounts of pMLBS. The *trans*-activating test plasmid was pHBV2836 and in the case of pBLcat2, the *trans*-activating plasmid was pHBV1982 (containing the *Bgl* II-*Bgl* II HBV DNA fragment from position 2836–1982 in the opposite orientation to pHBV2836)

one cellular promoter, the human metallothionein promoter (MTIIA), so far tested is also comparatively only poorly stimulated (less than three fold, data not shown). These results raise the possibility that in infected cells particular cellular genes may be activated or stimulated. It is very likely that alterations in gene expression contribute to pathogenesis and, as discussed below, may have implication in tumorigenesis.

## 8 X Product Stimulates Its Own Expression

The observation that a variety of viral enhancer/promoter elements are stimulated by the X protein suggests that its effect may be exerted via the enhancer rather than directly on the promoter and further that it does not bind directly to the DNA but more likely to other cellular proteins. This speculation is consistent with the observation that the X protein has been localized by immunofluorescence to the cytoplasm (SIDDIQUI et al. 1987). The possible binding of the X protein to cellular cytoplasmic protein(s) may activate it and cause limited amounts of such complexes to be transported to the nucleus where they function as a transcriptional *trans*-activator. The temporal and regulated expression of the X protein in infected cells is not understood partly because purified X protein and antibodies to it are not yet available. However, the studies to date (for a review, see GANEM and VARMUS 1987) indicate that the expression of this gene is well controlled. We studied the effect of the X protein on the HBV enhancer and X promoter (ZAHM et al. 1988; WOLLERSHEIM et al. 1988). Plasmid pHBVXcat, described above, comprising the CAT gene under the control of the HBV enhancer and X promoter, when transfected into CCL13 cells at low concentrations (i.e., 1 µg/100 mm plate) did not lead to CAT expression.

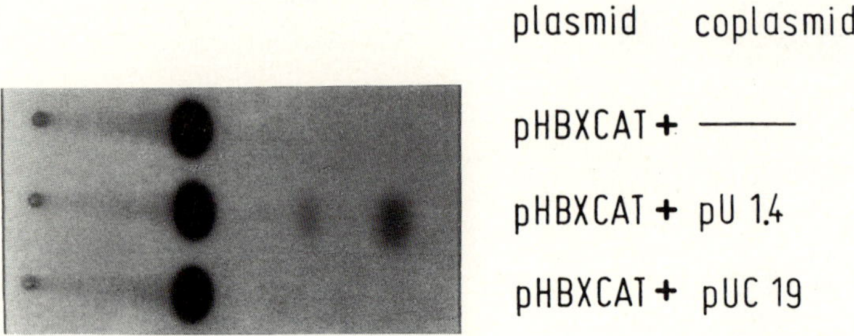



**Fig. 8.** The X product *trans*-activates the X promoter. Plasmid pHBXcat (described in legend to Fig. 3) was transfected into CCL13 cells at a concentration too low for it to be expressed (1.0 µg/plate). Cotransfection with an X-expressing plasmid, pU1-4 (see Fig. 10), resulted in CAT expression. Cotransfection with vector pUC19 did not have any effect

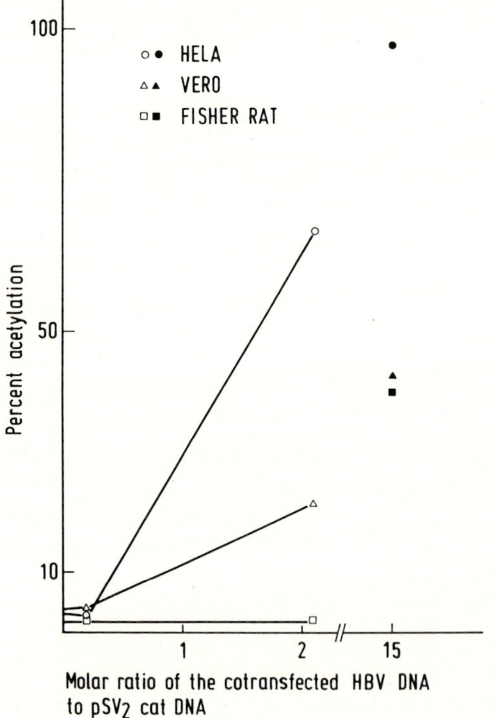



**Fig. 9.** *trans*-Activation of pSV2cat by pHBV2AG60 (*open symbols*) and pKKHBs34 (*closed symbols*) in cells of different species. Demonstration of differential cellular effects

However, when this plasmid was cotransfected with another plasmid, pU1.4 (see Fig. 10 for a description of this plasmid), expressing the X gene, there was stimulation of CAT expression (Fig. 8), indicating that the X product has a positive feedback on its own expression. This effect has been determined to occur via the enhancer as shown by the use of constructs in which the enhancer was deleted (see Fig. 10 and discussion below).

## 9 TRANS-Activation by HBV X-Gene Involves Cellular Factors

As discussed above, the action of the X protein appears to occur in concert with cellular proteins. These considerations and the fact that infection with HBV is tissue- and species-tropic led us to examine the properties of the X protein in cells of different species. For obvious reasons the experiments described thus far had been done in human liver-derived CCL13 cells. A variety of cells including HeLa (human), Fisher rat fibroblasts, CV-1 (monkey), Vero (monkey), GL2/2 (feline), BHK (hamster), and NIH 3T3 (mouse) were next used to determine the cellular effect on the function of the X-gene. The results of some of the experiments are shown in Fig. 9. It was clear that the best stimulation of CAT expression occurred in human liver cells. However, HeLa cells which are of human nonliver origin were also good expressers of the function. On the other hand Fisher rat cells permitted CAT stimulation to a much lower extent and then only at higher amounts of DNA transfection. The experiments with mouse, hamster, and cat cells confirmed these trends, thus suggesting that there are cellular factors required for optimum *trans*-activation and that these factors, though similar in function, are phylogenetically diverged.

## 10 Integrated HBV DNA from Tumors Retains TRANS-Activational Potential

The importance of a possible HBV *trans*-activating function in hepatocarcinogenesis has been suggested (KOSHY 1987). Therefore, having determined that such a function is indeed encoded by HBV DNA, we wanted to study this property in integrated HBV DNA. For this purpose, a copy of integrated viral DNA along with cellular flanking sequences was isolated by molecular cloning from cells of a primary hepatocellular carcinoma containing two integrated copies of HBV DNA. Restriction mapping of the cloned DNA (Fig. 10) shows that there are 2.1 kb of viral DNA with 12.6 kb of cellular DNA adjacent to it. The viral DNA present on the clone comprises only the surface (S) gene and most of the X-gene. The core (C) gene and the presurface (preS) gene are deleted, as is the S promoter. Sequence analyses (not shown) reveal that the last 28 nucleotides of the X ORF are deleted. This cloned DNA, pU 1.4, was tested in CCL13 cells and found to be extremely efficient in stimulating expression of cotransfected pSV2cat (Fig. 10). Because of the presence of a large amount of cellular sequences on the clone, it was necessary to reduce the

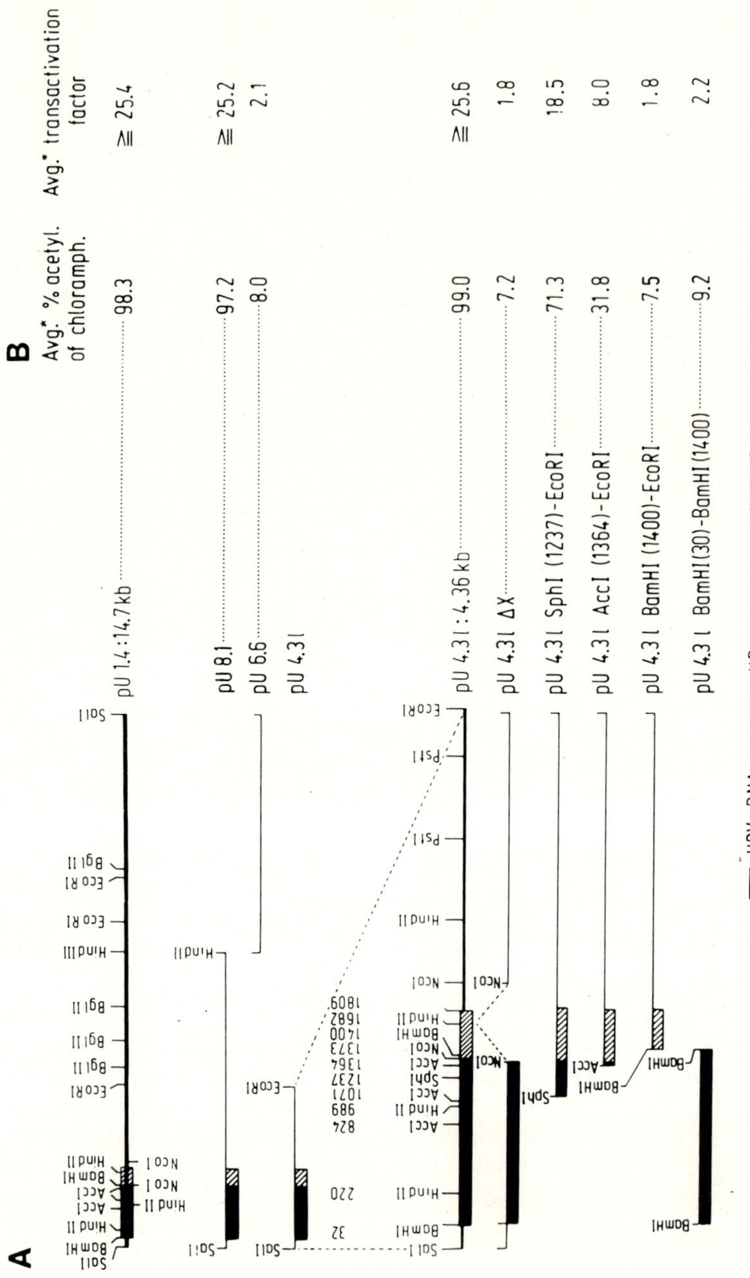


**Fig. 10A, B.** Restriction map of integrated HBV DNA clone pU1.4 and its deletion derivatives. The extent of deletions are indicated, and the stimulatory activities obtained upon transfection into CCL13 cells are presented to the *right* of each construct

tested DNA by progressive deletions in order to be sure that the stimulatory activity was a function of the X-sequences and to exclude any contributory effect of the cellular DNA. The plasmids thus produced were tested for their stimulatory capacity (Fig. 10). The results clearly indicate that a considerable part of the cellular DNA can be removed (pU4.31) without diminution of the activity. A deletion of part of the X sequences (pU431 X) results in total loss of activity. Removal of sequences upstream of the X-gene, i.e., enhancer and promoter (pU 4.31 Bam-Eco) also renders the construct inactive. Plasmid pU 4.31 Acc-Eco which retains a small region upstream of the X ORF but which is missing the enhancer and promoter shows a much diminished activity. Plasmid pU4.31 Sph-Eco containing the promoter but missing the enhancer is less active than the original. Thus, it can be concluded that the stimulatory activity of the plasmid pU 4.31 is encoded by the X sequences and furthermore, that the viral enhancer is necessary for optimum expression of the X-gene.

## 11 Expression of a Fusion X-Cellular RNA Transcript

The process of integration, as seen in plasmid pU 4.31, led to the deletion of 28 bases at the 3'-terminal of the X-gene as well as the viral polyadenylation signals normally required for production of viral mRNA. Therefore, in order to generate a functional X message, transcription would have to proceed though adjacent cellular DNA and terminate at a cellular polyadenylation site. We have examined RNA isolated from CCL13 cells transfected with pU 4.31 DNA. An

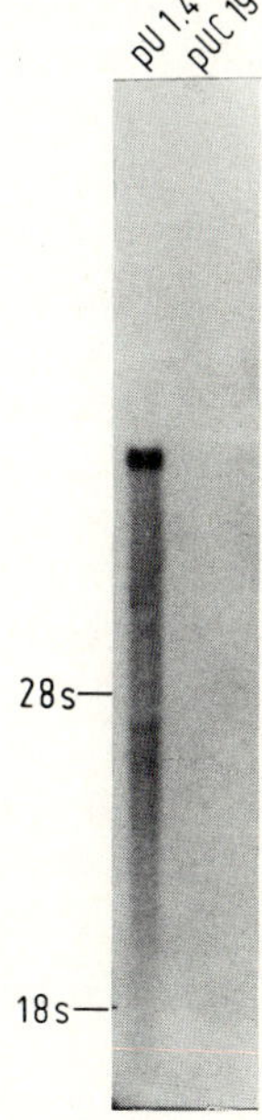


**Fig. 11.** Fusion mRNA consisting of X and cellular domains. Plasmid pU1.4 DNA was transfected into CCL13 cells. RNA was isolated 48 h after transfection and analyzed by Northern blotting. An X-specific probe was used for hybridization. There was no signal when the filter was strip-washed and re-hybridized to an S-specific probe (data not shown). Cells transfected with pUC19 vector DNA did not produce RNA hybridizing to the X probe. Ribosomal 28$S$ and 18$S$ RNA were used as size markers

unexpectedly long X transcript of about 10 kb is produced (Fig. 11). The failure of this transcript to hybridize to a probe derived from S sequences indicates that it could only have arisen from the X promoter, which means that the transcript contained up to 9 kb of cellular RNA. Nucleotide sequence analyses (not shown) indicate that there is a fused open reading frame between the viral X gene and at least 228 bases of cellular DNA, showing that a fusion protein is possible. While there is no direct evidence for the synthesis of such a protein, the results of the transcativation experiments strongly suggest that is is made. Lack of tissue from the original tumor precludes any effort to determine whether such a product was actually made in vivo. Future experiments with other tumors will provide important information on this question.

## 12  Discussion

The experiments described in this article establish that the X-gene of the HBV produces a protein with the ability to stimulate gene expression by a *trans* mechanism. This product increases the level of transcription of the activated gene. A frame-shift mutation in the X coding sequence abolishes the activity. In transfection experiments a promoter lying upstream of the X ORF is efficient in transcription of the gene, although this particular promoter has not yet been shown to be active in naturally infected cells. A number of viral enhancer-promoters tested show that the activity of the X product is rather broad, even though in certain cases there was little activity under the experimental conditions used. This indicates that the X protein may function indirectly by interacting with cellular proteins rather than directly with DNA. Experiments using a variety of cell lines strongly suggest the existence of cellular factors which influence the function of the X protein. The results indicate phylogenetically closer related species to be more supportive of X function, thus encouraging the speculation that X-gene function may have some role in determining the tropism of the virus.

The observation that various enhancer-promoters are activated by the X product also might suggest that the effect is exerted at the level of the enhancer. The efficient expression of the X-gene itself is dependet on the presence of the enhancer. This view is supported by the fact that when the enhancer is deleted as in plasmid pU 4.3 Sph-Eco, the level of activity is considerably reduced (Fig. 10). Similar conclusions have been recently reported by other investigators (SPANDAU and LEE 1988). In addition to the X promoter (TREININ and LAUB 1987), the surface (CHANG et al. 1987) and core (SHAUL et al. 1985) promoters have also been shown to be influenced by the HBV enhancer. The product of the X-gene in turn has a positive effect on the enhancer (Fig. 8). Thus, there may be a mechanism by which all the viral genes are regulated by the X-gene product via the enhancer. This possibility must be considered in attempting to define the viral function(s) of the X-gene. The conservation of the X-gene in mammalian hepadnaviruses (KODAMA et al. 1985) and the effects of its product on viral promoters would argue for an essential role. Recent experiments with frame-shift mutation of the X-gene led to the conclusion that it is not necessary for the replication of HBV (YAGINUMA et al.

1987). However, the system used in those studies was among the first ever established for HBV replication and, as such, perhaps not sensitive enough to monitor X-gene modulation of virus replication. It may well be that in the absence of a functional X product viral gene expression and virus replication are diminished. In the integrated HBV sequences from a tumor, described above, the X-gene was functional in transactivation despite the deletion of 28 bases at its 3'-end. This raises the possibility that a truncated X-gene could give rise to a functional product(s). This would also be consistent with the results of YAGINUMA et al. (1987) in whose experiments the frame-shift mutation was introduced in the 3'-portion of the X-gene.

Whatever the viral function of the X-gene, the ability of its product to transactive gene expression may have important implications in HBV-induced liver diseases particularly hepatocellular carcinoma. The need to consider the possible importance of transactivation in tumor development is also underscored by the lack of evidence for a common HBV *cis*-mediated role. The observation that the X function of integrated HBV in tumors is conserved in a number of cases studied (this report and unpublished observations) adds substance to this speculation. HBV X sequences are frequently present in tumor cells even when other HBV genes are not (KODAMA et al. 1985; NAGAYA et al. 1987). Because integration often involves the direct repeat (DR 2) DNA sequences which lie within the X-gene (NAGAYA et al. 1987), the gene is often truncated, and this leads to the expression of fusion transcripts of X-cellular composition. Such fusion transcripts could well have altered properties (FREYTAG VON LORINGHOVEN et al. 1985). The significance of such products for transformation is being investigated. In HTLV-related human T-cell leukemias, the HTLV I transactivator gene is shown to be able to activate the promoter of the gene encoding interleukin-2 receptor alpha chain (IL 2Rα, CROSS et al. 1987). Since IL 2Rα is specifically expressed in infected cells and is strongly implicated in T-lymphocyte transformation, there is good reason for considering transactivation to be very important to the process of viral transformation. Analogous mechanisms might be envisioned in HBV-related hepatocarcinogenesis. The potential of the HBV X-gene to influence the expression of cellular genes over prolonged periods of time via integration of viral DNA may be an important contributory factor in the progression to hepatocellular carcinoma. It is important to identify cellular genes involved in transformation and to study their possible activation by HBV. Other HBV genes may also have transactivational functions which remain to be evaluated in the future. It has been recently reported (TWU et al. 1988) that HBV represses the cellular β-interferon promoter by a *trans* mechanism and that the product of the HBc gene may be responsible for this effect. Such a situation in infected patients could promote the establishment of chronic infections which predispose to the development of cirrhosis and cancer.

*Acknowledgements.* We are thankful to Drs. P. Gruss, B. Howard, R. Renkawitz, H. Schaller, and H. Will for the kind gifts of plasmids. The excellent technical assistance of E. Bürgelt, S. Hankel, and A. Klem-Andres is gratefully acknowledged. The contributions of Drs. P. Zahm and M. Wollersheim to the work discussed in this article are recognised. This study was supported by funds from the Deutsche Stiftung für Krebsforschung.

# References


Beasley RP (1982) Hepatitis B virus as the etiologic agent in hepatocellular carcinoma–epidemiologic considerations. Hepatology 2: 21–26

Benbrook D, Lernhardt D, Pfahl M (1988) A new retinoic acid receptor identified from a hepatocellular carcinoma. Nature, 333: 669–672

Brand N, Petkovitch M, Krust A, Chambon P, deThe H, Marchio A, Tiollais P, Dejean A (1988) Identification of a second human retinoic acid receptor. Nature, 332: 850–853

Chang HK, Chou CK, Chang C, Su TS, Hu CP, Yoshida M, Ting LP (1987) The enhancer sequence of human hepatitis B virus can enhance the activity of its surface gene promoter. Nucleic Acids Res, 15: 2261–2268

Cross SL, Feinberg M, Wolf JB, Holbrook NJ, Wong-Staal F, Leonard WJ (1987) Regulation of the human interleukin-2-receptor alpha chain promoter: Activation of a nonfunctional promoter by the transactivator gene of HTLV-1. Cell, 49: 47–56

Dejean A, Bougueleret L, Grzeschik KH, Tiollais P (1986) Hepatitis B virus DNA integration in a sequence homologous to v-*erb*-A and steroid receptor genes in a hepatocellular carcinoma. Nature, 322: 70–72

deThe H, Marchio A, Tiollais P, Dejean A (1987) A novel steroid thyroid hormone receptor gene inappropriately expressed in human hepatocellular carcinoma. Nature, 330: 667–670

Freytag von Loringhoven A, Koch S, Hofschneider PH, Koshy R (1985) Co-transcribed 3′ host sequences augment expression of integrated hepatitis B virus DNA. EMBO J 4: 249–255

Ganem D, Varmus HE (1987) The molecular biology of the hepatitis B virus. Ann Rev Biochem 56: 651–693

Gorman C, Moffat L, Howard B (1982) Recombinant genomes which express chloramphenicol acetyltransferase in mammalian cells. Mol Cell Biol 29: 1046–1051

Gough NM (1983) Core and E antigen synthesis in rodent cells transformed with hepatitis B virus DNA is associated with greater than genome length viral messenger RNAs. J Mol Biol, 165: 683–699

Koch S, Freytag von Loringhoven A, Kahmann R, Hofschneider PH, Koshy R (1984) The genetic organisation of integrated hepatitis B virus DNA in the human hepatoma cell line PLC/PRF/5. Nucleic Acids Res 12: 6871–6886

Kodama K, Ogasawara N, Yoshikama H, Murakami S (1985) Nucleotide sequence of a cloned woodchuck hepatitis virus genome: evolutionary relationship between hepadnaviruses. J Virol, 56: 978–986

Koshy R (1987) Integrated hepatitis B virus genomes in human primary liver cancer: a consideration of molecular mechanisms of oncogenesis. In: Hofschneider PH, Munk K (eds) Viruses in human tumors. Contr Oncol, vol 24, pp 97–112

Miller RH, Robinson WS (1986) Common evolutionary origin of hepatitis B virus and retroviruses. Proc Nat Acad Sci USA, 83: 2531–2535

Nagaya T, Nakamura T, Tokino T, Tsurimoto T, Imai M, Mayumi T, Kamino K, Yamamura K, Matsubara K (1987). The mode of hepatitis B virus DNA integration in chromosomes of human hepatocellular carcinoma. Genes and Development, 1: 773–782

Rogler C, Hino O, Su C-Y (1987) Molecular aspects of persistent woodchuck hepatitis virus and hepatitis B virus infection and hepatocellular carcinoma. Hepatology, 7: 74s–78s

Roosinck MJ, Jameel S, Loukin SH, Siddiqui A (1986) Expression of hepatitis B viral core region in mammalian cells. Mol Cell Biol, 6: 1393–1400

Shaul Y, Rutter WJ, Laub O (1985) A human hepatitis B viral enhancer element. EMBO J 4: 427–430

Siddiqui A, Jameel S, Mapoles J (1987) Expression of hepatitis B virus X gene in mammalian cells. Proc Nat Acad Sci USA, 84: 2513–2517

Spandau FD, Lee C-H (1988) Transactivation of viral enhancers by the hepatitis B virus X protein. J Virol, 62: 427–434

Summers J, Mason WS (1982) Replication of the genome of a hepatitis B-like virus by reverse transcriptase of an RNA intermediate. Cell, 29: 403–415

Tiollais P, Pourcel C, Dejean A (1985) The hepatitis B virus. Nature, 317: 489–495

Tognoni A, Cattaneo R, Serfling E, Schaffner W (1985) A novel expression selection approach allows precise mapping of the hepatitis B virus enhancer. Nucleic Acids Res, 13: 7457–7472

Treinin M, Laub O (1987) Identification of a promoter element located upstream from the hepatitis B virus X gene. Mol Cell Biol, 7: 545–548

Twu JR-Shin, Lee C-H, Lin P-M, Schloemer RH (1988) Hepatitis B virus suppresses expression of human β-interferon. Proc. Nat. Acad. Sci. USA, 85: 252–256

Wollersheim M, Debelka U, Hofschneider PH (1988) A transactivating function encoded in the hepatitis B virus X gene is conserved in the integrated state. Oncogene, 3: 545–552

Yaginuma K, Shirakata Y, Kobayashi M, Koike K (1987) Hepatitis B virus (HBV) particles are produced in a cell culture system by transient expression of transfected HBV DNA. Proc. Nat Acad Sci USA, 84: 2678–2682

Zahm P, Hofschneider PH, Koshy R (1988) The HBV X-ORF encodes a transactivator: a potential factor in viral hepatocarcinogenesis. Oncogene, 3: 169–177

# Subject Index